Advanced Structured Materials

Series Editors: *Andreas Öchsner (Editor-in-chief),*
Lucas Filipe Martins da Silva, and
Holm Altenbach

Andreas Öchsner · Lucas Filipe Martins da Silva ·
Holm Altenbach
Editors

Materials with Complex Behaviour

Modelling, Simulation, Testing, and Applications

With 272 Figures

Springer

Editors
Prof. Dr.-Ing. Andreas Öchsner
Technical University of Malaysia
Fac. Mechanical Engineering
Dept. Applied Mechanics
91310 Skudai, Johor
Malaysia
andreas.oechsner@gmail.com

Prof. Dr.-Ing. habil. Holm Altenbach
Univ. Halle-Wittenberg
FB Ingenieurwissenschaften
LS Technische Mechanik
06099 Halle (Saale)
Germany
holm.altenbach@iw.uni-halle.de

Prof. Dr. Lucas Filipe Martins da Silva
Department of Mechanical Engineering
Faculty of Engineering of the
University of Porto
Rua Dr. Roberto Frias s/n
4200-465 Porto
Portugal
lucas@fe.up.pt

ISSN 1869-8433 e-ISSN 1869-8441
ISBN 978-3-642-12666-6 e-ISBN 978-3-642-12667-3
DOI 10.1007/978-3-642-12667-3
Springer Heidelberg Dordrecht London New York

Library of Congress Control Number: 2010932188

Cover design: WMXDesign GmbH, Heidelberg

Printed on acid-free paper

Springer is part of Springer Science+Business Media (www.springer.com)

Preface

Common engineering materials reach in many demanding applications such as automotive or aerospace their limits and new developments are required to fulfill increasing demands on performance and characteristics. The properties of materials can be increased for example by combining different materials to achieve better properties than a single constituent or by shaping the material or constituents in a specific structure. Many of these new materials reveal a much more complex behavior than traditional engineering materials due to their advanced structure or composition. Furthermore, the classical applications of many engineering materials are extended to new ranges of applications and to more demanding environmental conditions such as elevated temperatures. All these tendencies require in addition to the synthesis of new materials, proper methods for their manufacturing and extensive programs for their characterization. In many fields of application, the development of new methods and processes must be accomplished by accurate and reliable modeling and simulation techniques. Only the interaction between these new developments with regards to manufacturing, modeling, characterization, further processing and monitoring of materials will allow to meet all demands and to introduce these developments in safety-relevant applications.

The 3rd International Conference on Advanced Computational Engineering and Experimenting, ACE-X 2009, was held in Rome, Italy, from 22 to 23 June 2009 with a strong focus on the above mentioned developments. This conference served as an excellent platform for the engineering community to meet with each other and to exchange the latest ideas. This volume contains 23 revised and extended research articles written by experienced researchers participating in the conference. The book will offer the state-of-the-art of tremendous advances in engineering technologies of materials with complex behavior and also serve as an excellent reference volume for researchers and graduate students working with advanced materials. The covered topics are related to Material and Properties Part I, Modeling and Simulation of Non-classical Materials and Structures Part II, and New Technologies Part III.

The organizers and editors wish to thank all the authors for their participation and cooperation which made this volume possible. Finally, we would like to thank the team of Springer-Verlag, especially Dr. Christoph Baumann, for the excellent cooperation during the preparation of this volume.

Skudai, Malaysia
Porto, Portugal
Halle, Germany
January 2010

Andreas Öchsner
Lucas F.M. da Silva
Holm Altenbach

Contents

Part I Materials and Properties

A Statistical Study on Stress-Strain Relation of AISI 304 Stainless Steel Under Elevated Temperatures 3
S.H. Park, J.K. Kim, and J.H. Kim

Interaction Between Concrete Cylinders and Shape-Memory Wires in the Achievement of Active Confinement 19
Jean-François Destrebecq and Xavier Balandraud

Mathematic Modeling of the Osprey Process 35
M. Bodea, R. Muresan, and C.V. Prica

Cyclic-Bend-Over-Sheave Fatigue Testing of an Umbilical for Oil Production in Ultra-Deep Waters 47
Paula Ferreira Lépore and Miguel Angel Buelta Martinez

Dynamic Crack Propagation in Composite Structures 57
D. Bruno, F. Greco, and P. Lonetti

Computational Analysis of Loading–Unloading and Non-homogeneity Effects in Metallic Hollow Sphere Structures 83
Branca F. Oliveira, Luiz A.B. da Cunda, Andreas Öchsner, and Guillermo J. Creus

Dielectric Spectra Analysis: Reliable Parameter Estimation Using Interval Analysis 99
Adrien Brochier, Maëlenn Aufray, and Wulff Possart

Part II Modeling and Simulation of Non-classical Materials and Structures

Numerical Modeling of Complex Structures: Shells and Biological Cells 127
J.N. Reddy, R.A. Arciniega, G.U. Unnikrishnan, and V.U. Unnikrishnan

Free Vibration Characteristics of Thermally Loaded Cylindrical Shell 139
Byung-Hee Jeon, Hui-Won Kang, and Young-Shin Lee

Model of Large Displacements in Static Analysis of Shell 149
Domagoj Matešan, Jure Radnić, and Alen Harapin

Nonlinear Time-Dependent Analysis of Prestressed Concrete Shells . . 165
Domagoj Matešan and Jure Radnić

DBEM and FEM Analysis of an Extrusion Press Fatigue Failure 181
R. Citarella, G. Cricrì, M. Lepore, and M. Perrella

Damage Detections in Nonlinear Vibrating Thermally Loaded Plates . . 193
E. Manoach and I. Trendafilova

Macroscopic Stability Analysis in Periodic Composite Solids 213
Fabrizio Greco, Paolo Lonetti, Paolo Nevone Blasi, and Girolamo Sgambitterra

Finite Element Vibration Analysis of MHSS Based on 3D Tomography Image Processing 243
R. Winkler, J. Schulz, M. Merkel, and A. Öchsner

Computational Modelling and Experimental Characterisation of Heterogeneous Materials 257
A.J. Beveridge, M.A. Wheel, and D.H. Nash

Model Experiment and Numerical Modelling of Dynamic Soil-Structure Interaction 269
Noriko Kodama and Kazuhito Komiya

Part III New Technologies

The Laser Butt Welding Simulation of the Thin Sheet Metal 279
Takeji Arai

Laser Drilling Simulation Considering Multiple Reflection of Laser, Evaporation and Melt Flow 297
Etsuji Ohmura and Satoru Noguchi

Effect of Flight Spectrum Loads on the Damage Tolerance Evaluation of a Helicopter Frame 311
Marco Giglio and Andrea Manes

Effects of Manufacturing-Induced Residual Stresses and Strains on Hydrogen Embrittlement of Cold Drawn Steels 331
J. Toribio, M. Lorenzo, D. Vergara, and V. Kharin

Hybrid Bonding of Advanced High Strength Steels in the Lightweight Body Shell Design for the Automobile Manufacturing . . . 343
G. Weber, T. Bschorr, H. Cramer, O. Hahn, M. Rethmeier, and H. Thommes

Contributors

T. Arai Chuo University, Tokyo 112-8551, Japan, tarai@tamacc.chuo-u.ac.jp

R.A. Arciniega Department of Mechanical Engineering, Texas A&M University, College Station, TX 77843-3123, USA, rarciniegaa@hotmail.com

M. Aufray Chair Adhesion and Interphases in Polymers, University of Saarland, 66041 Saarbrücken, Germany, maelenn.aufray@ensiacet.fr

X. Balandraud Clermont Université, Institut Français de Mécanique Avancée, EA 3867, Laboratoire de Mécanique et Ingénierie, BP 10448, F-63000 Clermont-Ferrand, France, Xavier.Balandraud@ifma.fr

A.J. Beveridge Department of Mechanical Engineering, University of Strathclyde, Glasgow G1 1XJ, UK, andrew.beveridge@strath.ac.uk

P.N. Blasi Department of Structural Engineering, UNICAL, Cosenza, Italy, paolo.nevoneblasi@unical.it

M. Bodea Technical University of Cluj, Cluj-Napoca, Romania, mbodea@stm.utcluj.ro

A. Brochier Lehrstuhl Adhesion and Interphases in Polymers, Universität des Saarlandes, Gebäude C6-3, Postfach 15 11 50, 66041 Saarbrücken, Germany, brochier@math.unistra.fr

D. Bruno Department of Structural Engineering, University of Calabria, Via P. Bucci, Cubo 39-B, 87030, Rende, Cosenza, Italy, d.bruno@unical.it

T. Bschorr German Welding Institute SLV Munich, Munich, Germany, bschorr@slv-muenchen.de

M.A.B. Martinez Department of Structures Engineering, Engineering School, University of Sao Paulo, Sao Paulo, 05508-900, Brazil, buelta@usp.br

R. Citarella Department of Mechanical Engineering, University of Salerno, 84084 Fisciano (SA), Italy, rcitarella@unisa.it

H. Cramer German Welding Institute SLV Munich, Munich, Germany, cramer@slv-muenchen.de

G.J. Creus Cemacom, Universidade Federal do Rio Grande do Sul, 90035-190 Porto Alegre, RS, Brazil, creus@ufrgs.br

G. Cricrì Department of Mechanical Engineering, University of Salerno, 84084 Fisciano (SA), Italy, gcricri@unisa.it

L.A.B. da Cunda Escola de Engenharia, Universidade Federal do Rio Grande, 96201-900 Rio Grande, RS, Brazil, luizcunda@furg.br

J.-F. Destrebecq Clermont Université, Université Blaise Pascal, EA 3867, Laboratoire de Mécanique et Ingénieries, BP 10448, F-63000, Clermont-Ferrand, France, J-Francois.Destrebecq@univ-bpclermont.fr

M. Giglio Dipartimento di Meccanica, Politecnico di Milano, Milano, Italy, marco.giglio@polimi.it

F. Greco Department of Structural Engineering, UNICAL, Cosenza, Italy, f.greco@unical.it

O. Hahn University of Paderborn, Paderborn, Germany, ortwin.hahn@lwf.upb.de

A. Harapin Faculty of Civil Engineering and Architecture, 21000 Split, Croatia, alen.harapin@gradst.hr

B.-H. Jeon Agency for Defense Development, Daejeon 305-152, Korea

H.-W. Kang Agency for Defense Development, Daejeon 305-152, Korea

V. Kharin University of Salamanca, 49022 Zamora, Spain

J.K. Kim Defense Systems Test Center, Agency for Defense Development, Taeahn 357-900, Korea, kajun001@daum.net

J.H. Kim Department of Mechanical Design Engineering, Chungnam National University, Daejeon 305-764, Korea, kimjhoon@cnu.ac.kr

N. Kodama Waseda Institute for Advanced Study, Waseda University, Tokyo 169-8555, Japan, kodama@aoni.waseda.jp

K. Komiya Department of Civil Engineering and Architecture, Chiba Institute of Technology, Chiba 275-8588, Japan, komiya.kazuhito@it-chiba.ac.jp

Y.-S. Lee BK21 Mechatronics Group, Chungnam National University, Daejeon 305-764, Korea, leeys@cnu.ac.kr

M. Lepore Department of Mechanical Engineering, University of Salerno, 84084 Fisciano (SA), Italy, malepore@unisa.it

P.F. Lépore Department of Structures Engineering, Engineering School, University of Sao Paulo, Sao Paulo, 05508-900, Brazil, paula@mfx.com.br

P. Lonetti Department of Structural Engineering, UNICAL, Cosenza, Italy, lonetti@unical.it

M. Lorenzo University of Salamanca, 49022 Zamora, Spain

A. Manes Dipartimento di Meccanica, Politecnico di Milano, Milano, Italy, andrea.manes@polimi.it

E. Manoach Institute of Mechanics, Bulgarian Academy of Sciences, 1113 Sofia, Bulgaria, e.manoach@imbm.bas.bg

D. Matešan Faculty of Civil Engineering and Architecture, 21000 Split, Croatia, domagoj.matesan@gradst.hr

M. Merkel University of Applied Sciences Aalen, 73430 Aalen, Germany, markus.merkel@htw-aalen.de

R. Muresan Technical University of Cluj, Cluj-Napoca, Romania, radu.muresan@stm.utcluj.ro

D.H. Nash Department of Mechanical Engineering, University of Strathclyde, Glasgow G1 1XJ, UK, d.nash@strath.ac.uk

S. Noguchi Division of Materials and Manufacturing Science, Graduate School of Engineering, Osaka University, Osaka 565-0871, Japan

A. Öchsner Faculty of Mechanical Engineering, Technical University of Malaysia, UTM Skudai, 81310, Johor, Malaysia, andreas.oechsner@gmail.com

E. Ohmura Department of Management of Industry and Technology, Graduate School of Engineering, Osaka University, Osaka 565-0871, Japan, ohmura@mit.eng.osaka-u.ac.jp

B.F. Oliveira Virtual Design Group, Universidade Federal do Rio Grande do Sul, 90035-190 Porto Alegre, RS, Brazil, branca@ufrgs.br

S.H. Park Defense Systems Test Center, Agency for Defense Development, Taeahn 357-900, Korea, redfox89@dreamwiz.com

M. Perrella Department of Mechanical Engineering, University of Salerno, 84084 Fisciano (SA), Italy

W. Possart Saarland University, Saarbrücken, Germany, w.possart@mx.uni-saarland.de

C.V. Prica Technical University of Cluj, Cluj-Napoca, Romania, cprica@stm.utcluj.ro

J. Radnić Faculty of Civil Engineering and Architecture, 21000 Split, Croatia, jure.radnic@gradst.hr

J.N. Reddy Department of Mechanical Engineering, Texas A&M University, College Station, TX 77843-3123, USA, jnreddy@tamu.edu

M. Rethmeier Federal Institute for Materials Research and Testing (BAM), Berlin, Germany, Michael.Rethmeier@bam.de

J. Schulz Carl Zeiss MicroImaging GmbH, 81379 Munich, Germany, schulz@zeiss.de

G. Sgambitterra Department of Structural Engineering, UNICAL, Cosenza, Italy, gsgambitterra@libero.it

H. Thommes University of Paderborn, Paderborn, Germany

J. Toribio University of Salamanca, 49022 Zamora, Spain, toribio@usal.es

I. Trendafilova Department of Mechanical Engineering, University of Strathclyde, Glasgow G1 1XJ, UK, irina.trendafilova@strath.ac.uk

G.U. Unnikrishnan Department of Mechanical Engineering, Texas A&M University, College Station, TX 77843-3123, USA, ginu@tamu.edu

V.U. Unnikrishnan Department of Mechanical Engineering, Texas A&M University, College Station, TX 77843-3123, USA, unnithan@tamu.edu

D. Vergara University of Salamanca, 49022 Zamora, Spain

G. Weber Federal Institute for Materials Research and Testing (BAM), Berlin, Germany, gert.weber@bam.de

M.A. Wheel Department of Mechanical Engineering, University of Strathclyde, Glasgow G1 1XJ, UK, marcus.wheel@strath.ac.uk

R. Winkler University of Applied Sciences Aalen, 73430 Aalen, Germany, rolf.winkler@htw-aalen.de

Part I
Materials and Properties

A Statistical Study on Stress-Strain Relation of AISI 304 Stainless Steel Under Elevated Temperatures

Modelling, Simulation, Testing, and Applications

S.H. Park, J.K. Kim, and J.H. Kim

1 Introduction

Stainless steel members have been increasingly used in the construction industry, especially in architectural and structural applications because of high corrosion resistance and aesthetic appearance as well as ease of maintenance and ease of construction [1]. Currently available specifications for the design of stainless steel structural members are mainly for structures at room temperature. However, the material properties of stainless steel should be known at elevated temperatures for the fire resistant design of stainless steel structures [2–4]. A conventional design approach, which is based on somewhat arbitrary multipliers such as safety factors and safety margins, gives little indication of the failure probability of the component. The nature of mechanical behavior of materials and failure is probabilistic. The strength of materials has variations from size effect, surface finish, notch effect, etc. and the stress varies because of stress concentration, temperature factor, stress combinations, etc. So, the conventional design approach is not adequate from a reliability standpoint [5]. The statistical characteristics of tensile strength and yield strength should be known to use a reliable design approach at elevated temperatures. The material chosen for the present investigation was AISI 304 stainless steel (austenitic). The purpose of this study is to investigate the material's tensile properties at high temperatures, the distribution of the material's strength at high temperature and the use of normal statistical plot for a stress-strength interference model.

J.H. Kim (✉)
Department of Mechanical Design Engineering, Chungnam National University,
Daejeon, 305-764, Korea
e-mail: kimjhoon@cnu.ac.kr

A. Öchsner et al. (eds.), *Materials with Complex Behaviour*, Advanced Structured Materials 6, DOI 10.1007/978-3-642-12667-3_1,

2 Theory

2.1 Statistical Distribution of Strength

The allowable stress should be less than the material's strength to prevent failure. The material's strength is determined by standard test methods and specimens. But it has a statistical distribution from the deviation of manufacture, microstructure, test environment, etc. Moreover, there is variation of loads acting on the structural materials from load and environmental conditions. So, the test result of the material's strength is non-deterministic but statistical. It is important to investigate the distribution of the material's strength to analyze its characteristics statistically.

The statistical plot is a simple and easy way to determine the statistical distribution. If the data on the statistical plot are linearly pointed, the statistical plot matches the data. The correlation coefficient on some statistical plot shows the goodness of fit. On the statistical plot, to calculate the cumulative probability, the data should be arranged in ascending order. The mean rank is used to calculate the cumulative probability for ordered data. Generally, the tensile and yield strength comply with the normal distribution [6]. Therefore, the data were plotted on the normal probability plot and its correlation coefficient was checked for yield strength and tensile strength at each temperature to verify the linearity. Various characteristics of the normal distribution function can be used. The center of normal distribution curve corresponds to the average value. In the normal statistical plot, the probability of 50% corresponds to the average ($\bar{\mu}$) and 84.135% to the standard deviation added average ($\bar{\mu} + \bar{\sigma}$). Therefore, the standard deviation is a difference between two probabilities.

In U.S. Military Handbook [7], they determine the allowable stress from A-Basis or B-Basis for military planes and light alloy materials. The A-Basis means that 99% of the population has at least 95% reliability. The B-Basis means that 90% of the population has at least 95% reliability. A-Basis strength and B-Basis strength correspond to 1 and 10% probability respectively. A-Basis and B-Basis tensile and yield strengths can be read from normal statistical plot. Test data were plotted on the normal statistical plot by using a commercial data analysis program, OriginPro ver. 7.5. A linear regression curve was drawn on the normal statistical plot. A-Basis strength was chosen from the crossing point of 1% probability (y axis) and regression line. B-Basis strength was chosen from the crossing point of 10% probability (y axis) and regression line.

2.2 Stress-Strength Interference Model

The stress-strength interference model is used in the reliability design approach. In this model, the shaded portion in Fig. 1 shows the interference area, which is indicative of the probability of failure. In the conventional design approach, a safety factor is generally used. The safety factor (n) may be expressed as $n = \bar{\delta}/\bar{s}$. Here, $\bar{\delta}$ is a structural strength which is a material property and $\bar{s}$ is an allowable working stress. Generally, the ultimate strength is chosen to $\bar{\delta}$ and $\bar{s}$ is constrained by safety

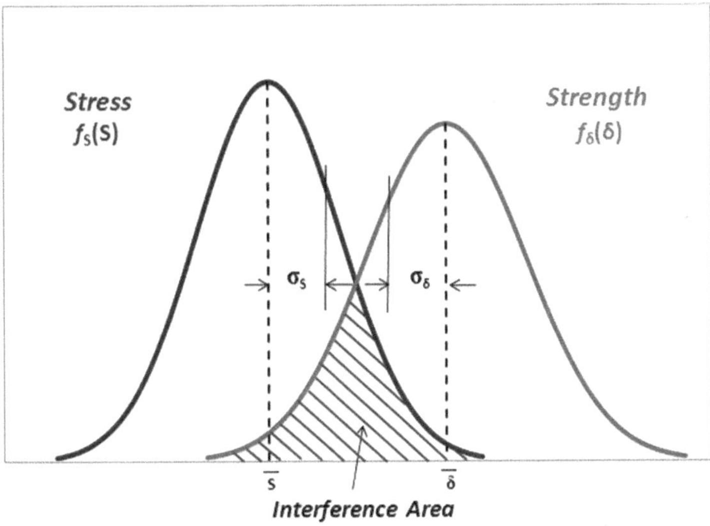

Fig. 1 Stress-strength interference model

factor which is determined by application. Though the strength of material and the stress acting on structure have statistical distribution, the design concept of safety factor cannot consider the variation of stress and strength. If the stress acts in the unexpected interference area the material may have failed as shown in Fig. 1.

Let the density function for the stress be denoted by $f_s(s)$ and that for strength by $f_\delta(\delta)$. Then, by definition the reliability is

$$R = P(\delta > s) = P(\delta - s > 0) \tag{1}$$

$$R = \int_{-\infty}^{\infty} f_\delta(\delta)\left[\int_{-\infty}^{\delta} f_s(s)ds\right]d\delta = \int_{-\infty}^{\infty} f_\delta(\delta)\left[F_s(\delta)\right]d\delta \tag{2}$$

The probability of failure (unreliability) is defined as

$$\bar{R} = 1 - R = P(\delta \leq s) = \int_{-\infty}^{\infty} [1 - F_s(\delta)] f_\delta(\delta) d\delta \tag{3}$$

If the strength of the material and the stress acting on the material are normally distributed, the probability density function is given by

$$f_s(s) = \frac{1}{\sigma_s\sqrt{2\pi}} \exp\left[-\frac{1}{2}\left(\frac{s-\mu_s}{\sigma_s}\right)^2\right], -\infty < s < \infty \tag{4}$$

$$f_\delta(\delta) = \frac{1}{\sigma_\delta\sqrt{2\pi}} \exp\left[-\frac{1}{2}\left(\frac{\delta-\mu_\delta}{\sigma_\delta}\right)^2\right], -\infty < \delta < \infty \tag{5}$$

Here, μ denotes the mean value and σ denotes the standard deviation.

If the random variable y is defined as $y = \delta - s$, y is normally distributed with a mean of $\mu_y = \mu_\delta - \mu_s$ and a standard deviation of $\sigma_y = \sqrt{\sigma_\delta^2 + \sigma_s^2}$. The reliability R can be expressed in terms of y as

$$R = P(y > 0) = \int_0^\infty \frac{1}{\sigma_y\sqrt{2\pi}} \exp\left[-\frac{1}{2}\left(\frac{y-\mu_y}{\sigma_y}\right)^2\right] dy \tag{6}$$

If we let z be defined by $z = (y - \mu_y)/\sigma_y$, the reliability is as follows and the random variable z is the standard normal variable.

$$R = \frac{1}{\sqrt{2\pi}} \int_{-\frac{\mu_\delta-\mu_s}{\sqrt{\sigma_\delta^2+\sigma_s^2}}}^{\infty} e^{-z^2/2} dz \tag{7}$$

Equation (7) may be rewritten as

$$R = 1 - F\left(-\frac{\mu_\delta - \mu_s}{\sqrt{\sigma_\delta^2 + \sigma_s^2}}\right) \tag{8}$$

Although abnormally large stress acts on the component, the material should endure this allowable stress S_a. Then the probability of failure is

$$P_f = F\left(\frac{S_a - \mu}{\sigma}\right) \tag{9}$$

We should determine S_a such the probability of failure (P_f) should be lower that the limit of some value of P. So, the allowable stress S_a is as

$$S_a = \mu - F^{-1}(1-P)\sigma = \mu - \mu_p\sigma \tag{10}$$

Here, $\mu_p = F^{-1}(1-P)$ is the upper tail probability of the standard normal distribution. As an example, for the confidence level of 99% ($P = 0.01$), the upper tail probability is 2.326 and the allowable stress is $S_a = \mu - 2.326\sigma$. For the confidence level of 90% ($P = 0.10$), the upper tail probability is 1.282 and the allowable stress is $S_a = \mu - 1.282\sigma$.

3 Experimental Procedure

3.1 Material

The material used in this study is AISI type 304 stainless steel. Table 1 summarizes the chemical composition.

Table 1 Chemical composition for AISI type 304

Element	C	Si	Mn	P	S	Ni	Cr
Composition (max, %)	0.08	1.0	2.0	0.045	0.03	8~10.5	18~20

3.2 Tensile Tests

The tensile tests were conducted in accordance with ASTM E8 [8] and E21 [9] using an MTS 810 servo-hydraulic test machine as shown in Fig. 2. Figure 3 is the geometry of the test specimen. The thermocouples (K type) were attached at the center of it. The test was run in strain control mode at a strain rate of 2 mm/min. 10 specimens were tested at each temperatures of 21, 100, 200, 350, 500, 650 and 800°C. The heating speed was 1.9–2.0°C/s and the temperature was maintained for 30 min. An extensometer with a total gauge length of 25 mm was used.

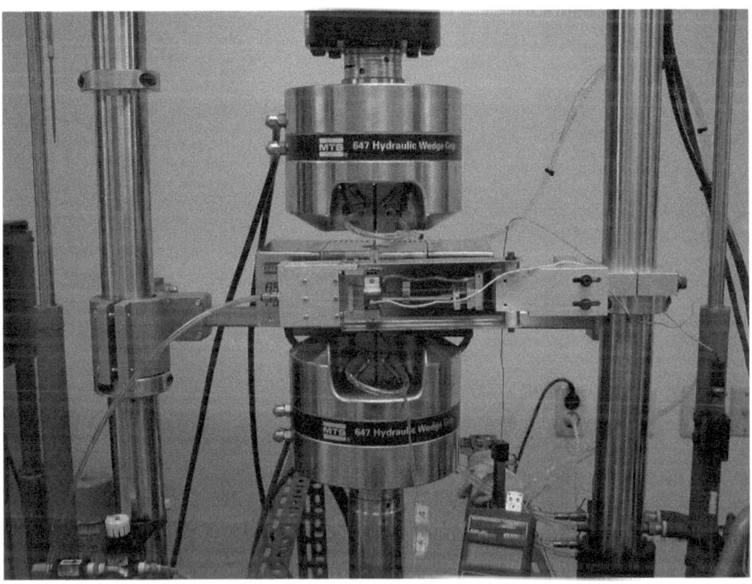

Fig. 2 Experimental setup for uniaxial tension tests under various temperatures

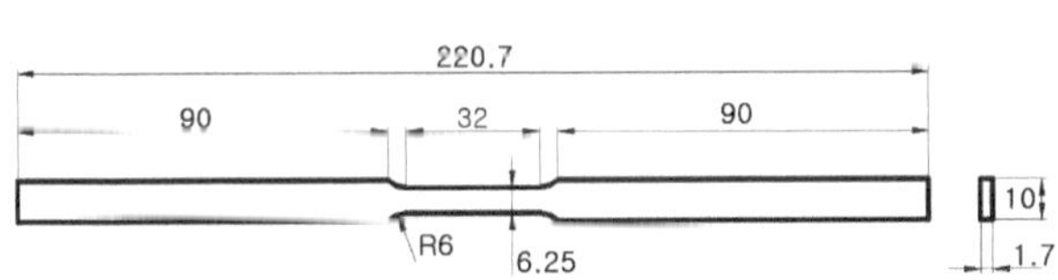

Fig. 3 Test specimen geometry (dimensions in mm)

4 Results and Discussion

4.1 Tensile Characteristics

Load-displacement curves and stress-strain curves at room and elevated temperatures are plotted in Figs. 4 and 5 respectively. The mechanical properties of AISI 304 stainless steel like ultimate tensile strength, yield strength, Young's modulus and elongation were obtained from the tensile test. Table 2 shows their average ($\bar{\mu}$) and standard deviation ($\bar{c}$), and the A-Basis and B-Basis strength calculated by Eq. (10) (A-Basis at $P = 0.01$, B-Basis at $P = 0.10$).

The tensile strength, yield strength and elastic modulus decrease as the temperature increases. Figure 6 shows the tensile strength as a function of temperature. As shown in Fig. 7, the elongation decreases as the temperature increases to 650°C,

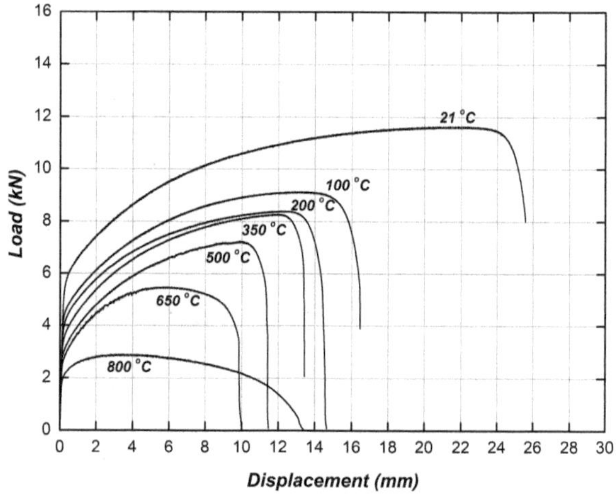

Fig. 4 Load-displacement curves at various temperatures

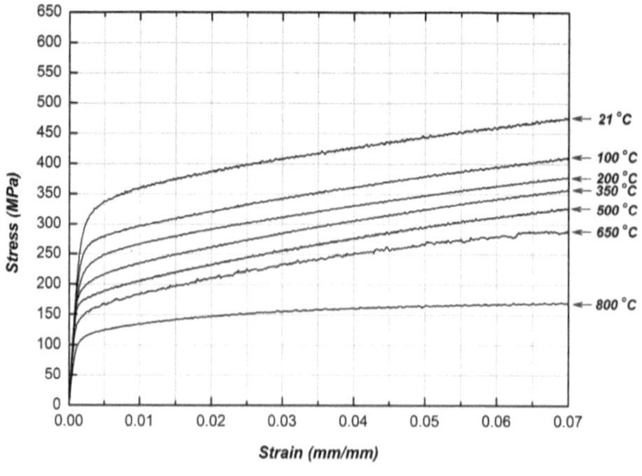

Fig. 5 Stress-strain curves at various temperatures

Table 2 Tensile characteristics for AISI 304 stainless steel at various temperatures

	Ultimate tensile strength (MPa)				Yield strength (MPa)				Young's modulus (GPa)		Elongation (%)	
Temp. (°C)	$\bar{\mu}$	$\bar{\sigma}$	A-basis	B-basis	$\bar{\mu}$	$\bar{\sigma}$	A-basis	B-basis	$\bar{\mu}$	$\bar{\sigma}$	$\bar{\mu}$	$\bar{\sigma}$
21	691.6	7.97	673.1	681.4	322.7	7.49	305.3	313.1	199.0	8.98	78.4	0.84
100	562.2	3.52	554.0	557.7	268.9	3.45	260.9	264.5	187.7	7.02	52.0	0.49
200	502.0	3.17	494.6	497.9	233.0	4.91	221.6	226.7	183.2	3.84	44.5	1.12
350	492.9	10.81	467.8	479.0	202.4	3.16	195.0	198.4	173.1	7.83	42.9	0.93
500	448.2	3.70	439.6	443.5	182.0	3.56	173.7	177.4	167.4	3.72	36.1	0.27
650	327.4	10.92	302.0	313.4	156.4	4.91	145.0	150.1	151.9	5.79	31.5	0.55
800	170.0	7.04	153.6	161.0	113.0	5.57	100.0	105.9	111.4	5.16	43.1	1.80

Fig. 6 Tensile strength at various temperatures

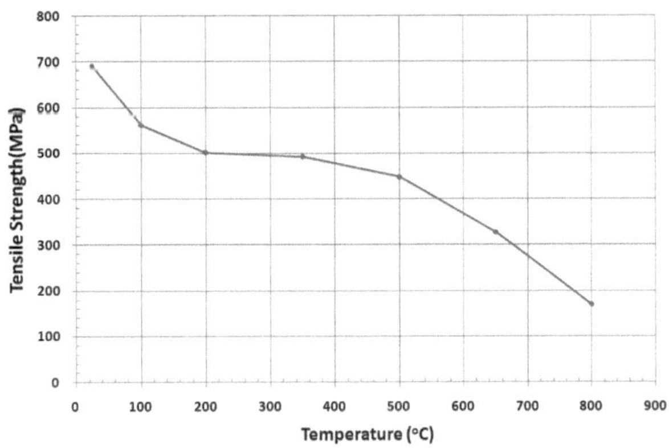

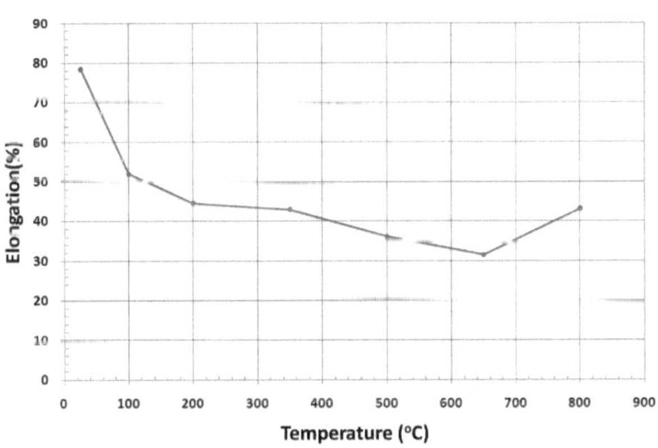

Fig. 7 Elongation at various temperatures

but increases at 650–800°C. Rhines and Wray [10] have pointed out that the minimum ductility occurs in an intermediate temperature range for all ductile metals and alloys. At low temperatures, fracture occurs by the usual transgranular crack propagation mechanism and ductility is high. At temperatures near minimum, deformation occurs by grain boundary sliding, causing the formation of intergranular cavities at triple junctions. At high temperatures, recrystallization occurs simultaneously with intergranular cavity formation as a result of which intergranular crack propagation is retarded. The ductility therefore increases at high temperatures. Sikka et al. [11] have reported elevated temperature tensile ductility minima in AISI 304 and 316 stainless steels and their metallographic findings were found to be consistent with the model proposed by Rhines and Wray. That is, the ductility minimum was associated with the temperature, the strain rate and the metallurgical condition under which intergranular crack propagation was not inhibited [12]. The variation of ductility from different fracture mechanism is the cause of variation of the elongation with temperature.

After the tensile test, the fractured specimens were analyzed by EDX (Energy Dispersive X-ray) microanalysis. The wt% of carbon and chrome is plotted as a function of temperature in Figs. 8 and 9 respectively. The AISI 304 stainless steel is an austenitic type. An austenitic stainless steel has the characteristic of being harder and more brittle at the temperature between 500 and 800°C. This is due to the precipitation of α phase chrome carbide ($Cr_{23}C_6$) in this temperature range. The density of chrome in the material decreases as a result of the precipitation. From the EDX analysis, it can be seen in Figs. 8 and 9 that the wt% of carbon and chrome decrease above 500°C.

SEM (Scanning Electron Microscope) images are shown in Figs. 10, 11, 12, and 13. There are dimples associated with transgranular ductile fracture. As the temperature increases, the size of the dimples increases. Up to 800°C, evidence of any intergranular failure was not found from the SEM images. The tensile ductility of AISI 316 stainless steel is influenced by temperature as well as grain size. The temperature corresponding to the ductility minimum increases with an increase in grain size which is not considered in this study.

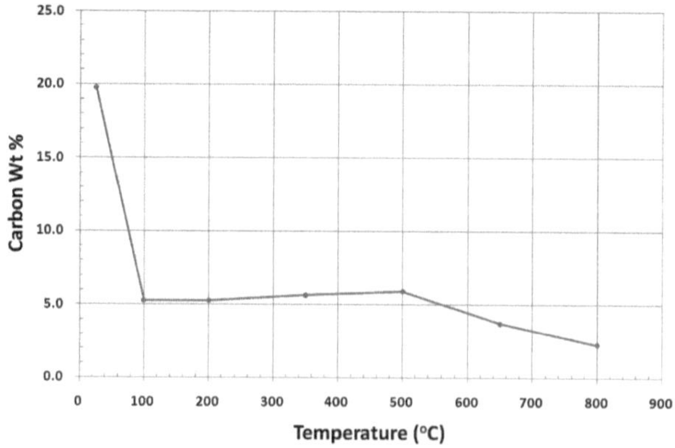

Fig. 8 Carbon wt% at various temperatures

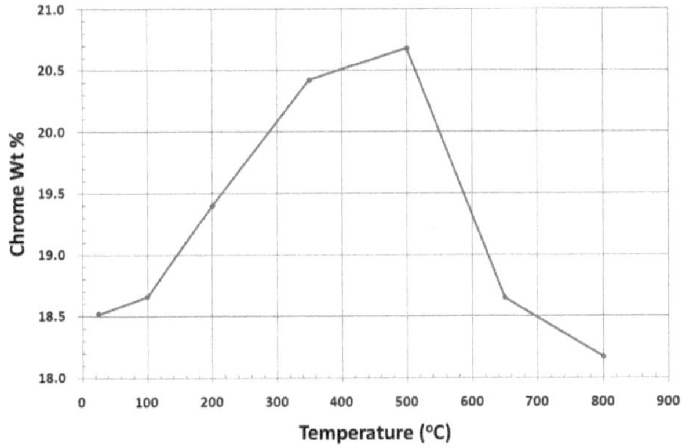

Fig. 9 Chrome wt% at various temperatures

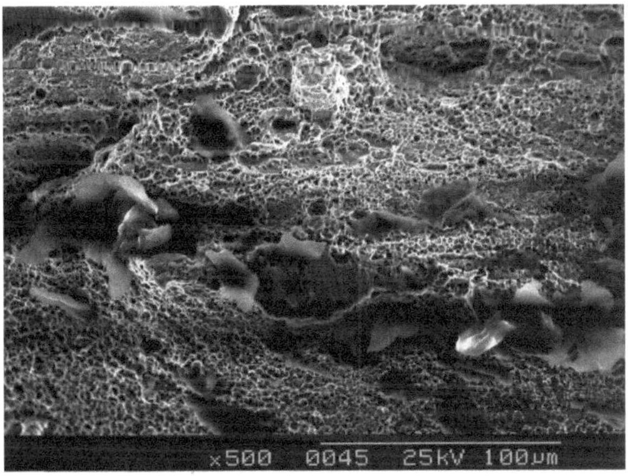

Fig. 10 SEM image after test at 21°C

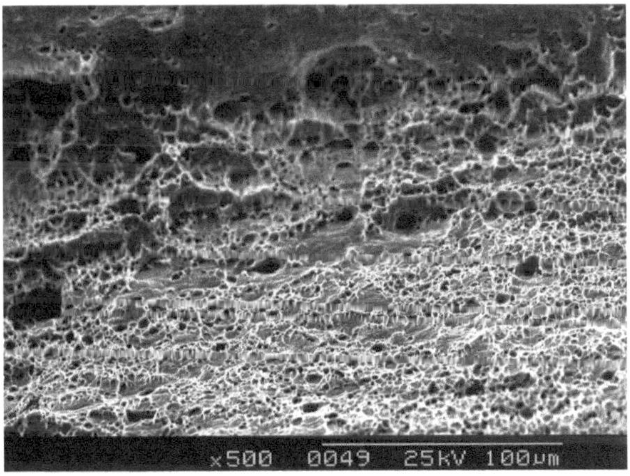

Fig. 11 SEM image after test at 200°C

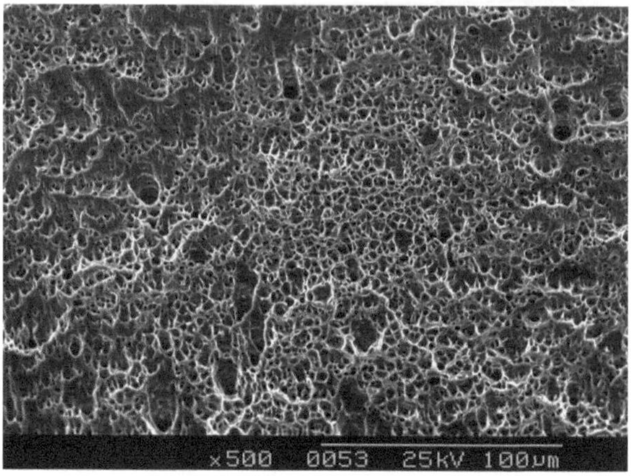

Fig. 12 SEM image after test at 500°C

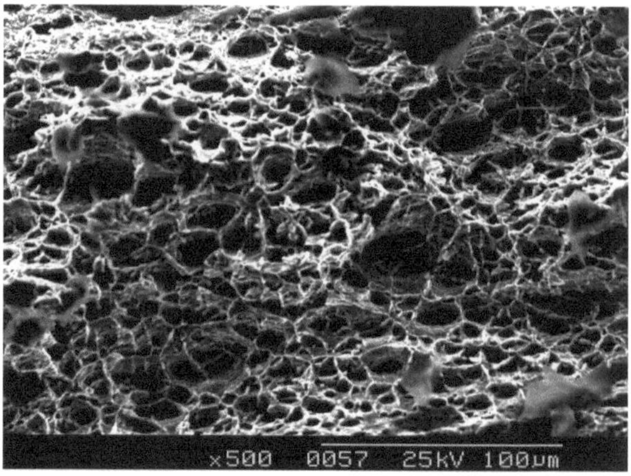

Fig. 13 SEM image after test at 800°C

4.2 Statistical Test Results at Elevated Temperatures

The probability of yield strength and tensile strength at each temperature was plotted on the normal statistical plot, as shown in Figs. 14, 15, 16, 17, 18, 19, and 20. At each temperature, the normal statistical plot shows that the data are linearly plotted and the r-values (correlation coefficients) are nearly 1.0. This means that the normal distribution describes the variation of data in a good manner. The normal distribution describes well the tensile strength and the yield strength at room temperature as well as at elevated temperature. The A-Basis and B-Basis tensile and yield strength are included in the normal statistical plot.

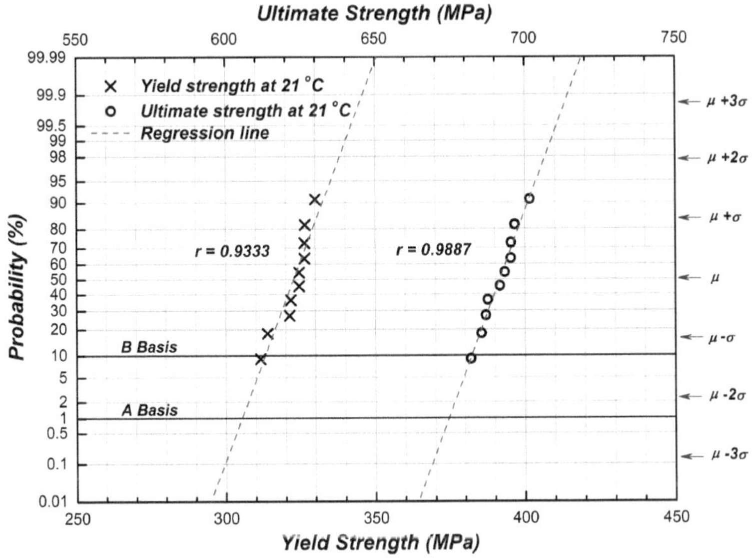

Fig. 14 Normal statistical plot at 21°C

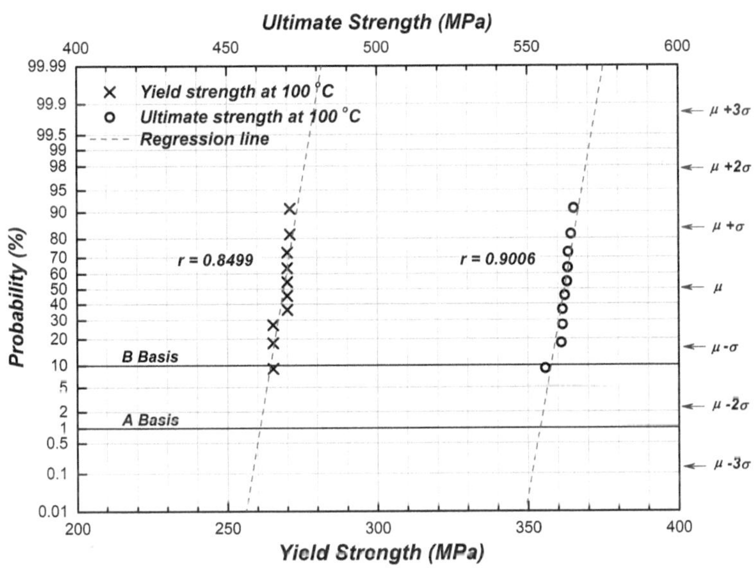

Fig. 15 Normal statistical plot at 100°C

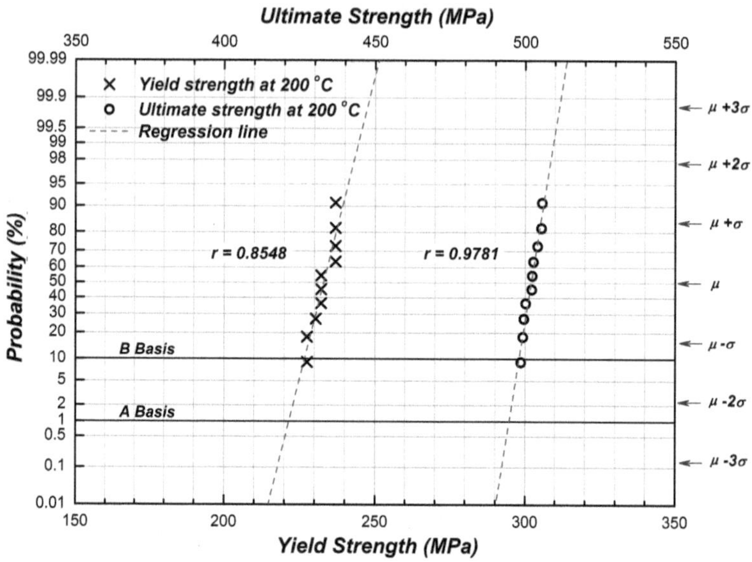

Fig. 16 Normal statistical plot at 200°C

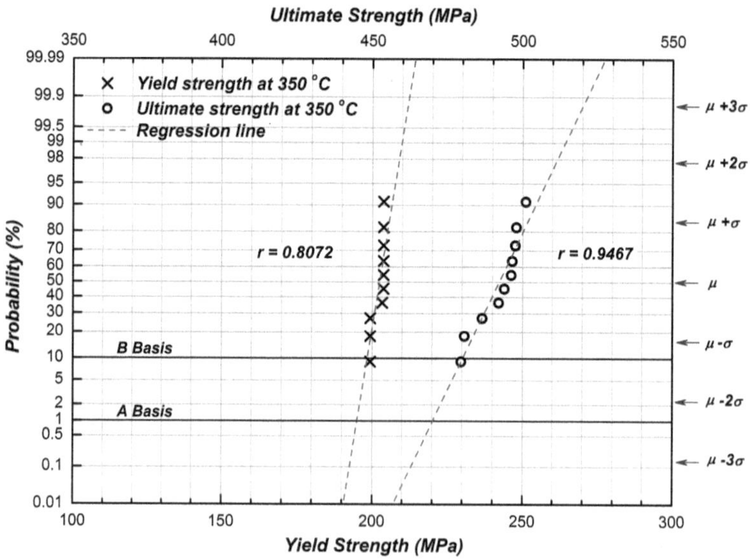

Fig. 17 Normal statistical plot at 350°C

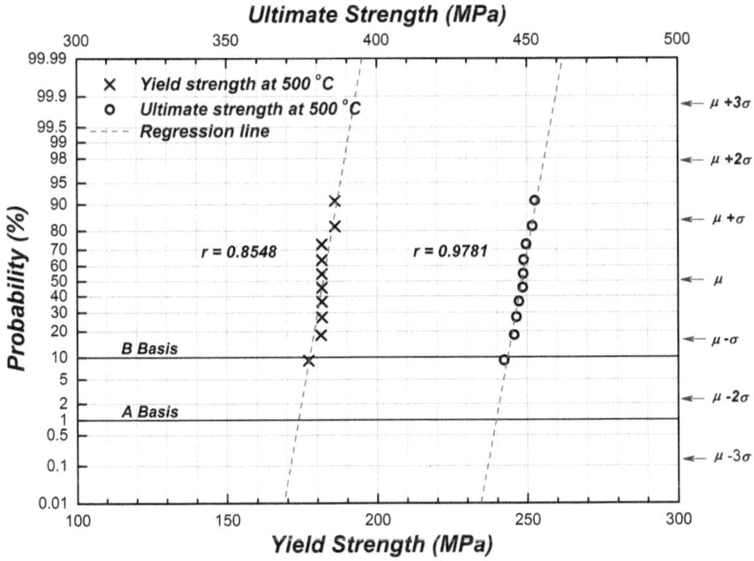

Fig. 18 Normal statistical plot at 500°C

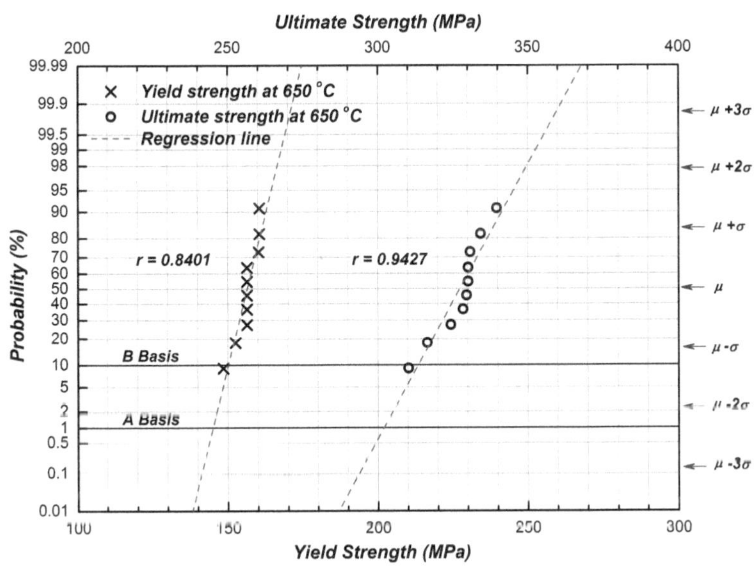

Fig. 19 Normal statistical plot at 650°C

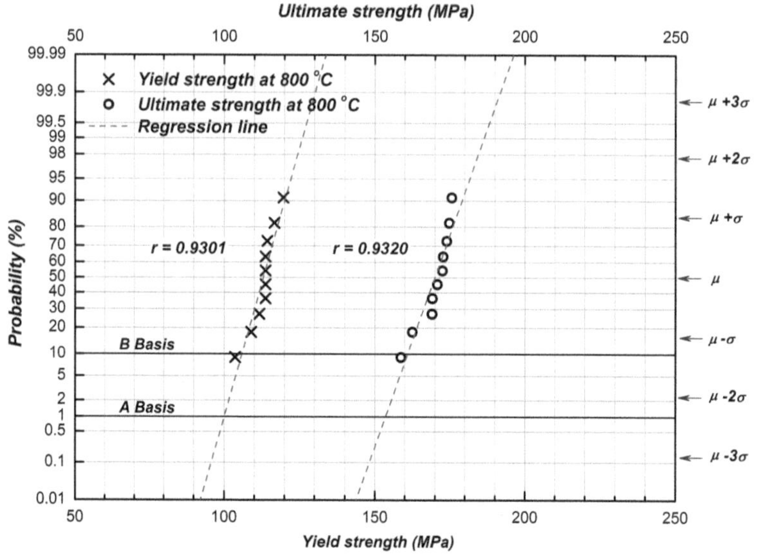

Fig. 20 Normal statistical plot at 800°C

4.3 Use of Statistical Plot for Design

Figure 21 is a normal statistical plot and can explain the use of a stress-strength model for statistical design. The mean and standard deviation can be read from the normal statistical plot. The inclination of the points in normal statistical plot is the variation of data. When the inclination increases, the standard deviation decreases. If the stress acts as the dotted line, the material may be expected to fail. Though the mean value of stress is less than the material's tensile strength, the acting stress has maximum 10% probability failure in Fig. 21. It corresponds to the interference area of stress-strength model in Fig. 1. If we consider the A-Basis strength as the minimum required level, the safe stress area is the shaded area. So, it is important to know the distribution of strength in the lab test as well as the expected stress distribution. Tables 3 and 4 shows that the difference between the value calculated by Eq. (10) and the value picked on the normal statistical plot is almost the same. The statistical plot is a very useful tool to guess the allowable strength.

5 Conclusions

This paper aimed to statistically analyze the strength data of AISI 304 stainless steel at room and elevated temperatures. From tensile tests, material constants such as ultimate tensile strength, yield strength, Young's modulus and elongation are acquired. The material characteristics at room and elevated temperatures were

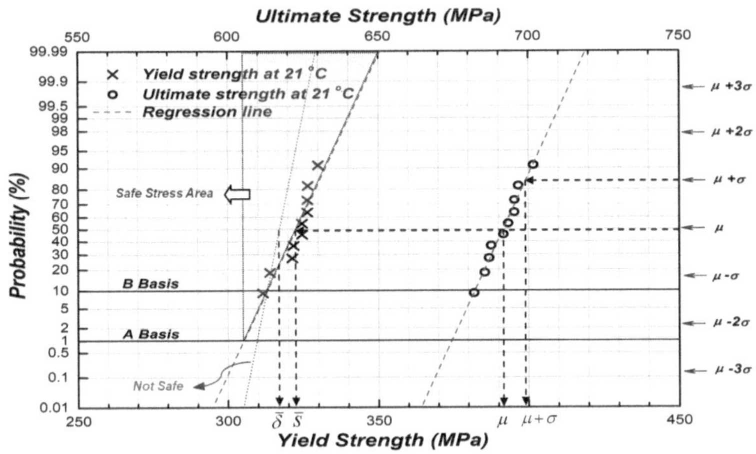

Fig. 21 Use of normal statistical plot for statistical design

Table 3 Comparison of A-basis and B-basis tensile strength

			A-basis		B-basis	
Temp. (°C)	$\widehat{\mu}$	$\widehat{\sigma}$	Plot	Eq. (10)	Plot	Eq. (10)
21	691.6	7.97	674.5	673.1	682.2	681.4
100	562.2	3.52	554.0	554.0	557.6	557.7
200	502.0	3.17	494.7	494.6	497.9	497.9
350	492.9	10.81	470.2	467.8	480.1	479.0
500	448.2	3.70	439.6	439.6	443.5	443.5
650	327.4	10.92	302.0	302.0	313.4	313.4
800	170.0	7.04	153.6	153.6	160.9	161.0

Table 4 Comparison of A-basis and B-basis yield strength

			A-basis		B-basis	
Temp. (°C)	$\widehat{\mu}$	$\widehat{\sigma}$	Plot	Eq. (10)	Plot	Eq. (10)
21	322.7	7.49	305.3	305.3	313.1	313.1
100	268.9	3.45	260.9	260.9	264.5	264.5
200	233.0	4.91	221.6	221.6	226.7	226.7
350	202.4	3.16	195.0	195.0	198.4	198.4
500	182.0	3.56	173.7	173.7	177.4	177.4
650	156.4	4.91	145.0	145.0	150.2	150.1
800	113.0	5.57	100.0	100.0	105.9	105.9

analyzed by using EDX analysis and the distribution of strength data was investigated by plotting on the normal statistical plot. The use of normal statistical plot to express a stress-strength inference model graphically was investigated. The following conclusions were obtained.

(1) The tensile strength, yield strength and elastic modulus decrease as the temperature increases. But the elongation decreases as the temperature increases below 650°C, but increases at 650–800°C. The EDX analysis showed that the precipitation of α phase chrome carbide ($Cr_{23}C_6$) at this temperature range was the cause of hardening and embrittlement of this material. As the temperature increases, the size of dimples increases.
(2) The tensile strength and the yield strength at elevated temperatures vary according to a normal distribution. It can be confirmed on the normal statistical plot. The normal statistical plot can be usefully utilized for statistical design just by picking points. It gives various information such as the average, the standard deviation, the safe design area, the allowable stress for the material, etc.

References

1. J. Chen, B. Young, Stress-strain curves for stainless steel at elevated temperatures. J. Eng. Struct. **28**, 229–239 (2006)
2. AS/NZS 4673, Cold-formed stainless steel structures (Sydney, 2001)
3. EC3. Eurocode 3, *Design of Steel Structures – Part 1.4: General Rules – Supplementary Rules for Stainless Steels* (European Committee for Standardization, Brussels, 1996)
4. SEI/ASCE-8-02, Specification for the design of cold-formed stainless steel structural members (Reston, VA, 2002)
5. K.C. Kapur, L.R. Lamberson, *Reliability in Engineering Design* (John Wiley & Sons, New York, NY, 1977)
6. J.H. Song, J.H. Park, *Introduction to Reliability Engineering* (Intervision, Korea, 2007)
7. MIL-HDBK-5H, *Metallic Materials and Elements for Aerospace Vehicle Structures* (DOD, Washington, DC, 1998)
8. ASTM E8, *Standard Test Methods for Tension Testing of Metallic Materials* (ASTM International, West Conshohocken, PA, 2004)
9. STM E21, *Standard Test Methods for Elevated Temperature Tension Tests of Metallic Materials* (ASTM International, West Conshohocken, PA, 2005)
10. F.N. Rhines, P.J. Wray, Trans. Am. Soc. Met. **54**, 117 (1961)
11. V.K. Sikka, R.W. Swindeman, C.R. Brinkman, Proceedings of the 4th International Conference on Fracture, vol 2 (University of Waterloo Press, Waterloo, Ontario, 1977), p. 561
12. S.L. Mannan, K.G. Samuel, P. Rodriguez, Influence of temperature and grain size on the tensile ductility of AISI 316 stainless steel. Mater. Sci. Eng. **68**, 143–149 (1984)

Interaction Between Concrete Cylinders and Shape-Memory Wires in the Achievement of Active Confinement

Jean-François Destrebecq and Xavier Balandraud

Abbreviations

A	austenite phase
M	martensite phase
M_i	martensite *i*-variant
As	austenite start transformation temperature
Af	austenite finish transformation temperature
Ms	martensite start transformation temperature
Mf	martensite finish transformation temperature
SMA(s)	shape memory alloy(s)
Ni-Ti	nickel-titanium

1 Introduction

The mechanical behaviour of structural concrete is governed by a process of damage which results from the progress of macro or microcracks in the loaded material [1]. The damaging process can be delayed by applying a uni- or multiaxial compression in order to counterbalance local tensile stresses in the material. This can be achieved by means of passive or active reinforcement, external or internal to the concrete component.

Shape memory alloys (SMAs) are known as smart materials due to their peculiar thermomechanical properties, like the pseudoelasticity and the memory effect [2]. These properties take their origin from an austenite-to-martensite phase transition occurring when stress or temperature changes. In particular, SMAs have the ability to "freeze" the deformation in the martensitic state, then to recover their initial shape in the austenitic state.

J.-F. Destrebecq (✉)
Clermont Université, Université Blaise Pascal, EA 3867, Laboratoire de Mécanique et Ingénieries, BP 10448, F-63000, Clermont-Ferrand, France
e-mail: J-Francois.Destrebecq@univ-bpclermont.fr

A. Öchsner et al. (eds.), *Materials with Complex Behaviour*, Advanced Structured Materials 6, DOI 10.1007/978-3-642-12667-3_2, © Springer-Verlag Berlin Heidelberg 2010

Several reported experiments confirmed the feasibility of creating stresses in concrete components using SMA wires as external or embedded active reinforcement [3–5]. Also, recovery forces resulting from the activation of shape memory effect in SMA rods were used to reduce crack opening in a damaged structure [6]. In all cases, the prestressing was achieved by thermal activation of the shape memory effect in the SMAs. Difficulties of achieving permanent prestressing in the concrete specimens were reported, due to unwanted stress loss upon cooling down of the SMA wires [3]. According to Motavalli et al. [7], the loss is most likely caused by improper choice of the SMA transformation temperatures. Another potential reason is the stress induced austenite-to-martensite phase transition of the SMA [8].

Recently reported experiments showed that SMAs could be used to create confinement stresses in concrete cylinders in order to improve their performances in uniaxial compression. Choi et al. [9] used austenitic and martensitic SMA wires as external active reinforcement. The existence of a confinement effect was estimated indirectly by looking at the concrete behaviour during crush tests. Cylinders wrapped with prestrained martensitic wires showed an increase in their ductility and ultimate compressive strength. On the contrary, cylinders wrapped with austenitic wires at ambient temperature showed a ductility increase without increase in the ultimate strength. The authors concluded in the existence of confinement stresses due to the shape memory effect in the martensitic wires. This conclusion agreed with experimental results previously reported by Krstulovic-Opara and Thiedeman [10] who observed a substantial increase in the ultimate compressive strength of plain concrete cylinders wrapped with SMA wires.

Attempts were done of estimating the progressive stressing of small-scale SMA/concrete specimens caused by the shape recovery process in the SMA [11, 12]. When the shape memory effect was activated by the heating of the specimen, the prestressing effects were usually estimated by comparing strain measurements in the SMA/concrete specimen with strain measurements in a plain concrete companion specimen. By means of this method, compressive prestressing up to 6 MPa has been estimated in mortar samples reinforced with SMA short fibres [12].

The aim of the present work is to investigate the process of confinement of concrete cylinders combined with wrapped SMA wires used as external active reinforcement. The objective is to get a precise description of the mechanisms involved in the development of the confinement stresses due to the thermal activation of the shape memory effect in the SMA. For this purpose, preliminary tests are carried out in order to precisely determine thermomechanical properties of the SMA wire used in the study. The experimental procedure consists in wrapping SMA wires in different martensitic states around a series of concrete cylinders. The circumferential strains in the cylinders are monitored during the thermal activation of the memory effect in the SMA in order to estimate the confinement stress state induced in the concrete by the martensite-to-austenite transformation of the SMA. The transformation temperatures have been chosen in order to highlight the interaction between temperature and recovery stress in the SMA during the transformation process, leading in some cases to partial stress loss due to austenite-to-martensite phase transition of the SMA. The value of the prestrain of the martensitic wire is a key point in the resulting amplitude of the shape memory effect. Having in mind that the wrapping of the wire on the cylinder induces additional strains and stresses in the wire, valuable information is expected from the tests about the influence of the wire curvature on the effective confinement effect.

2 Creating Forces with SMA

Depending on the thermomechanical loading, a given SMA can exhibit different mechanical behaviours. The present section briefly describes the physical phenomena which explain how a SMA component may be used to create large forces taking advantage of the shape memory effect.

The macroscopic properties of SMAs take their origin from a solid-solid phase transition mechanism occurring when stress or temperature changes [2]. The parent phase is called *austenite* (referred to as A below). It has a bold-centered-cubic structure in all known SMAs. The product phase is called *martensite* (referred to as M below). It exhibits different types of structures which mainly depend on the composition. Due to the symmetry breaking involved in the transformation, the martensite phase exhibits multiple *variants* (referred to as M_i below), which correspond to the same crystal with different orientations in space with respect to the parent phase. The atom displacements involved in a transformation $A\rightarrow M_i$ lead to a so-called transformation strain. When the different variants of martensite exist in equal proportion, the macroscopic strain is nearly equal to zero, and the martensite is said to be "self-accommodated" [13].

Figure 1a, b present the phase diagram indicating the material state as a function of temperature and stress. The material can be either purely austenite, or purely martensite (with various proportions of martensite variants), or a mixture of austenite and martensite. Let us describe the evolution of the material state as a function of the thermomechanical loading.

Starting from austenite, the martensite phase can be obtained by decreasing the temperature or by increasing the stress. Figure 1a shows the two lines in the temperature-stress plane which correspond to the beginning and the end of the transformation $A\rightarrow M$:

- The temperature at which the transformation $A\rightarrow M$ starts by cooling is named Ms. The Mf value is the temperature at which the transformation $A\rightarrow M$ finishes. At zero stress, a self-accommodated martensite is obtained, leading to no change in the specimen shape between the austenite state and the martensite state.
- Martensite appears also when stress increases at a given temperature. In this case, the martensite obtained is not self-accommodated. As a result, large strains are obtained.

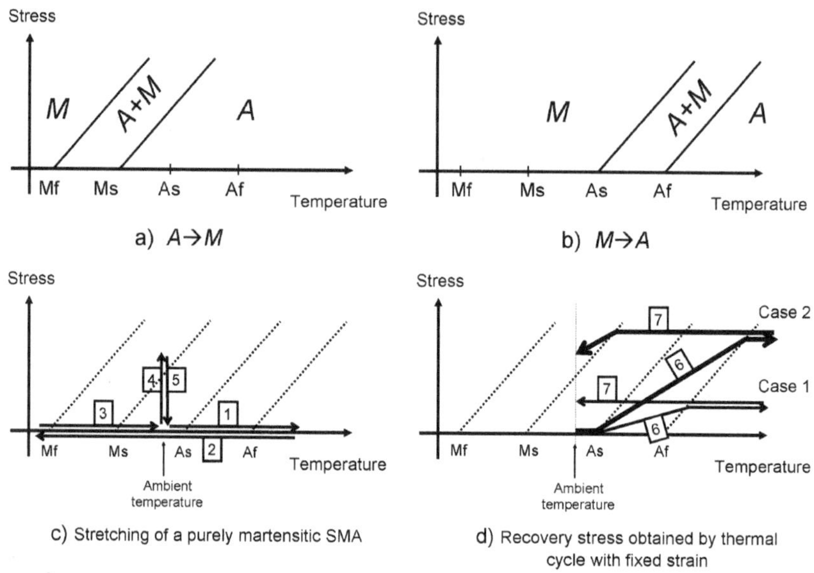

Fig. 1 (**a**) and (**b**) Phase diagrams in the temperature-stress plane. (**c**) and (**d**) Thermomechanical loading steps to create forces in a SMA specimen

Starting now from the martensite phase, changes in the material state can be obtained by increasing the temperature or by stretching the specimen (see Fig. 1b):

- A martensitic specimen returns to austenite state by increasing the temperature. At zero stress, the temperature at which the transformation *M*→ *A* starts is named As. The Af value is the temperature at which the transformation *M*→*A* finishes.
- The stretching of a purely martensitic specimen at a given temperature leads to a change in its proportions of the martensite variants. The martensite is said to be "reoriented" and the deformation is preserved upon unloading. As a result, different shapes of a given martensitic specimen can be obtained at zero stress.

For Ni-Ti (nickel-titanium) SMAs, it is possible to have Ms < As as shown in Fig. 1 [2]. Let us describe now the capability of creating forces with a Ni-Ti specimen in the case of an ambient temperature between Ms and As. Fig. 1c, d illustrate the thermomechanical loading.

First stage: the objective is to stretch a purely martensitic specimen in order to create a residual strain (prestrain). The following procedure enables to precisely define the level of martensite reorientation (see Fig. 1c):

- step 1: increase the temperature to a value higher than Af in order to obtain a purely austenite specimen.
- step 2: decrease the temperature to a value lower than Mf in order to obtain a purely self-accommodated martensite specimen.

- step 3: return to ambient temperature. The specimen keeps its self-accommodated martensite state (since the ambient temperature is lower than As). No change has occurred in the specimen shape since step 1.
- step 4: apply a mechanical load. The martensite is then reoriented. The shape of the specimen is modified during this loading.
- step 5: return to zero stress. The martensite keeps its reorientation level reached at the end of the previous step. The actual shape of the specimen is different from the initial shape, due to the prestrain caused by the martensite reorientation.

Second stage: the objective is to create active forces at ambient temperature. Starting from step 5, the following procedure is applied (see Fig. 1d):

- step 6: increase the temperature in order to return an austenite state while the strain is fixed. The return to austenite leads to stresses, as the specimen would like to recover its initial shape [14]. Indeed, the free-stress state of the austenite corresponds to a strain equal to zero. As the strain is kept constant to a non-zero value, stresses are created in the specimen. Note that the higher the imposed prestrain obtained by reorientation, the higher the stress obtained.
- step 7: return to ambient temperature. Two cases must be distinguished:

- Case 1: during the temperature decrease, no martensite is created. The corresponding final point in the temperature-stress plane is on the right-hand side of the transformation starting line $A \rightarrow M$ (line including the Ms point). As a consequence, the stress obtained at the end of step 6 is retained at the end of the procedure.
- Case 2: during the temperature decrease, the corresponding point in the temperature-stress plane crosses the transformation starting line $A \rightarrow M$, which means that martensite is created. As a consequence, a decrease in stress is expected because the material optimizes the proportions of martensite variants to relax the stresses in the specimen.

It can be noted that the classic effect of thermal expansion has not been included in this presentation for the sake of simplicity. The next section describes the experimental set-up to perform such a phenomenon with SMA wires wrapped around concrete cylinders.

3 Experimental Set-Up

3.1 Materials

A nickel-titanium ($Ni_{50.8}$-$Ti_{49.2}$, at.%) wire 1 mm in diameter was used in the present study. The choice was motivated by the performance of this type of SMA regarding memory effects. The transformation temperatures were chosen in order to have the austenite-start temperature at about 23°C. The effective transformation temperatures were found as follows:

- martensite to austenite : As = 23°C and Af = 28°C,
- austenite to martensite: Ms = –10°C and Mf = –25°C.

It can be noted that Ms < 20°C < As. This constraint is specially required for the present application involving an ambient temperature at about 20°C. This was achieved by the supplier thanks to a specific thermal treatment.

A series of concrete cylinders were prepared for the tests, 74.4 mm in diameter and 180 mm in height. The moulds were precisely cut from a polyvinyl chloride (PVC) plastic pipe, such that their inside surface was smooth and they were all very similar in geometry. The concrete mix was designed, aiming at a compressive strength about 25 MPa at 28 days. The proportions used for the concrete preparation were as follows (in kg/m^3): gravel 1039.2, sand 782.7, cement 324.5, and water 176.1. Owing to the size of the samples, the maximum size of the coarse aggregate (gravel) was limited to 12.5 mm.

For every cylinder, two pairs of metallic thread inserts were fixed inside the mould before cast, in such a way to have them embedded in the concrete after hardening (see Fig. 2). For each pair, the two inserts were positioned diametrically opposite to each other at a distance of 20 mm from the nearest cylinder end (top and bottom). The inserts were aimed at fixing the SMA wire ends on the cylinder without damaging the concrete.

After cast, the cylinders were kept in the moulds and sealed with plastic foils in order to avoid moisture loss. Foils and moulds were removed after 28 days of hardening. Then the cylinders were kept in ambient in-door atmosphere until testing.

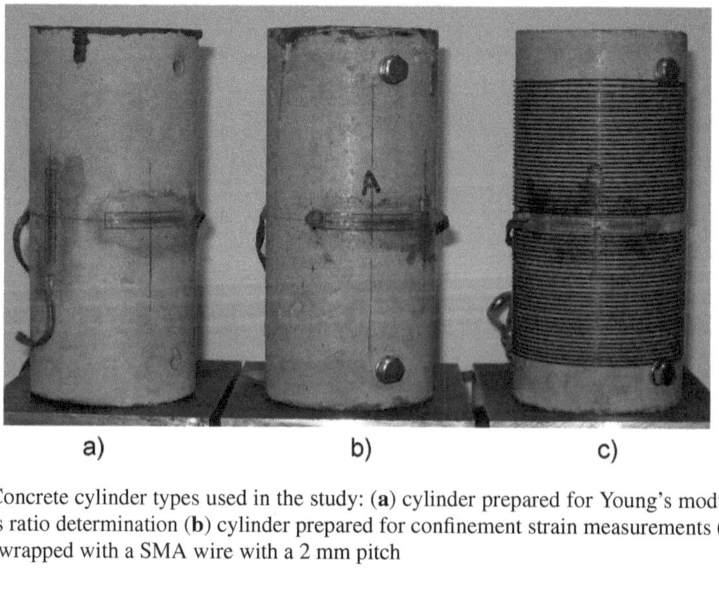

a) b) c)

Fig. 2 Concrete cylinder types used in the study: (**a**) cylinder prepared for Young's modulus and Poisson's ratio determination (**b**) cylinder prepared for confinement strain measurements (**c**) same cylinder wrapped with a SMA wire with a 2 mm pitch

3.2 Specimen Preparation

Four cylinders were prepared from the same batch:

- One of them was aimed at measuring the mechanical properties of the concrete (Fig. 2a). It was equipped with two pairs of strain gauges. They were bonded onto the concrete surface at mid-height of the cylinder: two gauges were bonded opposite to each other parallel to its longitudinal axis; the other two were bonded opposite to each other along its mid-height circumference.
- The other three were prepared for the active confinement tests (Fig. 2b). Each cylinder was equipped with two strain gauges bonded opposite to each other along its mid-height circumference. The gauges were aimed at measuring the circumferential strain at the concrete surface of the cylinder in order to estimate the confinement stress induced in the concrete by the SMA wire during the test.

All gauges were 30 mm in length. According to the supplier specifications, the operational temperature is 80°C for the gauges and the connecting leads, 120°C for the adhesive.

In order to get a close to uniform confinement stress over the cylinder lateral surface, it was felt necessary to wrap the wire around the cylinder with a small constant pitch. Preliminary calculations showed that the choice of a 1 mm diameter wire with a constant 2 mm pitch would be a satisfying compromise. Owing to the cylinder dimensions, the total wire length required for wrapping one cylinder between the two thread inserts was about 19.9 m.

The preparation of each specimen consisted in a sequence of operations in order to prepare the SMA wire and to wrap it around the concrete cylinder under controlled temperature conditions:

- transforming the SMA wire in the self-accommodated martensite state – this was done by heating the wire up at a temperature over Af, then cooling it down at a temperature under Mf (Fig. 1c, steps 1 and 2);
- stretching the SMA wire at ambient temperature (under As) to obtain a specified prestrain in the reoriented martensite state (Fig. 1c, steps 4 and 5)
- wrapping the SMA wire around the concrete cylinder at ambient temperature (under As) with a constant pitch equal to 2 mm.

Practically, each SMA wire was coiled round a 350 mm diameter drum and kept in an oven at a temperature of 60°C for 30 min. Then, it was cooled down and kept in a freezer at a temperature of –30°C. After that, the wire was ready for stretching at ambient temperature. Despite its length (19.9 m), it was decided to stretch each wire as a whole in order to ensure a uniform prestrain. This was done by setting the wire horizontally and pulling its other end with a jack to a prescribed value. The vertical sag of the wire was taken into account for this operation. After stretching, the wire was wrapped around the concrete cylinder using a device specially designed to ensure a constant 2 mm pitch (Fig. 2c). All the operations were done at an ambient temperature between 18 and 21°C.

3.3 Thermal Loading and Strain Measurement

Thermal activation of the shape memory effect in the SMA wire was achieved by heating the wrapped cylinders at a temperature above Af (i.e. the $M \rightarrow A$ final transformation temperature).

For this purpose, the specimens were placed in an oven. The temperature was increased at a constant rate of 1.6°C/min up to 60°C, and then decreased to 23°C at a rate about –1.2°C/min. The total test duration was about 75 min. Values given by the circumferential strain gauges were monitored and recorded during the whole test for each cylinder.

The experiments were organized in two stages. In the former stage, the plain cylinders (i.e. without SMA wires) were tested under the conditions described above in order to record the influence of the thermal loading on the gauges and the concrete strains. In the later stage, the wrapped cylinders were tested under the same thermal conditions. For each specimen, the difference between the strain measurements obtained for the wrapped and the plain cylinder was assumed to reflect the net deformation of the concrete caused by the activation of the shape memory effect in the SMA, i.e. the net strains in the wrapped cylinder freed from the influence of temperature on the concrete and the gauge measurements.

4 Preliminary Tests

4.1 Determination of Concrete Characteristics

The mechanical characteristics required for the concrete in the present study are its Young's modulus E_c and its Poisson's ratio ν. For this purpose, compression tests were carried out 105 days after cast on the cylinder equipped with four strain gauges (Fig. 2a). The compressive stress was applied to the concrete in the range 0–2.5 MPa, which corresponds to the stress range expected in the wrapped cylinders during the confinement tests. Average values of the Young's modulus and the Poisson's ratio are calculated from the measured strains. The average values obtained are as follows: $E_c = 25.2$ GPa and $\nu = 0.193$.

4.2 Determination of SMA Wire Characteristics

The different steps described in Sect. 2 to create forces in SMAs were experimentally performed on a Ni-Ti wire coupon 150 mm in length. The aim was to determine the relationship between the tensile prestrain in the SMA wire in the martensite state and the tensile stress obtained in the wire after thermal activation of the shape recovery. Some points are here detailed:

- The ambient temperature was the same for all tests: $19 \pm 1.5°C$.
- The loading and unloading (steps 4 and 5) were applied with a MTS uniaxial testing machine. Tests differ in the stretching amplitude imposed to the wire at the end of step 4.

Figure 3 presents a typical strain-stress curve, corresponding here to the case 2 illustrated in Fig. 1d. Three quantities are defined:

- ε_{mar} is the strain obtained at the end of step 5, which is kept constant during the final thermal loading (steps 6 and 7),
- σ_{aus} is the maximum stress obtained in the austenite state (end of step 6),
- σ_{rec} is the residual stress so-called "recovery stress" at the end of the procedure.

Figure 4 presents the values obtained for σ_{aus} and σ_{rec} as a function of ε_{mar}. As predicted from Fig. 1d, the higher ε_{mar}, the higher σ_{aus}. It is also confirmed that high values of ε_{mar} lead to recovery stress values σ_{rec} lower than σ_{aus} (see case 2 in Fig. 1d). Indeed, σ_{rec} shows a constant value (205.3 MPa) for $\varepsilon_{mar} \geq 5.79 \cdot 10^{-3}$.

Finally, the Young's modulus of the austenite phase needs to be determined. To be sure that the tested wire coupon would be in the austenite state, it was first heated to a temperature higher than Af to obtain a purely austenite state. The return to room temperature did not produce martensite because the room temperature was higher than Ms. Then, the austenitic wire coupon was tensioned at ambient temperature. Finally, the Young's modulus of the SMA wire in the austenite state was determined by comparing the applied tensile stress with the longitudinal strain measured during the test. It was found to be equal to 62.9 GPa.

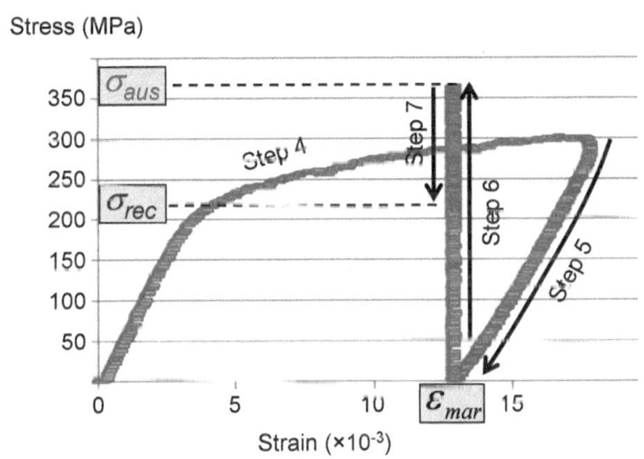

Fig. 3 Preliminary tests on Ni-Ti wires: typical strain-stress curve corresponding to the case 2 of Fig. 1d

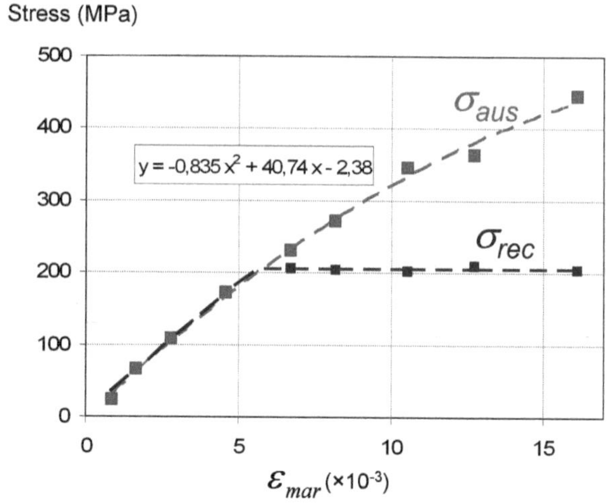

Fig. 4 Preliminary tests on Ni-Ti wires: results obtained as a function of ε_{mar} (see Fig. 3 for the meaning of ε_{mar}, σ_{aus} and σ_{rec})

5 Testing of Wrapped Cylinders

5.1 Characteristics of Tested Specimens

Two series of three specimens were prepared according to the procedure described in Sect. 3.2. The specimens are called C1 to C6, by order of increasing prestrain in the SMA wire. Data concerning the prestrain ε_{mar} and the corresponding recovery stresses σ_{aus} and σ_{rec} are listed in Table 1 for the six specimens. The letter (a) or (b) printed after every specimen name indicates which series the specimen belongs to.

Table 1 Effective prestrains (ε_{mar}) and recovery stresses (σ_{aus} at high temperature, σ_{rec} at ambient temperature) in the SMA wires for the tested specimens

	C1 (b)	C2 (a)	C3 (a)	C4 (b)	C5 (a)	C6 (b)
ε_{mar} (10^{-3})	2.08	3.76	4.95	6.22	8.84	10.71
σ_{aus} (N/mm^2)	78.8	139.1	178.9	218.9	292.4	338.2
σ_{rec} (N/mm^2)	78.8	139.1	178.9	205.3	205.3	205.3

5.2 Test Results

The cylinders were tested after ageing between 100 and 105 days after cast. All three specimens of the same series were placed together in the oven in order to achieve the prestressing by thermal activation of the memory shape effect in the SMA. As explained in Sect. 3.3, the net circumferential strains were obtained for each cylinder by correcting the values given by the gauges by the corresponding values previously recorded for the same cylinder previously tested without SMA wire (plain cylinder).

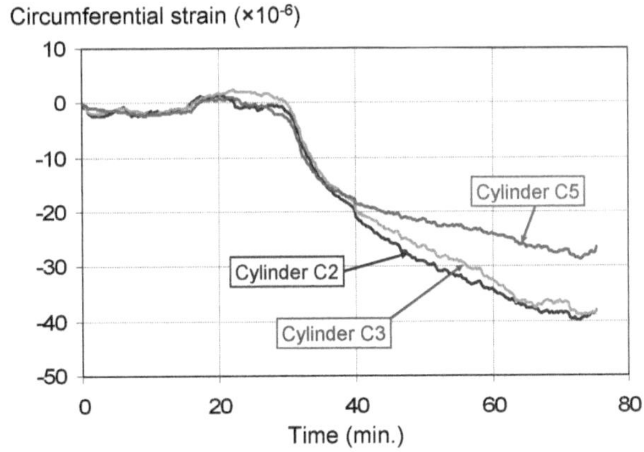

Fig. 5 Development of circumferential strains in cylinders C2, C3 and C5 (series a) during the thermal cycle

As an example, the net circumferential strains obtained for specimens C2, C3 and C5 (series a) are shown in Fig. 5. The competition between the shape recovery of the SMA wire due to the memory effect (transformation to austenite), and the thermal expansion in the wire and concrete materials results in small circumferential strains during the heating phase of the test, i.e. during the first 30 min of the thermal cycle. A change is observed when the temperature begins to decrease. The progressive development of a negative circumferential strain in the concrete observed in Fig. 5 shows the existence of a confinement effect caused by the developing recovery stress in the SMA wire. The strain evolution is pretty similar for the three samples, until the temperature reaches 50°C (i.e. 40 min after test beginning). Under 50°C, the circumferential concrete strains evolve separately in the three specimens. Finally, they stabilize when the temperature reaches about 23°C.

It can be observed that the final values of the circumferential strains are similar for specimens C2 and C3. A smaller strain value is obtained for specimen C5. This trend is in contradiction with the values expected for the recovery stress σ_{rec} in the three specimens (see Table 1): according to the respective prestrain ε_{mar} in the wire, the largest circumferential strain would be expected in specimen C5 and the smallest in specimen C2, with an intermediate value in specimen C3. The same trend is observed in series b (not presented here). This point will be discussed in Sect. 6.

5.3 Post-processing of Strain Gauge Measurements

Based on the hypothesis of isotropic plane state of stresses, the circumferential strain ε_{c0} caused by a uniform confinement stress σ_{cc} applied to a plain cylinder can be expressed as follows:

$$\varepsilon_{c0} = (1 - \nu)\frac{\sigma_{cc}}{E_c}. \quad (1)$$

There is no SMA wire wrapped within the zone where the circumferential gauges are bonded to the concrete at the mid-height of the cylinder (see Fig. 2c). As a consequence, no confinement force is applied to the concrete, leading to lower concrete strains in this zone. Thus, the circumferential strain ε_{cc} to be expected at mid-height of a wrapped cylinder can be expressed as follows:

$$\varepsilon_{cc} = \frac{\varepsilon_{c0}}{\xi}, \tag{2}$$

where ξ is a correction coefficient ($\xi > 1$). A finite element analysis was carried out under the hypothesis of axi-symmetry in order to determine the appropriate value for ξ. Precise values were considered for the geometry and the mechanical characteristics of the wrapped cylinder. A value of 1.464 was obtained for the correction coefficient ξ.

From Eqs. (1) and (2), the confinement stress σ_{cc} corresponding to the circumferential concrete strain ε_{cc} measured at mid-height of a wrapped cylinder can be expressed as follows:

$$\sigma_{cc} = \xi \frac{E_c \varepsilon_{cc}}{(1-\nu)}. \tag{3}$$

Based on local static equilibrium in the wrapped zone of the cylinder, the expression of the tensile force F in the SMA wire corresponding to the confinement stress can be easily derived:

$$F = \frac{pD}{2} \sigma_{cc}, \tag{4}$$

where p is the pitch and D is the diameter of the concrete cylinder.

The effective tensile stress σ_{SMA} in the SMA wire is obtained by dividing the force F by the cross-section area A_W of the SMA wire with account of Eq. (3), hence:

$$\sigma_{SMA} = \frac{\xi pD}{2(1-\nu)A_W} E_c \varepsilon_{cc}. \tag{5}$$

Finally, the expected confinement stress $\sigma_{cc,exp}$ induced by the recovery stress σ_{rec} given in Table 1 for every tested specimen, is easily derived taking $F = A_W \sigma_{rec}$ in Eq. (4):

$$\sigma_{cc,exp} = \frac{2A_W}{pD} \sigma_{rec}. \tag{6}$$

Equations (3), (5) and (6) will allow us to derive the effective confinement stress σ_{cc} in the concrete and the effective tensile stress σ_{SMA} in the SMA wire from the values of the circumferential strain ε_{cc} measured during the thermal cycle, and to estimate the expected confinement stress $\sigma_{cc,exp}$ in terms of the expected recovery stress for each tested specimen.

5.4 Effective Confinement Stress

The values obtained for the net circumferential concrete strains ε_{cc} after thermal activation of the SMA memory effect and cooling down to 23°C are listed in Table 2 for the six tested specimens. The corresponding values of the confinement stresses σ_{cc} in the concrete are processed from the measured circumferential strain values ε_{cc} by means of Eq. (3). The expected values of the confinement stresses $\sigma_{cc,exp}$ are derived from Eq. (6) for the expected values of the recovery stress σ_{rec} given in Table 1. In order to compare the expected values with the effective values of the confinement stress derived from tests, the $\sigma_{cc}/\sigma_{cc,exp}$ ratio was calculated and is plotted in Fig. 6 in terms of the initial prestrain ε_{mar} in the SMA wire.

Surprisingly, the confinement effect is much higher than expected ($\sigma_{cc}/\sigma_{cc,exp} > 1$) when the initial prestrain is low. On the contrary, the confinement effect is much lower than expected ($\sigma_{cc}/\sigma_{cc,exp} < 1$) when the prestrain in the SMA wire is high. In between, the $\sigma_{cc}/\sigma_{cc,exp}$ ratio sharply decreases when the SMA prestrain is

Table 2 Results for the tested specimens

	C1 (b)	C2 (a)	C3 (a)	C4 (b)	C5 (a)	C6 (b)
ε_{cc} (10^{-6})	–39.4	–39.8	–39.0	–41.7	–28.6	–24.9
σ_{cc} (N/mm^2)	–1.80	–1.82	–1.78	–1.91	–1.31	–1.14
$\sigma_{cc,exp}$ (N/mm^2)	–0.83	–1.47	–1.89	–2.17	–2.17	–2.17
σ_{SMA} (N/mm^2)	170.6	172.4	168.9	180.6	123.9	107.8
$\sigma_{cc}/\sigma_{cc,exp}$ (–)	2.16	1.24	0.94	0.88	0.60	0.53

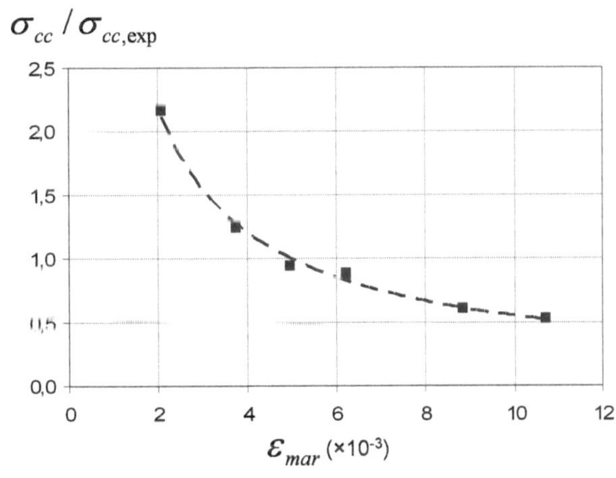

Fig. 6 Effective to expected confinement stresses ratio in terms of the wire prestrain

increased. The balance ($\sigma_{cc}/\sigma_{cc,exp} = 1$) is obtained for prestrain in the wire of about $5{\cdot}10^{-3}$. This result highlights the complex interaction between cylinder and wire responses. It can be concluded that the confinement effect in the cylinder is not resulting only from the recovery of the tensile prestrain in the wire. An additional phenomenon should be considered involving a more complex interaction between the SMA wire and the concrete cylinder. In the next section, the hypothesis of an influence of the curvature caused by the wrapping of the wire around the cylinder is investigated.

6 Effect of the Curvature on the Recovery Stress

It is obvious that the effective tensile stresses σ_{SMA} obtained in the wrapped SMA wires (see Table 2) are in no agreement with the recovery stresses σ_{rec} expected from the prestrains ε_{mar} produced by stretching of the martensitic wires before wrapping (Table 1). The two sets of values are plotted in Fig. 7 in terms of ε_{mar}. On the left hand side of the figure ('low' ε_{mar} values), the effective tensile stress σ_{SMA} has a constant value, whereas the expected values σ_{rec} increase with increasing values of ε_{mar}. On the right hand side of the figure ('high' ε_{mar} values), the effective tensile stress σ_{SMA} decreases, with a maximum 'loss' about 37%.

Based on these observations, the issue of an additional interaction phenomenon, due to the effect of the wrapping of the wire on the cylinder can be addressed. As a matter of fact, the wrapping induces an additional deformation of the wire in order to fit the geometry of the cylinder. Indeed, the additional deformation corresponds

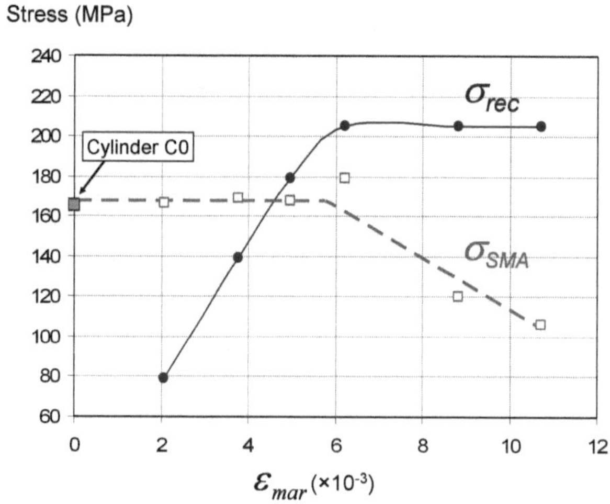

Fig. 7 Comparison between the expected recovery stress and the effective residual stress in the wire in terms of the wire prestrain

to a bending curvature with a fixed value. Taking the cylinder and the wire radii into account, the curvature is found equal to 26.5 m^{-1}. The corresponding maximum strain in the martensitic wire is equal to the curvature multiplied by the wire radius. It is found equal to 13.3 10^{-3}, which is higher than the highest value of the prestrain ε_{mar} in the martensitic wire (see Table 1). For low ε_{mar} values, the resulting strains in the martensitic wire are dominated by the curvature effect, leading to about constant effective stresses in the austenitic wire after the thermal cycle. For high ε_{mar} values, the strains and stresses induced by the curvature combine with the tensile prestrains, leading to decreasing effective stresses in the SMA wire. The fact that the effective values of the wire stresses are lower than the maximum value (205.3 MPa) obtained for the recovery stress can be a result of the gradient in the additional strain through the wire diameter.

Finally, in order to confirm the influence of the wire curvature on the confinement effect, one additional test was carried out with a cylinder wrapped with a non-stretched martensitic wire, i.e. $\varepsilon_{mar} = 0$ and $\sigma_{rec} = 0$. The final effective confinement stress obtained in the concrete was found equal to 1.77 MPa, which is pretty close to the values obtained for specimens C1 to C3 (see Table 2). The effective stress in the SMA wire was found equal to 167.6 MPa. This value is plotted in Fig. 7 (cylinder C0). It can be concluded that the confinement effect is significantly modified by the curvature strains due to the wrapping of the SMA wire on the cylinder.

7 Conclusion

The confinement of concrete cylinders wrapped with Ni-Ti SMA wires was investigated. The main results are as follows:

1. An active confinement at ambient temperature is possible using a SMA wire with suitable transformation temperatures. The martensite start (Ms) and the austenite start (As) temperatures must be chosen to be lower and higher respectively than the ambient temperature. So it is possible to stretch the wires in their martensitic state before wrapping around the cylinders. The recovery forces are developed by heating and they are retained when returning to ambient temperature.
2. The influence of the initial stretch of the wires on the final confinement is evidenced. Surprisingly, it is observed that the confinement level is constant for low-stretched wires and decrease for higher values of wire stretch. This behaviour is not predictable from results obtained with straight wire coupons tested under uniaxial tension.
3. A complex interaction between cylinder and wire responses may explain this result. An additional test with a non-previously stretched wire shows that an active confinement is also obtained. This result illustrates that the curvature due to the wrapping around the cylinders has an effect on the confinement.

References

1. G. Pijaudier-Cabot, J.-M. Reynouard, J.-M. Torrenti, *Mechanical Behavior of Concrete* (ISTE & Wiley, London, 2010)
2. K. Otsuka, C.M. Wayman, Shape Memory Materials (Cambridge University Press, Cambridge, 1999)
3. S. El-Tawil, J. Ortega-Rosales, Prestressing concrete using shape memory alloy tendons. ACI Struct. J. **101**, 846–851 (2004)
4. Z. Deng, Q. Li, H. Sun, Behavior of concrete beam with embedded shape memory alloy wires. Eng. Struct. **28**, 1691–1697 (2006)
5. H. Li, Z. Liu, J. Ou, Behavior of a simple concrete beam driven by shape memory alloy wires. Smart Mater. Struct. **15**, 1039–1046 (2006)
6. P. Soroushian, K. Ostawari, A. Nossoni, H. Chowdhury, Repair and strengthening of concrete structures through application of corrective posttensionning forces with shape memory alloys. Transp. Res. Rec. **1770**, 20–26 (2001)
7. M. Motavalli, C. Czaderski, A. Bergamini, L. Janke, Shape Memory Alloys for Civil Engineering Structures – On the Way from Vision to Reality. AMCM'2008 Lodz, Poland, pp. 91–108
8. L. Janke, C. Czaderski, M. Motavalli, J. Ruth, Applications of shape memory alloys in civil engineering structures: overviews, limits and new ideas. Mater. Struct. **38**, 578–592 (2005)
9. E. Choi, Y.S. Chung, B.S. Cho, T.H. Nam, Confining concrete cylinders using shape memory alloy wires. Eur. Phys. J. **158**, 255–259 (2008)
10. N. Krstulovic-Opara, P.D. Thiedeman, Active confinement of concrete members with self-stressing composites. ACI Mater. J. **97**, 297–308 (2000)
11. N. Krstulovic-Opara, A.E. Naaman, Self-stressing fiber composites. ACI Struct. J. **97**, 335–345 (2000)
12. K. Moser, A. Bergamini, R. Christen, C. Czaderski, Feasibility of concrete prestressed by shape memory alloy short fibers. Mater. Struct. **38**, 593–600 (2005)
13. K. Bhattacharya, Microstructure of Martensite: Why It Forms and How It Gives Rise to the Shape-Memory Effect (Oxford University Press, Oxford, 2003)
14. P.H. Lin, H. Tobushi, K. Tanaka, C. Lexcellent, A. Akai, Recovery stress of TiNi shape memory alloy under constant strain. Arch. Mech. **47**, 281–293 (1995)

Mathematic Modeling of the Osprey Process

M. Bodea, R. Muresan, and C.V. Prica

1 Introduction

Spray forming is a flexible process which can be used to manufacture a wide range of materials that are difficult to be produced by other methods such as: metal matrix composites (MMCs), Al–Si alloys or high speed steels with a relatively fine-scale microstructure in large cross-sections [1].

The process benefits arise from the single-step operation of converting molten alloy directly into a semifinished product. The high solidification rate in spray forming promotes the microstructural refinement and allows to obtain a wide range of alloys free of macrosegregation. The range of products currently being manufactured covers a broad spectrum of shapes, alloys and markets, including round billets, tubes, rings, and clad rolls, for automotive, military, electronic and aerospace applications [1, 2].

Spray deposition involves atomization of a molten material by high velocity gas jets into a spray of micron-sized droplets which are subsequently propelled and deposited onto a substrate to produce preforms of desired shapes [3–5]. The main process variables which influence the thermal state of the spray during flight and the deposition stage are: the nozzle assembly, the atomization gas pressure, melt superheat and the nozzle to substrate distance [3, 4, 6].

In the past, many attempts have been made for modeling the various steps involved in spray forming [5–7] that have evolved today in complex studies trying to combine all aspects involved in the process. Some of them have tried to simplify the mathematic model [8] preserving however good correlations between the outcome and the experimental observations.

Meanwhile new technologies were developed using new materials and widening the range products. An example is the spray rolling which is a relatively new strip-casting technology, being developed, that combines elements of spray forming and twin-roll casting [4, 5].

M. Bodea (✉)
Technical University of Cluj, Cluj-Napoca, Romania
e-mail: mbodea@stm.utcluj.ro

A. Öchsner et al. (eds.), *Materials with Complex Behaviour*, Advanced Structured Materials 6, DOI 10.1007/978-3-642-12667-3_3, © Springer-Verlag Berlin Heidelberg 2010

This work tries to contribute to overall knowledge in the field taking in account the phenomena involved in the spray forming, starting with the melt flow study from the crucible through the nozzle. Then, the gas flow, droplets velocity and the heat transfer mechanisms are modeled using numerical methods implemented in integrated software developed by the authors, called MetLAB [9].

2 Model Formulation

2.1 Modeling of the Melt Flow Rate

The melt flow rate represents a key variable that has a direct influence on the atomized droplets and on the deposition process. The head loss was calculated using the Colebrook implicit equation [10] that was solved by Ridder's numerical method [11]. A software program called MetLAB [9] was realized for this study, and the iterations were performed tacking in consideration the variation of the metal head or liquid delivery nozzle diameter.

The exit liquid velocity through the delivery nozzle was calculated using the Bernoulli equation [12, 13], Eq. (1), for the system shown in Fig. 1:

$$z_1 + \frac{u_{m1}^2}{2\alpha_1 \cdot g} + \frac{p_1}{\gamma_m} = z_2 + \frac{u_{m2}^2}{2\alpha_2 \cdot g} + \frac{p_2}{\gamma_m} + \sum h_L \tag{1}$$

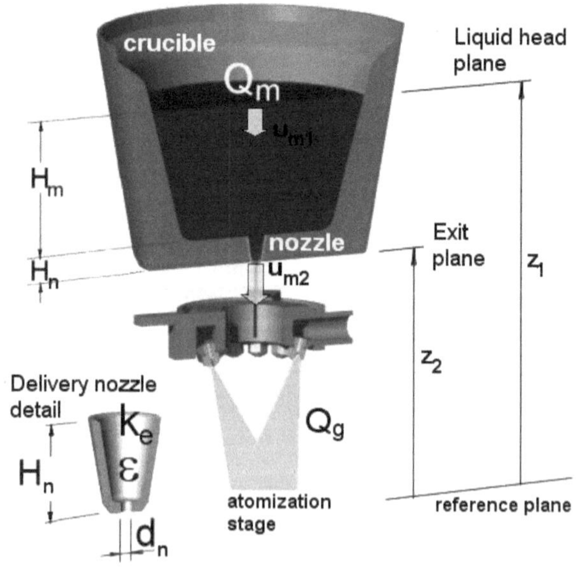

Fig. 1 Variables used in the melt flow rate mathematical model

where $z_{1,2}$ (m) represents the height of the liquid head from the exit section of the liquid delivery nozzle measured from a reference plane; $u_{m1,2}$ (m/s) represents the melt flow velocity related to the $z_{1,2}$ liquid head sections; $\alpha_{1,2}$ is the velocity distribution coefficient which is approximately 1 for turbulence flow and 0.5 for laminar flow [10]; $p_{1,2}$ (Pa) is the medium pressure related to $z_{1,2}$ liquid head sections and Σh_L(m) represents the total losses due to friction and section variations.

The energetic loss h_f (m) due to friction in the delivery nozzle is expressed by the Darcy–Weisbach formula [1, 13], Eq. (2), where H_n (m) stands for liquid nozzle height and d_n (m) for liquid nozzle diameter:

$$h_f = f \cdot \frac{H_n \cdot u_{m2}^2}{d_n \cdot 2g} \tag{2}$$

The friction factor *f* from Darcy–Weisbach formula, Eq. (2) could be determined using an empirical formula proposed by Colebrook [13] expresed in Eq. (3), where ε represents the rugosity factor of the delivery nozzle surface and *Re* is the Reynolds number.

$$\frac{1}{\sqrt{f}} = -2 \cdot \log\left(\frac{\varepsilon}{d_n \cdot 3.7} + \frac{2.51}{\mathrm{Re} \cdot \sqrt{f}}\right) \tag{3}$$

Because the Colebrook formula is an implicit equation, the solution could be found using a numerical method. In this paper we have used the Ridder's method [11] which was implemented in MetLAB program [9] to solve the Colebrook implicit equation.

The energetic loss h_e (m) due to the geometric changes at the entrance of delivery nozzle depends mainly on the nozzle entrance angle (Fig. 1), quantified by the entrance loss coefficient k_e as expressed in Eq. (4):

$$h_e = k_e \frac{u_{m2}^2}{2g} \tag{4}$$

The metal flow expression Q_m ($m^3 \cdot s^{-1}$) given in Eq. (5) is obtained from Eq. (1) after replacing the expressions: $(z_1 - z_2) = H_m + H_n$ and the melt stream velocity $u_{m1} = u_{m2} \cdot S_2 / S_1$ and neglecting the fraction $(1/S_1)^2$ because the crucible section S_1 is much larger than the exit section S_2 of the delivery nozzle.

$$Q_m = S_2 \cdot \sqrt{\frac{2g\left(H_m + H_n - \frac{\Delta p}{\gamma_m}\right)}{\left(f\frac{H_n}{d_n} + k_e + 1\right)}} \tag{5}$$

In Fig. 2 is shown the experimental and predicted values of the melt flow rates for different alloys and nozzle sizes. The height of metal head in crucible was maintained aproximately to 300 mm and the melt was overheated about 100°C above liquidus temperature.

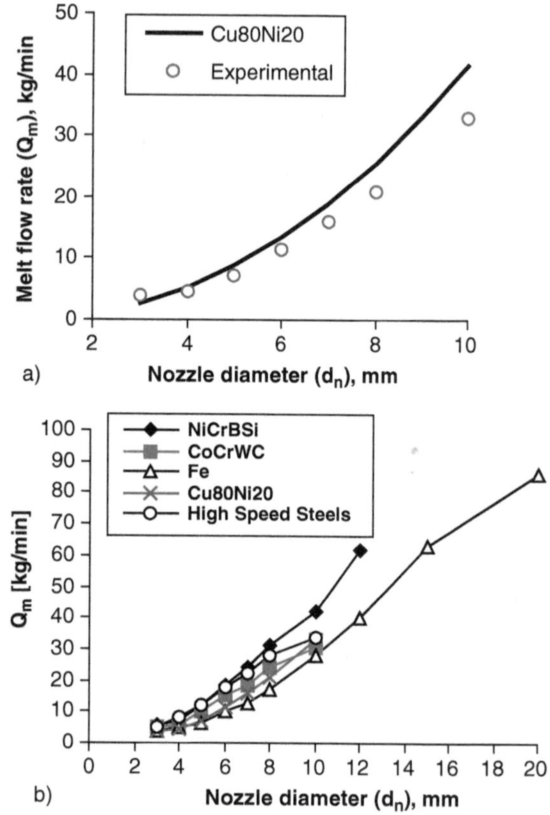

Fig. 2 Experimental and modeling of the melt flow for different alloys (**a**) Cu80Ni20 (**b**) others different alloys

It can be observed a resonable correlation between modeled and experimental results for the melt flow rate, especially for the smaller liquid delivery nozzle diameters. The deviation from the predicted values that occurs in case of large melt flow rates can be interpreted as a result of crucible metal head variation and also due to the incorrect assessment of the entrance loss coefficient, or relative roughness ε of the delivery nozzle.

2.2 Modeling of Gas Flow

The dynamics of the gas flow has direct influence on the first stage of the spray forming. The momentum and the heat transfer of the molten droplets will determine the microstructure and the preform properties. For that reason, the modeling of the gas flow represents also one of the key processes for optimization purposes.

The gas velocity is maximum at the exit section of the atomizer nozzle and is decreasing with the increasing distance Z (m) from the head atomizer, as expressed in Eq. (6), [6, 8].

$$\frac{u_g(Z)}{u_0} = \left[1 + \left(\frac{Z}{\lambda}\right)^{20}\right]^{-0.05} \tag{6}$$

$$\lambda = 7.414\sqrt{S_n} \tag{7}$$

where $u_g(Z)$ is the axial gas velocity (m·s^{-1}), u_o (m·s^{-1}) the initial gas velocity from the atomizer nozzle, Z (m) the axial distance from the atomizer, λ the exponential gas velocity decay, S_n (m^2) the exit area of the atomizer nozzle and r (m) the radial distance from the axe of the atomizer nozzle. The gas velocity in a point described by the coordinates (Z,r) is given by Eq. (8) [7]:

$$u_g(Z, r) = u_g(Z, 0) \cdot \frac{r}{\frac{0.004}{0.268} + Z} \tag{8}$$

The software program realized by the authors called MetLAB [9], was developed in Borland C++ Builder, could represent the 2D distribution of the gas velocity according to initial conditions: the gas properties, atomiser head geometry and the gas pressure, Fig. 3.

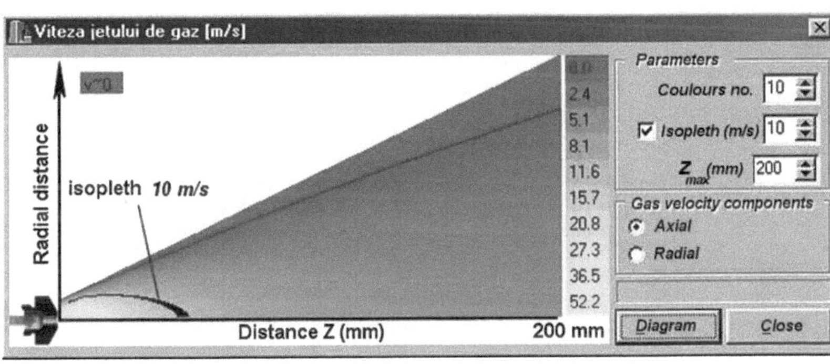

Fig. 3 Two-dimensional map of gas velocity definition in the MetLAB program

2.3 Modeling of the Droplets Velocity

The molten droplets and the ligaments after the secondary disintegration are assumed to have a spherical shape. These droplets are accelerated or decelerated during their motion into atomization chamber, accordingly to the 2nd Newton's law that has the form expressed in Eq. (9), [7, 6].

$$\frac{4}{3}\pi r_p^3 \rho_m \frac{d(u_p)}{d\tau} = \frac{4}{3}\pi r_p^3(\rho_m - \rho_g) - \frac{C_D \pi r_p^2}{2}\rho_g \left(u_p - u_g\right)\left|u_p - u_g\right| \tag{9}$$

where r_p(m) is the droplet radius, ρ_m, ρ_g (kg·m^{-3}) are the melt and the gas density, respectively u_p, u_g (m·s^{-1}) are the droplet and the gas velocity.

Equation (9) takes in consideration the inertial force, gravitational force and the droplet drag force that depends on the relative speed between the droplets and the surrounding gas, where the drag coefficient C_D is related to the Reynolds number by Eq. (10), which is defined in Eq. (11).

$$C_D = 0.28 + \frac{6\sqrt{\mathrm{Re}} + 21}{\mathrm{Re}} \tag{10}$$

$$\mathrm{Re} = \frac{\rho_g \cdot |u_p - u_g| \cdot d_p}{\eta_g} \tag{11}$$

Because Eq. (9) is an implicit ordinary differential equation that has not an analytical solution, a numerical integration is required. In order to solve this set of equations the fourth-order Runge–Kutta method [11] has used and implemented in the MetLAB program.

Studying the data presented in Fig. 4, we observe that the finer droplets are subjected to higher accelerations and decelerations relative to the coarse particles. This behavior is a consequence of the inertial forces that are related to the droplets mass. The Reynolds number becomes zero when the relative speed between the droplets and the gas is also zero. After this point, the droplets suffer a deceleration influence from the gas stream.

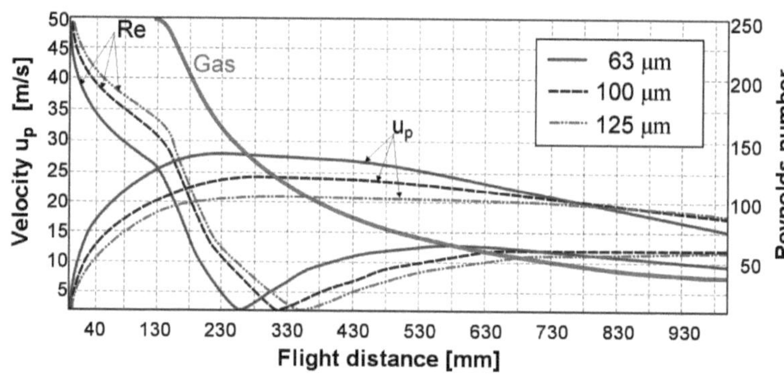

Fig. 4 Droplets and gas velocity vs. flight distance calculated with the MetLAB program

2.4 Modeling of the Droplets Heat Transfer

The heat extraction from the droplet is realised by convective and radiation mechanisms transfer. The heat conduction within the droplet is neglected considering that the temperature of the droplet is uniform. Assuming a spherical shapes for particles, we have Eq. (12) for the heat balance, [7, 6], where ε is the emissivity of the particle surface, k_{SB} is the Stefan-Boltzmann constant and h_c the heat transfer coefficient for a particle given by Eq. (13):

$$-\rho_p \frac{4\pi {r_p}^3}{3} C_{pL} \frac{dT_p}{d\tau} = h_c \left(T_p - T_g\right) 4\pi r_p^2 + \varepsilon k_{SB} \left(T_p^4 - T_g^4\right) 4\pi r_p^2 \quad (12)$$

$$h_c = \frac{k_g}{d_p} \left(2 + 0.6 \cdot \mathrm{Re}^{0.5} \cdot \mathrm{Pr}^{0.33}\right) \quad (13)$$

$$\mathrm{Pr} = \eta_g \frac{C_{pG}}{k_g} \quad (14)$$

The Prandtl number is expressed in Eq. (14), where $k_g$is the gas thermal conductivity, η_g the dynamic gas viscosity and C_{pg} the gas specific heat. The solidification process occurs in distinctive stages: in the 1st stage we have cooling in the liquid phase, in the 2nd stage the recalescence of particle is produced, in the 3rd stage we have segregated solidification followed by possible reactions at constant temperature (i.e. peritectic or eutectic reaction, corresponding to the 4th stage) and finally in the 5th stage we have cooling in the solid state. Each stage of the solidification process is described by a balance heat equation. The first stage is completed when the droplet temperature reach the nucleation temperature T_N, the degree of droplet undercooling being calculated using Eq. (15):

$$\Delta T_{\text{actual undercooling}} = \Delta T_{\text{homogeneous}} \cdot \exp(-2.2 \cdot 10^{12} \cdot V_p) \quad (15)$$

In the 2nd stage the particle recalescence is produced, which is a rapid heating of an undercooled droplet due to the nucleation of the solid phase. The recalescence process is considered complete when the rate of heat release is equal to the rate of heat extraction from the droplet. The recalescence is completed in less than 10^{-6} s.

$$T_R = \begin{cases} T_L & \text{if} \quad \frac{H_f}{C_{pS}} + T_N > T_L \\ \frac{H_f}{C_{pS}} + T_N & \text{if} \quad \frac{H_f}{C_{pS}} + T_N \leq T_L \end{cases} \quad (16)$$

The solidified fraction is calculated based on the thermal energy balance between the latent heat released and the thermal energy which is needed to accommodate the temperature rise in the droplet, Eq. (17).

$$H_f \cdot f_r = (T_R - T_N)\left[C_{pL}(1 - f_r) + C_{pS} \cdot f_r\right] \quad (17)$$

$$f_r = \frac{C_{pL} \cdot (T_R - T_N)}{H_f + \left(C_{pL} - C_{pS}\right)(T_R - T_N)} \quad (18)$$

In the next stage, the segregated solidification, the completion of the recalescence process is described by the Scheil equation [6–8]:

$$(c_L - c_S)\, df = (1 - f)\, dc_L \tag{19}$$

which after integration becomes:

$$f = 1 - (1 - f_r)\left(\frac{c_L - c_{ref}}{c_0 - c_{ref}}\right)^{1/(k_e - 1)} \tag{20}$$

In terms of temperature variables, we get the final equation that express the fraction solidified in this stage, Eq. (21):

$$\frac{df}{dT} = \frac{(1 - f_r)}{(k_e - 1)\left(T_{ref} - T_L\right)}\left(\frac{T_{ref} - T_p}{T_{ref} - T_L}\right)^{(2-k_e)/(k_e-1)} \tag{21}$$

For the final stage (cooling in the solid state), we use the Eq. (12) where C_{pL} was substituted by C_{pS}.

The finer droplets are cooled at higher rates and also the undercooling degree is greater for the smaller particles, Fig. 5. The solidification process occurs, however, in a shorter time relative to the coarse ones. The solidified fraction shows us the evolution of the solidification process vs. the droplet flight distance in order to establish the optimum distance of the preform relative to the atomization head. That distance is affected also by the atomization gas, as can be seen in Fig. 6, where the same droplet has a different thermal history and the solidification occurs more rapidly in the nitrogen atomization case.

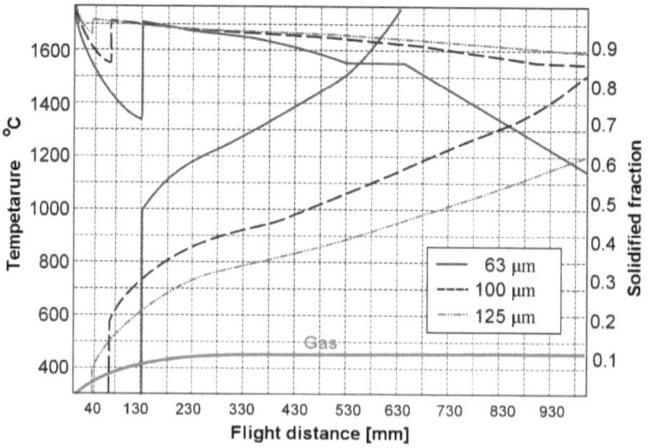

Fig. 5 Droplets temperature history and solidified fraction vs. flight distance calculated with the MetLAB program

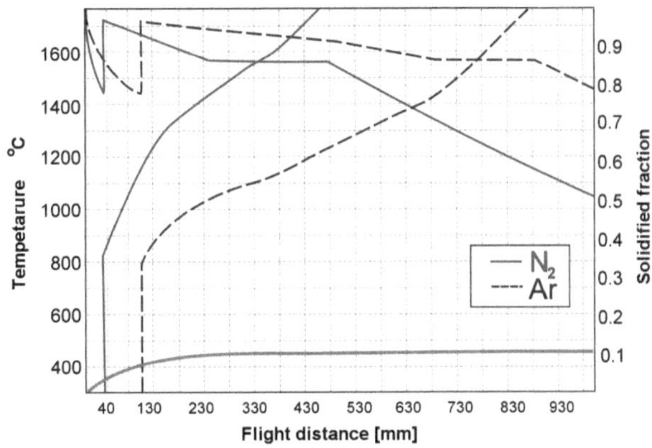

Fig. 6 Temperature history and solidified fraction vs. flight distance for a 80 μm droplet's size in a N_2 and Ar gas atomization

3 Conclusions

1. The atomising gas mass flow rate to molten metal mass flow rate ratio is a key parameter in controlling the droplet diameter and hence the cooling rate, billet temperature and resulting solid particle nucleant density.
2. The droplet diameters, are in the range 40–400 μm diameter with a mean diameter of ~160 μm. The droplet diameter governs the dynamic behaviour of the droplet in flight which in turn determines the time available for in-flight cooling which is critical in controlling the resulting billet microstructure.
3. The droplet velocities are in the range of 30–40 (ms^{-1}) for droplets diameters in the range 80–125 μm and at distance of up to 300 mm from the atomization head the droplets are accelerated by the gas.
4. The heat extraction from the droplet is realised mainly by convective and radiation mechanisms transfers. The smaler droplets can experience undercoolings of up to 250°C prior to nucleation.
5. Not all droplets that impact the billet surface are incorporated. Some solid droplets will splash-off the billet top surface or will be directed out of the deposition region by turbulent gas movement in the atomization chamber.

List of Symbols

C_{pL}	specific heat of the liquid droplets	$J\cdot kg^{-1}\cdot K^{-1}$
C_{ps}	specific heat of the solidified droplets	$J\cdot kg^{-1}\cdot K^{-1}$
C_{pg}	specific heat of the gas	$J\cdot kg^{-1}\cdot K^{-1}$
C_D	drag coefficient	–
d_n	liquid nozzle diameter	m
f	Colebrook friction factor	–
f_r	solidified fraction of the droplet	–

H_f	enthalpy of fusion	$J{\cdot}kg^{-1}$
H_m	liquid metal head	m
H_n	liquid nozzle height	m
k_g	thermal conductivity of the gas	$W{\cdot}m^{-1}{\cdot}K^{-1}$
k_{SB}	is the Stefan-Boltzmann constant	$W{\cdot}m^{-2}{\cdot}K^{-4}$
M	atomic weight of alloy	kg mol^{-1}
Q_m	melt flow rate	$m^3{\cdot}s^{-1}$
$p_{1,2}$	medium pressure related to $z_{1,2}$ liquid head sections	Pa
Pr	Prandtl number	–
Re	Reynolds criteria	–
R	universal gas constant	$J{\cdot}mol^{-1}{\cdot}K^{-1}$
r_p	droplet radius	m
T	melt temperature	K
T_L	alloy liquidus temperature	K
T_N	nucleation temperature	K
T_R	recalescence temperature	K
$u_{m1,2}$	represents the melt flow velocity related to $z_{1,2}$ liquid head sections	$m{\cdot}s^{-1}$
u_p	droplet velocity	$m{\cdot}s^{-1}$
u_g	gas velocity	$m{\cdot}s^{-1}$
Z	distance from the atomizer	m
Z_1	liquid head height	m
Z_2	exit section height of liquid delivery nozzle	m
V_p	droplet volume	m^3
Σh_L	total losses due to friction and section variations	m
$\alpha_{1,2}$	velocity distribution coefficient	–
μ_m	melt viscosity	$Pa{\cdot}s$
η_g	dynamic gas viscosity	$kg{\cdot}m^{-1}{\cdot}s^{-1}$
σ_m	surface tension of melt	$N{\cdot}m^{-1}$
ρ_m	melt density	$kg{\cdot}m^{-3}$
ρ_g	gas density	$kg{\cdot}m^{-3}$

References

1. A. Leatham, Spray forming: Alloys, products, and markets. JOM **51**(4), (1999).
2. A. Moran, C. Madden et al., Spray forming technology for military applications. JTTEE **3**, 197–198 (1994). doi:10.1007/BF02648278
3. P.S. Grant, Solidification in spray forming. Metall. Mater. Trans. A **38A**, 1520–1529 (2007). doi:10.1007/s11661-006-9015-3
4. S.B. Johnson, J.P. Delplanque et al., Numerical Simulation and Experimental Characterization of a Binary Al Alloy Spray – Application to the Spray Rolling Process. TMS 2005 – 134th Annual Meeting and Exhibition, San Francisco (2005)
5. K.M. McHugh, Y. Lin et al., Microstructure evolution during spray rolling and heat treatment of 2124 Al. Mater. Sci. Eng. A **477**(1–2), 26–34 (2008)
6. N.S. Mahesh, J. Mendonca et al., Modeling of droplet dynamic and thermal behaviour during spray deposition. Bull. Mater. Sci. **26**(3), 355–364 (2003)
7. B. Li, X. Liang, J.C. Earthman, E.J. Lavernia, Two dimensional modeling of momentum and thermal behavior during spray atomization of γ-TiAl. Acta Mater. **44**(6), 2409–2420 (1996)
8. G. Vedovato, A. Zambon, E. Ramous, A simplified model for gas atomization. Mater. Sci. Eng. **A304–306**, 235–239 (2001)

9. M. Bodea, MetLAB – A software program for modeling of the atomization process (2005), http://zeus.east.utcluj.ro/sim/bodea/MetLAB.zip
10. Y.M. Chen, Y.H. Su et al., Modeling of atomization rate during gas atomization. Acta Mater. **46**, 1011–1023 (1998)
11. W.H. Press, S.A. Teukolski et al., Numerical recipes in C (Cambridge University Press, Cambridge, 1992)
12. B. Bergquist, Influence of operating conditions related to physical properties of the iron melt during water atomization. Powder Metall. **42**(4), 331–342 (1999)
13. Le Tuyet, R. Stefaniuk et al., Measurement and analysis of melt flowrate in gas atomization. Int. J. Powder Metall. **35**(1), 51–59 (1999)
14. B. Bergquist, T. Ericsson, A robustness simulation of water atomization. Powder Metall. **43**(1), 37–42 (2000)
15. P. Shukla, R.K. Mandal, S.N. Ojha, Modeling of heat flow and solidification during spray deposition process. Trans. Indian Inst. Met. **57**(3), 283–296 (2004)

Cyclic-Bend-Over-Sheave Fatigue Testing of an Umbilical for Oil Production in Ultra-Deep Waters

Paula Ferreira Lépore and Miguel Angel Buelta Martinez

1 Introduction

During the last two decades, technological qualification for oil production and gas fields located up to 1,000 m water depths developed, consolidating the existing operational experience. However, a series of challenges and difficulties will have to be faced and overcome to increase the oil production for fields located in ultradeep waters (beyond 2,000 m). A much more compreensive analysis of this case scenario discloses different technological demands. Among those, there is the development of a brand new technology regarding the submarine umbilicals used for control and base connections. That aproach shall consider all the structure life span: design, manufacure, prototype testing, installation and *in loco* operation.

Basically, the structural basic conception of an umbilical consists in overlapping plastic and metalllic layers that supply internal and external tightness, and structural strenght, keeping its functional core intact. The different characteristics of those materials, their mechanical properties, as well as their behavior under different kinds of loads, make it rather difficult for any type of simplified modeling approach.

It is proven that this kind of numerical and mathematical modeling might become very laborious due to the umbilcal constructive conception, with different materials and built-in layers. The assembly of the transversal section of an umbilical is based on the principle of a core with functional components that must be protected from external loads by a series of concentric layers (Fig. 1). The components are normally the following: electrical and/or optic cables; hydraulic hoses; chemical injection hoses. The umbilical major characteristic is that it is flexible so that the line can be easily spooled over transport reels to smooth the progress of the installation operation. Also, the flexibility of the line is an important factor as it shall not restrict the movements of the floating unit which the umbilical is attached to during the operation process.

P.F. Lépore (✉)
Department of Structures Engineering, Engineering School, University of Sao Paulo,
Sao Paulo, 05508-900, Brazil
e-mail: paula@mfx.com.br

A. Öchsner et al. (eds.), *Materials with Complex Behaviour*, Advanced Structured Materials 6, DOI 10.1007/978-3-642-12667-3_4, © Springer-Verlag Berlin Heidelberg 2010

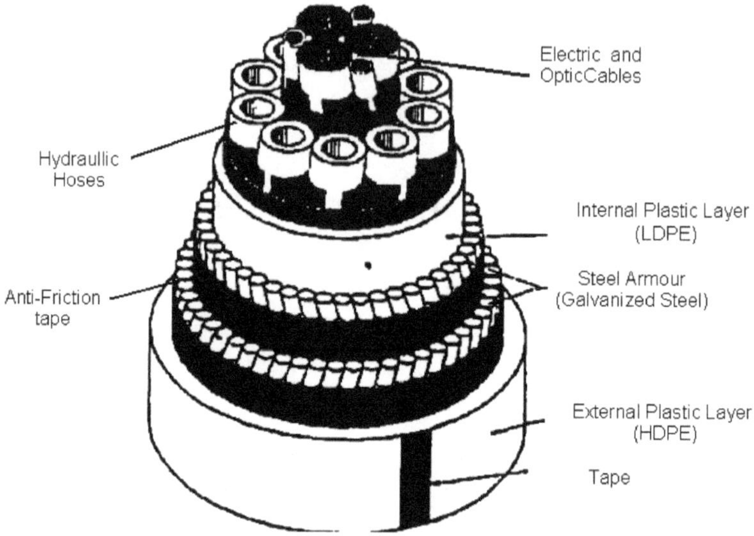

Fig. 1 Typical umbilical section

As a basic step for the design of an umbilical line, its utilization and ultimate strength limits and inner stresses and strains have to be well defined. Those limits, when exceeded, may cause comprehensive damages to the functional elements core.

The evaluation of the strength limit of an umbilical line, during the design phase of a specific production system consists basically of a general analysis of the interaction among the different material layers.

That task must be optimized in order to establish the correct dimensions and materials, considering critical operational and installation loads. The traditional methods applied to describe an umbilical structure are bi-dimensional models, considering only the axi-symmetrical loads, such as tensile and pressure loads [1]. Those cannot adequately describe the three-dimensional 'Stress x Strain' state of this multi-layered structure.

So as to improve those deficiencies, a three-dimensional numerical model for a reduced longitudinal portion of an umbilical cross-section was developed. In order to create a comprehensive basis of experimental data, a series of umbilical samples were submitted to different load conditions, simulating its operation and installation conditions. The functional core and the strength elements were monitored in order to feed the numerical model. For this purpose, two test rigs were assembled.

With the data of both the physical and numerical experimental analysis at hand, a comparative study was performed, aiming at presenting an optimized tool for designing an umbilical cross section.

One of the main purposes of this work was to add detailed information and to define some aspects that would provide a more practical tool for the thecnical qualification requirements presented to umbilical manufacturers, during the bid phase. A specific technical regulation for great depths umbilicals that does not exist: usually, the manufacturer would inform the utilization factor of the structure (life span, minimum bend radius for operation and installation, maximum pull in straight line, etc).

Therefore, considering the growing need to optimize the structure design, the simulation through experimental tests of the structure behavior of a representative type of umbilical during its phase of installation and operation was proposed.

The specifications [2, 3] provide details on design tools, manufacturing, as well as the homologation of those phases for multi-function umbilicals, for both static and dynamic applications.

Each type of experimental test proposed at this phase had its specific objectives, such as fatigue analysis simulating the installation and operations conditions.

2 Fatigue Tests

This test (Fig. 2) simulates the installation operation of an umbilical line (critical load condition), during which a combination of tensile and bending loads occurs.

Three samples of a typical cross-section umbilical were submitted to a combination of f bending cycles under tension (Table 1). This test program is meant to simulate the interruption of the installation process during a certain period, and the flexible line will be exposed to the inherent movements of the vessel.

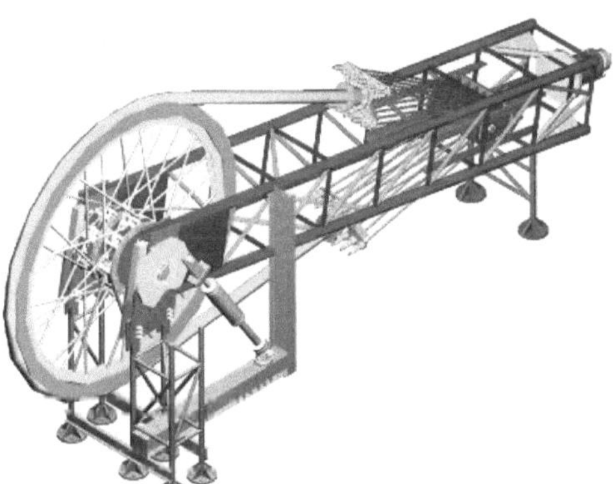

Fig. 2 Fatigue test ig

Table 1 Test program

Sample	Tension (tf)	Cycles	"f" (cycles/min)	"T" (s)
CP1	60	30,000	10	6.0
CP2	30	30,000	8	7.5
CP3	30	100,000	10	6.0

Those values were supplied by the manufacturer and represent the percentages of the tension during the installation phase. In this case, the highest value adopted was the maximum installation load obtained trough a global dynamic analysis performed by the manufacturer for the electrical-optical cable umbilical: $3 \times 240\,\text{mm}^2 + 12\text{CE}$ (RONCADOR Field) (Fig. 3).

Two positions of the umbilical were monitored with strain gauge transducers (Fig. 4).

At one of those positions, the tensile stress was predominant (section "T"), and at the other, the bending combined with tensile (section "F"). Position "F" is at a region in which the umbilical was always in contact with the wheel. Because of that, it was submitted to constant bending combined to tensile stress.

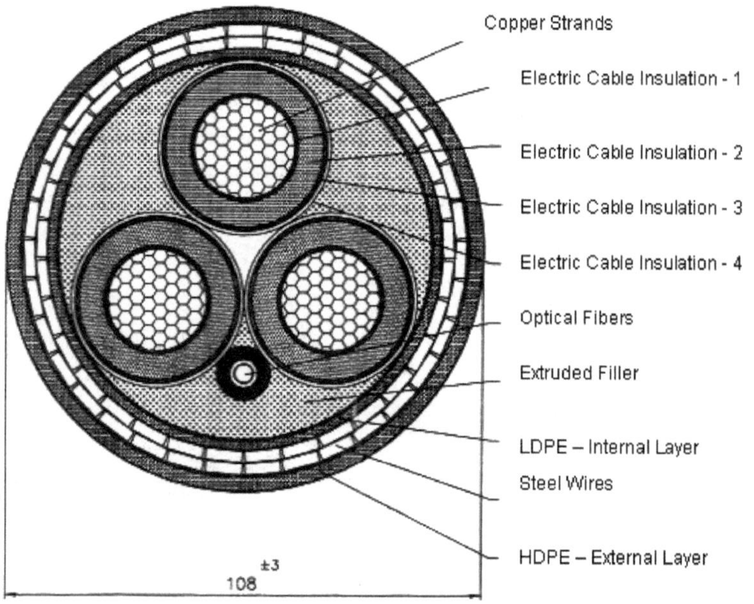

Fig. 3 Electro-hydraulic umbilical: $3 \times 240\,\text{mm}^2$ 12/20 kV + 12 single mode optical fibers

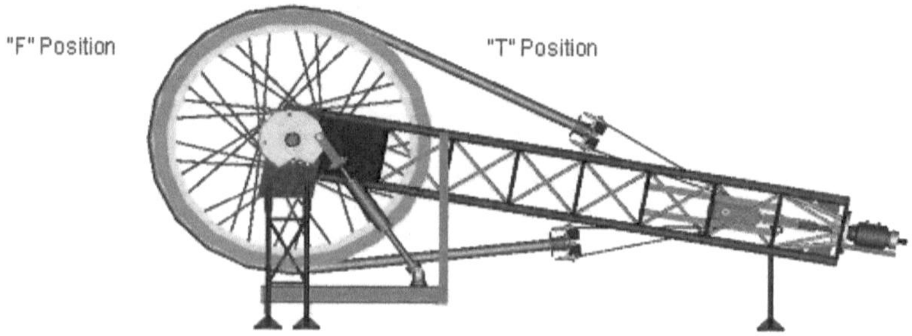

Fig. 4 Strain gauges positioning

The "T" position is at a region not in contact with the wheel and only submitted to tensile stresses.

At both positions, strain gauges were positioned over the external steel armor.

3 Numerical Analysis

In general, the usual numerical analysis methods for flexible lines consider only two-dimensional models and axi-symmetric loads, not being able to simulate the particularities of some of the critical conditions of the umbilical life operation. Using the premises of the FEM (Finite Element Method), an optimization analysis of the interaction among the varius types of layers of an umbilical under critical load conditions was performed. A pre-processor GERFLEX [1] was used for the preparation of the entry data, such as geometry and material caracteristics,. That tool was important in order to smooth the progress of the activities of modeling and adjusting the umbilical typical cross-section. The phase of mathematically processing was performed by ANSYS which, associated with the pre-processor GERFLEX, supplied the data for the results complete analysis.

3.1 Calculus Hypothesis

3.1.1 Plastic Layers

The external plastic layer (high density polyethylene – HDPE), is normally extruded over the external tensile armor, providing waterproof and a protective element for the internal core against corrosion, abrasion and impact. It also helps to keep the wires of the tensile armor in their correct position.

The internal plastic layer manufactured of low density polyethylene (LDPE), is extruded over the filler (elastomer) acting as an isolating and protection element

Non-linear Properties
Stress specified as Normalised

Strain	Stress (kN/m²)	Stress / E
0,00	0	0,000
0,02	4200	0,020
0,04	7350	0,035
0,06	9450	0,045
0,08	10710	0,051
0,10	11760	0,056
0,12	12390	0,059
0,14	13020	0,062
0,16	13440	0,064
0,18	13000	0,066
0,20	14280	0,068

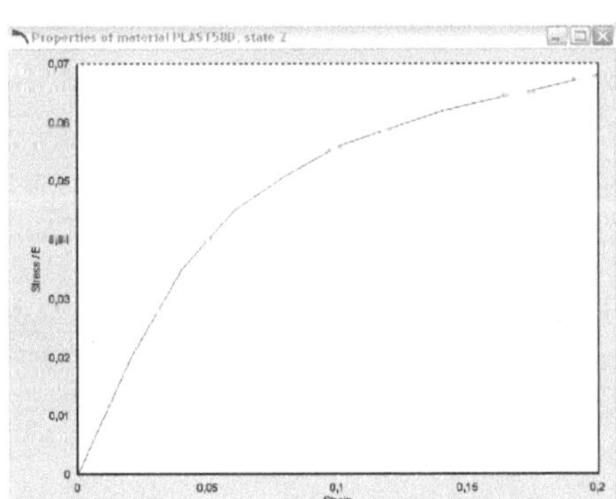

Fig. 5 Material properties

for the copper conductors and the optic fiber. It is also capable of transmitting the contact loads between the metallic layer and the adjacent rubber filler.

For the modeling of those two plastic layers, the inherent material non-linearities and the viscoelastic properties of the polyethylene material were considered (Fig. 5). Therefore, a solid isoparametric type element was chosen: 8-point brick type, with capacity up to 20 intermediate points and with 3 degrees of freedom for each point.

3.1.2 Steel Armors

The tensile armor is composed by two internal layers of steel wires conformed around the core with a helical shape. Those two complementary layers are positioned with opposing angles between them (the layers are placed so as to directly resist the axial loads applied to the line, converting this tension into a contact pressure among the internal layers [4]. The geometry non-linearities (system with large deformations) were taken into account by adopting elements joined continuously in several helical springs, representing groups of neighboring wires.

3.1.3 Electric-Optic Nucleus

The electro-optic crore is composed of electric and optic cables, extruded over an elastomer of propilene-ethylene. To model this set of elements, the average properties of the electrical core were evaluated with the data supplied by the manufacturer. The representative element type of the functional core was chosen taking into account the fact that the largest percentage of its composition (more than 80%) consists of extruded rubber (EPR). The same type of element used for the plastic layers was adopted for this case.

3.1.4 Contact

The steel-plastic contact is represented by an axi-symmetric surface contact (the contact elements are contained into a surface while the target elements are positioned at the other one). The plastic layers were defined as the contact surface, while the metallic layers were assigned as a target surface. After the definition of the element type, the correct set of attributes were selected (the same for the contact surface and for the target surface). Thus, the ANSYS program uses the material properties at the surfaces to calculate an equivalent stiffness. Moreover, the program also automatically defines a value for the tangent contact stiffeness, proportional to the value of the axial stiffness.

The steel-steel interaction was simulated as a contact surface between flexible elements using small elements (beam type) with small torsional stiffness, set radially between the wires of the armor.

3.1.5 Friction

The friction between the plastic and metallic layers was taken into account, because during the slide between those materials, many times, a mechanism known as 'film deposition' can occur. The tribologic properties of HDPE (material component of the external layer) in contact with the galvanized steel (material component of the helical wires) were studied and the data obtained through experimental tests [5], determined the wear rate and the friction coefficient between the plastic and metallic layers. Those values were adopted to describe the friction between external and internal layers. The steel-steel contact was not modeled (presence of anti-friction layers between the wires).

3.1.6 Model Geometry

Figure 6 show the installation process of a umbilical. The experimental tests and numerical models proposes a simulation of this installation and reproduce the umbilical contact with the reel (also the floating buoys during its operation).

3.1.7 Model Mesh

The perimeter of the umbilical was divided circumferentially in N parts forming a polygon of identical sides (Fig. 7). In the longitudinal direction, each fraction of its length corresponds to one the steel armors layers (also divided into N parts). This process guarantees the generation of a regular mesh (by GERFLEX).

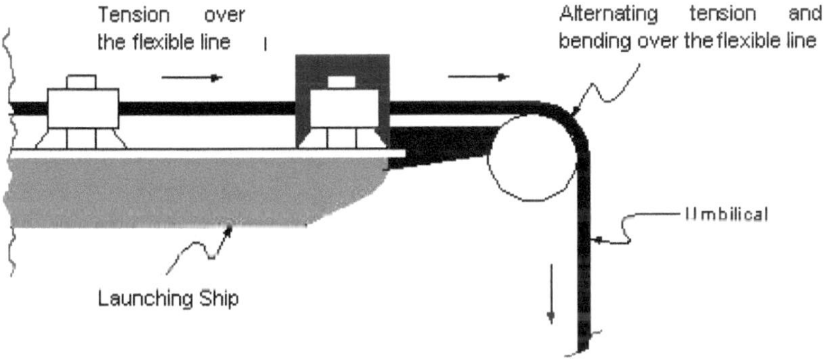

Fig. 6 Loading conditions

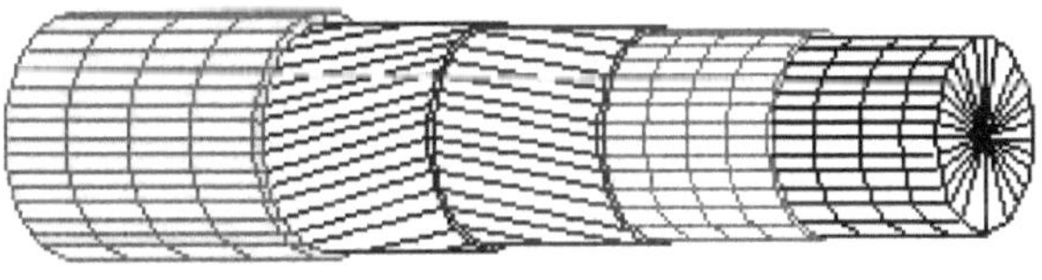

Fig. 7 Model mesh

3.1.8 Model Loading Conditions

Regarding the model symmetry, the umbilical was pinned to one of its ends where it was in direct contact with the wheel. The structure that simulates the wheel was fixed in its central point with freedom to rotate. Tension and bending loads were applied to the umbilical in order to reproduce the experimental fatigue tests.

The two models generated are : MOD1-A (60 tons tension + bending) and MOD1-B (30 tons tension + bending).

4 Results

Only the stresses results are presented. The comparative studies proved that the numerical model was able to represent the geometry of the layers under some of the critical conditions faced by an umbilical during the installation and operation processes. A good approximation was verified between the numerical and experimental results for strains and stresses values (external steel armor). Figure 8 presents the stress distribuition over several different layers of the umbilical. Tables 2, 3, 4, and 5 presents the numeric results in detail.

Good correlation was found between the experimental and the numerical results.

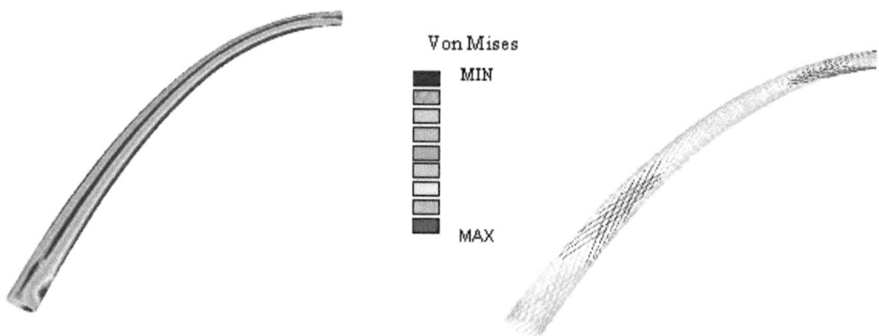

Steel Plastic

Fig. 8 Von mises stress: external steel and plastic layers.

Table 2 Maximum stress values (model1-A)

Layer	Tension load (ton)	Maximum stress[1] (MPa)	Nominal stress (MPa)	Safety factor
Internal steel armor		199.04	1040	5.23
External steel armor	60	231.44	1040	4.49
Internal plastic layer		11.57	18	1.56
External plastic layer		25.46	32	1.26

[1]Maximum stress evaluated trough numerical model

Table 3 Maximum stress values (model1-B)

Layer	Tension load (ton)	Maximum stress[1] (MPa)	Nominalstress (MPa)	Safety factor
Internal steel armor		117.76	1040	8.45
External steel armor	30	130.33	1040	7.98
Internal plastic layer		3.84	18	4.68
External plastic layer		10.24	32	3.12

[1]Maximum stress evaluated trough numerical model

Table 4 Comparison between experimental x numerical model (CP1)– external steel armor

MOD1-A – 60 ton	Comparative stress data: experimental x numerical		
Positioning	Experimental (MPa)	Numerical (MPa)	Error (%)
F_s – bending+tensile	247	231.44	6.3
F_i – bending+tensile	206	194.05	5.8
T_s – tensile	186	182.09	2.1
T_i – tensile	167	160.65	3.8

Table 5 Comparison between experimental x numerical (CP2) – external steel armor

MOD-1B – 30 ton	Comparative stress data: experimental x numerical		
Positioning	Experimental (MPa)	Numerical (MPa)	Error (%)
F_s – bending+tensile	93	94.07	1.15
F_i – bending+tensile	88	83.60	5.0
T_s – tensile	125	128.00	2.4
T_i – tensile	123	120.29	2.2

5 Conclusions

The level of stresses and strain is a critical point for a flexible line design under critical load conditions. The data generated from this study can be used as a complementary fatigue study which may become a more realistic forecast about the structure service life.

Therefore, this computational analysis model viewed to supply a validation process for design procedures of a brand new umbilical line. This numerical model gives the most critical points of the flexible line. That local analysis will determine in advance if the structure global analysis is correct and evaluate the capacity of a specific type of umbilical to hold different load combinations.

The hypotesis adopted, as well as some of the analysis proposals for the different types of operational conditions, were presented aiming to perform a comparative study of the adopted experimental prerogatives. The results were very good, with a positive confluence between the experimental and numerical values.

References

1. F.T. Lopes, Análise estrutural de linhas flexíveis. Máster Thesis, Escola Politécnica da USP, São Paulo, 1996
2. American Petroleum Institute, Specification for Sub Sea Production Control Umbilicals - API Specification 17-E, Sept 1998
3. American Petroleum Institute, Specification for Unbonded Flexible Pipe – API Specification 17–J, Mar 1997
4. M.H. Patel, J.A. Witz, Z. Tan, *A Flexible Riser Design Manual*, 2nd edn. (London, Betham Press, 1994)
5. C.H. Silva, C.P Pesce, D.K. Tanaka, A. Sinatora, Desgaste de Polietileno de Alta Densidade empregados na fabricação de tubos para extração de petróleo. COBEM, 15th Brazilian Congress of Mechanical Engineering, Águas de Lindóia, São Paulo, 22–26 Nov 1999.

Dynamic Crack Propagation in Composite Structures

D. Bruno, F. Greco, and P. Lonetti

1 Introduction

Composite materials in the form of laminated structures are affected by interface damage mechanisms, known in the literature as delamination phenomena, which typically produce high stiffness and strength reductions with catastrophic failure modes [1]. From an experimental point of view, many observations have shown that the evolution of such interface damage mechanisms is strongly influenced by dynamic effects related to the loading rates, the inertial forces and the wave propagation phenomena [2–4]. The full characterization of the dynamic crack propagation requires specialized techniques to predict the physical quantities, which govern the crack growth. Computational analyses are frequently conceived to explain the main features of the dynamic crack growth phenomena, which are really complex if investigated from an experimental point of view. As a matter of this fact, the progress toward a complete understanding of dynamic fracture behavior has been limited by the intrinsic complexities of the growth phenomena, such as time dependence, high speeds in the crack propagation with multiple cracks and branching mechanisms [5].

A brief literature review denotes that the interface cracking mechanisms are frequently analyzed by means of simplified models developed in the context of static analysis or of a steady state advance of the crack. In the former formulation, the time dependent behavior is practically annihilated and thus the solution is based on a quasi-static evolution of the crack, neglecting a priori the inertial effects in the crack growth [6]. The latter modelling describes the crack growth enforcing the crack tip to evolve at constant speed. In this context, typically, a moving reference system fixed crack tip is utilized to describe the time dependence of the solution in terms of spatial variables only, leading to crack propagation equations involving only ordinary instead of partial differential equations [7, 8].

P. Lonetti (✉)
Department of Structural Engineering, University of Calaberia, Via P. Bucci,
Cubo 39-B, 87030 Rende, Cosenza, Italy
e-mail: lonetti@unical.it

A. Öchsner et al. (eds.), *Materials with Complex Behaviour*, Advanced Structured Materials 6, DOI 10.1007/978-3-642-12667-3_5, © Springer-Verlag Berlin Heidelberg 2010

In the framework of the dynamic crack propagation, extended analyses are developed to predict the behavior of propagating cracks for monolithic structures, in which limiting crack speeds and crack arrest phenomena are evaluated for different kind of materials [9, 10]. The literature on the dynamic fracture mechanics dealing with composite structures is relatively limited. Most research efforts were confined to analyze static or low velocity crack propagation conditions [1], whereas the dynamic delamination phenomena are not completely investigated. Typically, the composite structures are formed by weak interfaces, in which the interfacial cracks are constrained to propagate along preferred paths, suppressing any tendency to branch or kink out from the weak planes. Therefore, the evolution of the interfacial cracks is characterized by high speeds and thus, in order to describe accurately the dynamic crack behavior, it is necessary to take into account for the time dependence effects in the definition of the energy release rate and the crack growth behavior [5, 11].

From a computational point of view, the analysis of delamination phenomena requires a detailed description of the growth area since, typically, the interfacial cracks propagate very rapidly, i.e. at such speeds close to those of the material waves. Numerical methods are frequently preferred to analyze dynamic propagation phenomena, since analytical solutions of the growth phenomena are very difficult to be extracted, unless for simple loading and geometric conditions [7]. In order to predict the crack growth conditions, several approaches have been proposed in the literature. The node release technique [12, 13] is based on the assumption that the crack growth is described by uncoupling nodes at the crack faces, whose acting tractions are reduced as far as the crack opens. In this approach, the evolution of the crack is strictly dependent from the size of the element mesh around the crack tip, since it governs the amount of the crack advance. Moreover, the advancing process is not really continuous since a proper iteration scheme is necessary to evaluate accurately the dynamic crack growth during the time integration [14]. Models based on the virtual crack closure method (VCCM) calculate the energy release rate as the work performed by the internal traction forces at the crack faces during a virtual crack advance of the tip. In dynamic fracture mechanics, the VCCM is applied by using the modified form, in which the ERR, during the time integration, is evaluated by the product between the reaction forces and the relative displacements at the crack tip and at the nodes closer to the crack tip front, respectively [15]. The prediction of the energy release rate is strictly dependent from the mesh discretization of the crack tip. Analogously to the node release technique, the crack growth is not continuous, since the crack is able to advance only of a length equal to that adopted in the definition of the mesh element size at the crack tip. Moreover, in order to evaluate correctly the crack evolution during the time integration, an accurate discretization of the crack tip is needed and thus the computational cost of the analysis may increase notably.

Crack growth phenomena can be predicted also by means of damage formulation making the use of interface cohesive elements. In this framework, strain softening interface elements with a damaged constitutive relationship are introduced between crack faces. In order to predict the crack growth phenomena, accurately, a detailed finite element mesh at the crack tip front is often required [16, 17].

Crack growth phenomena should be analyzed by means of the moving mesh methodology, which is able to simulate the motion of an initial defect by changing the position of the mesh elements by means of prescribed displacements or velocities. Typically, moving mesh techniques are utilized in the field of fluid mechanics, since large displacements are involved in the deformed configuration. However, some applications can be recovered in solid mechanics also in the context of crack propagation. In particular, moving element procedures for non-singular finite element methods can be recovered in [18], in which the entire mesh is moved with the crack tip. Contrarily, local mesh update procedures based on the motion of a small portion of the crack tip are developed by Nishoka and Atluri [19] and by Nishioka [20], in which the singular fields are reproduced as far as the finite elements move with the crack. Moving mesh strategy can be developed also in the framework of Arbitrary Lagrangian-Eulerian (ALE) methods. The mesh elements configuration is decoupled by the material motion introducing a fictitious reference coordinate system, in which the position of the mesh points does not introduce any mesh distortion during the geometry variation [21].

In the present paper, dynamic effects related to crack propagation phenomena in composite structures are analyzed. The proposed modeling is developed according to a moving mesh strategy based on Arbitrary Lagrangian-Eulerian formulation. This choice is motivated by the fact that in composite laminated structures, the crack path is constrained to advance at the interface planes and thus the moving mesh technique becomes a suitable tool to modify the geometry of the model and to account for the evolution of preexisting interfacial cracks. In order to ensure an accurate prediction of the crack growth, the dynamic energy release rate mode components are calculated by means of the decomposition methodology of the J-integral [8, 22], which is proposed in the framework of an unsteady crack growth. It is worth noting the integration path utilized to calculate the J-integral and its mode components moves with the crack tip, ensuring a high discretization of the mesh only on a small region around the crack tip. However, as far as distorted mesh is achieved due to mesh motion of the tip, a remeshing algorithm is utilized to ensure accuracy in the finite element results. The crack tip motion is based on an explicit dynamic criterion, in which the relation between ERR mode components and velocity of the crack tip is assumed to be a material property of the composite laminate. Comparisons with experimental results are reported to validate the proposed modeling. Moreover, a parametric study is developed to show some characteristic phenomena of the dynamic crack growth.

2 Formulation of the Damage Model

The proposed modeling is based on the combination of Fracture Mechanics and moving mesh methodology. The former predicts the crack growth, by the use of the ERR concept and a proper crack advance criterion, whereas the latter is introduced to account for the changes of the geometry during the crack evolution. The general formulation of the structural model is consistent to a 2D plane stress approach, in which the behavior of each lamina is homogeneous, linear and elastic. The interfacial defects are assumed to propagate along the interfaces between the laminas, which are basically weak planes in which the delamination defects are able to growth, producing high stiffness and strength reductions [1]. This assumption can be motivated from a physical point of view, since many experimental observations have shown that the evolution of such interfacial defects proceeds along a prescribed path almost fixed in the interface zones, leading to measured crack speeds ranging also in the framework of intersonic crack propagation [10]. From an engineering point of view, it is reasonable to assume that the crack position along the thickness direction for any interfacial cracks is fixed. This assumption leads to high simplifications in the crack growth prediction, since crack branching or kinking phenomena are excluded from the simulation.

The crack growth is predicted by the use of the moving mesh methodology based on an arbitrary Lagrangian-Eulerian formulation. This technique despite the conventional Lagragian or Eulerian approaches provides a better way to take into account mesh movements. As a matter of this fact, Lagrangian moving mesh strategy provides good accuracy in the prediction of the motion of free surfaces and interfaces between different materials. However, during the mesh movements, the mesh geometry is affected by large distortions and the recourse of remeshing operations is strongly required. Eulerian moving mesh algorithms predict easily large distortions of the continuum, but, typically, numerical difficulties arise in the simulation of material interfaces or, in the solving procedure, due to the unsymmetrical character of the convection operator. The ALE technique combines the best features of the two methods, since the mesh motion is disconnected by the material points and thus it can be arbitrary chosen.

3 ALE Formulations: Description of the Motion and Notation

In this section, the main features on ALE formulation are summarized, whereas more details on this topic can be recovered in Donea et al. [1]. In solid mechanics, the description of the body motion can be developed by means of a Lagrangian approach, introducing a family of mappings φ, which associates at each point $\underset{\sim}{X}$ of the material or Lagrangian configuration B_L the position of the particle $\underset{\sim}{x}$ in the current configuration, B_E at a generic t, as:

$$\varphi : B_L \rightarrow B_E \quad \underset{\sim}{x} = \varphi\left(\underset{\sim}{X}, t\right) \tag{1}$$

In the framework of the ALE formulation, the finite element mesh movement is described introducing a fictitious reference system, known as "referential coordinates", whose motion is typically arbitrary and it does not coincide neither with that of the material points neither with current configuration. In the referential coordinate, the motion of the continuum is described by means of a family of mapping $\chi : B_R \rightarrow B_L$ between the material and referential configurations:

$$\underset{\sim}{X} = \chi\left(\underset{\sim}{r}, t\right) \tag{2}$$

where $\underset{\sim}{r}$ is the referential coordinate introduced to identify the grid points (Fig. 1). It is assumed that χ is a homeomorphism and is differentiable everywhere. In view of Eqs. (1) and (2), the material ($\underset{\sim}{u}$) and referential displacements ($\underset{\sim}{\psi}$) are defined as:

$$\underset{\sim}{u} = \underset{\sim}{x} - \underset{\sim}{X} = \varphi\left(\underset{\sim}{X}, t\right) - \underset{\sim}{X}, \quad \underset{\sim}{\psi} = \underset{\sim}{X} - \underset{\sim}{r} = \chi\left(\underset{\sim}{r}, t\right) - \underset{\sim}{r} \tag{3}$$

Therefore, the corresponding differentiation with respect to the vector $\underset{\sim}{X}$ and $\underset{\sim}{r}$ are:

$$\underset{\sim}{\nabla}_{\underset{\sim}{X}} \underset{\sim}{u} = \frac{d\underset{\sim}{u}}{d\underset{\sim}{X}}, \quad \underset{\sim}{\nabla}_{\underset{\sim}{r}} \underset{\sim}{\psi} = \frac{d\underset{\sim}{\psi}}{d\underset{\sim}{r}} \tag{4}$$

where $\underset{\sim}{\nabla}_{\underset{\sim}{X}}$ and $\nabla_{\underset{\sim}{r}}$ are the notations for the gradient operators in the material and referential configurations, respectively. Taking into account of Eqs. (1) and (2), the

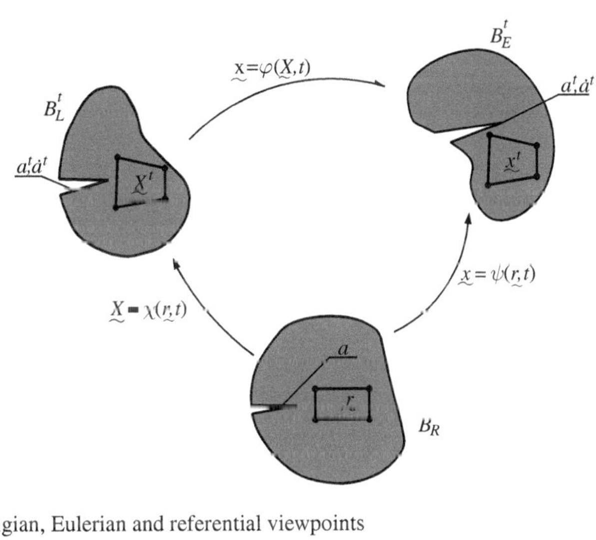

Fig. 1 Lagrangian, Eulerian and referential viewpoints

gradient operators are related to each other by introducing the Jacobian ($\underset{\sim}{J}$) of the transformation between the material and the referential configurations as:

$$\underset{\sim}{\nabla}_{\underset{\sim}{X}} \underset{\sim}{u} = \underset{\sim}{\nabla}_{\underset{\sim}{r}} \underset{\sim}{u} \underset{\sim}{J}^{-1}, \quad \underset{\sim}{J} = \frac{d\underset{\sim}{X}}{d\underset{\sim}{r}}. \tag{5}$$

In order to be consistent the transformation between material and reference coordinates systems, a one-to-one relationship must be enforced during the motion, i.e. $\det \underset{\sim}{J} \neq 0$. Since the proposed modeling is developed in dynamic framework, the governing equations require the computation of the time derivatives. Consistently to ALE formulation, the material and referential derivatives are defined as the rate of change of the function with $\underset{\sim}{X}$ and $\underset{\sim}{r}$ fixed:

$$\dot{f} = \frac{d}{dt} f\left(\underset{\sim}{X}, t\right)\Big|_{\underset{\sim}{X}}, \quad f' = \frac{d}{dt} f\left(\underset{\sim}{r}, t\right)\Big|_{\underset{\sim}{r}} \tag{6}$$

where the *dot* and the *prime* are used here and in the following to represent material and referential time rates, respectively. Material and the grid point velocities in the Lagrangian and referential configurations, i.e. $\underset{\sim}{v}$ and $\underset{\sim}{X}'$ respectively, in view of Eqs. (1) and (2), are defined as:

$$\underset{\sim}{v} = \frac{d}{dt} \varphi\left(\underset{\sim}{X}, t\right)\Big|_{\underset{\sim}{X}}, \quad \underset{\sim}{X}' = \frac{d}{dt} \chi\left(\underset{\sim}{r}, t\right)\Big|_{\underset{\sim}{r}} \tag{7}$$

According to ALE formulation, the time derivatives of a generic physical field, in referential and material configurations can be related by the following relationship [21]:

$$\dot{f} = f' - \underset{\sim}{X}' \frac{d}{d\underset{\sim}{X}} f\left(\underset{\sim}{X}, t\right) \tag{8}$$

where $\underset{\sim}{X}'$ represents the relative velocity of the grid points in the material reference system.

4 Governing Equations for Laminated Structures

The governing equations are presented for a 2D continuum model based on plane stress formulation in small deformations. In order to account for the behavior of a laminated structure, a multilayer formulation is adopted. In particular, the composite structure is modeled as an assembly of orthotropic elastic layers, which are

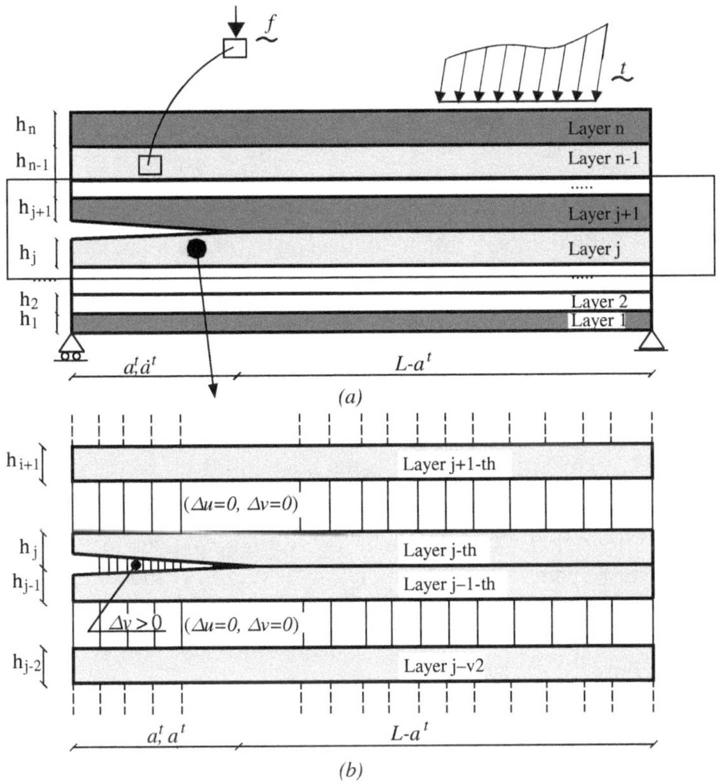

Fig. 2 (**a**) Schematic illustration of the delaminated model subjected to general loading conditions, (**b**) compatibility conditions at the interfaces

connected each other by perfect and imperfect interfaces, reproducing the continuity between each lamina or the presence of initial interfacial defects (Fig. 2a). Therefore, assuming that the laminate is formed by n laminas, the constitutive equations for the generic i-th lamina present the following form:

$$\sigma_{11}^i = c_{11}^i \frac{\partial u_1}{\partial X_1} + c_{12}^i \frac{\partial u_2}{\partial X_2},$$

$$\sigma_{22}^i = c_{12}^i \frac{\partial u_1}{\partial X_1} + c_{22}^i \frac{\partial u_2}{\partial X_2}, i = 1 \ldots n \tag{9}$$

$$\sigma_{66}^i = c_{66}^i \left[\frac{\partial u_1}{\partial X_2} + \frac{\partial u_2}{\partial X_1} \right],$$

where $c_{11}^i, c_{12}^i, c_{22}^i$ and c_{66}^i are the standard elastic moduli of the material. According to the Lagrangian multiplier method, compatibility conditions between each lamina are introduced to simulate the perfect adhesion or contact phenomena at the undelaminated or the delaminated interfaces, respectively (Fig. 2b):

$$\begin{array}{ll} \Delta u_i = u_{i+1} - u_i = 0, \quad \Delta v_i = v_{i+1} - v_i = 0, & \text{(undelaminated interfaces)} \\ \Delta v_i = v_{i+1} - v_i \geq 0, & \text{(delaminated interfaces)} \end{array} \tag{10}$$

The governing equations in the material configuration can be written by means of the principle of d'Alembert, taking into account, for each lamina, virtual works of inertial, external and internal forces:

$$\sum_{i=1}^{n} \int_{V_i} \underset{\sim}{\sigma} \delta \nabla \underset{\sim}{u} dV + \sum_{i=1}^{n} \int_{V_i} \rho \ddot{\underset{\sim}{u}} \delta \underset{\sim}{u} dV = \sum_{i=1}^{n} \int_{\Omega_i} \underset{\sim}{t} \delta \underset{\sim}{u} dA + \sum_{i=1}^{n} \int_{V_i} \underset{\sim}{f} \delta \underset{\sim}{u} dV \tag{11}$$

where $\underset{\sim}{\sigma}$ is the Cauchy stress tensor, $\underset{\sim}{t}$ is the traction forces vector on the free surfaces, $\underset{\sim}{f}$ is the volume forces vector, dV and dA are the volume and the loaded area in the material configuration. Consistently to ALE formulation, the motion of the body is described in the referential configuration and thus Eq. (11) should be reformulated in such a way to take into account for the transformation rule between Lagrangian and referential coordinate systems (Fig. 3). In particular, the first terms at LHS can be expressed by using Eq. (5):

$$\sum_{i=1}^{n} \int_{V_i} \underset{\sim}{\sigma} \delta \nabla \underset{\sim}{u} dV = \sum_{i=1}^{n} \int_{V_{ri}} \underset{\sim}{C} \left(\underset{\sim}{\nabla}_{r} \underset{\sim}{u} \underset{\sim}{J}^{-1} \right) \delta \left(\underset{\sim}{\nabla}_{r} \underset{\sim}{u} \underset{\sim}{J}^{-1} \right) \det(J)\, dV_r \tag{12}$$

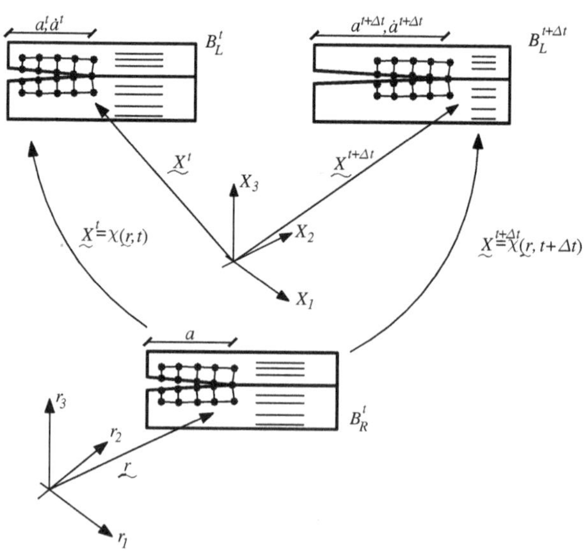

Fig. 3 ALE representation for fracture mechanics problem

where V_r is the volume in the referential configuration and $\underset{\sim}{C}$ is the elastic moduli matrix collecting stiffness coefficients of Eq. (9) for the i-th lamina. The second term at LHS requires the computation of the time second derivative of the displacement vector, which is evaluated by using Eq. (8) as:

$$\ddot{\underset{\sim}{u}} = \underset{\sim}{u}'' - 2\nabla_x \underset{\sim}{u}' \cdot \underset{\sim}{X}' - \nabla_x \underset{\sim}{u}\underset{\sim}{X}'' + \nabla_x \left(\nabla_x \underset{\sim}{u} \right) \underset{\sim}{X}' \underset{\sim}{X}' + \nabla_x \underset{\sim}{u} \nabla_x \underset{\sim}{X}' \underset{\sim}{X}' \tag{13}$$

Introducing the transformation rule provided by Eq. (5), Eq. (13) can be written in the referential configuration as:

$$\begin{aligned} \ddot{\underset{\sim}{u}} &= \underset{\sim}{u}'' - 2\nabla_{\underset{\sim}{r}} \underset{\sim}{u}' \underset{\sim}{J}^{-1} \cdot \underset{\sim}{X}' - \left(\nabla_{\underset{\sim}{r}} \underset{\sim}{u}\, \underset{\sim}{J}^{-1} \right) \cdot \underset{\sim}{X}'' \\ &+ \nabla_{\underset{\sim}{r}} \left(\nabla_{\underset{\sim}{r}} \underset{\sim}{u}\, \underset{\sim}{J}^{-1} \right) \underset{\sim}{J}^{-1} \underset{\sim}{X}' \underset{\sim}{X}' + \nabla_{\underset{\sim}{r}} \underset{\sim}{u}\, \underset{\sim}{J}^{-1} \cdot \left(\nabla_{\underset{\sim}{r}} \underset{\sim}{X}' \underset{\sim}{J}^{-1} \right) \underset{\sim}{X}' \end{aligned} \tag{14}$$

Therefore, the second term at LHS of Eq. (11), representing the work performed by the inertial forces in view of Eq. (14) is:

$$\begin{aligned} \sum_{i=1}^{n} \int_{V_i} \rho \ddot{\underset{\sim}{u}} \delta \underset{\sim}{u} dV &= \sum_{i=1}^{n} \int_{V_{ri}} \rho \Bigg[\underset{\sim}{u}'' - 2\nabla_{\underset{\sim}{r}} \underset{\sim}{u}' \underset{\sim}{J}^{-1} \cdot \underset{\sim}{X}' - \left(\nabla_{\underset{\sim}{r}} \underset{\sim}{u}\, \underset{\sim}{J}^{-1} \right) \cdot \underset{\sim}{X}'' \\ &+ \nabla_{\underset{\sim}{r}} \left(\nabla_{\underset{\sim}{r}} \underset{\sim}{u}\, \underset{\sim}{J}^{-1} \right) \underset{\sim}{J}^{-1} \underset{\sim}{X}' \underset{\sim}{X}' + \nabla_{\underset{\sim}{r}} \underset{\sim}{u}\, \underset{\sim}{J}^{-1} \cdot \left(\nabla_{\underset{\sim}{r}} \underset{\sim}{X}' \underset{\sim}{J}^{-1} \right) \underset{\sim}{X}' \Bigg] \delta \underset{\sim}{u} \det(J)\, dV_r \end{aligned} \tag{15}$$

The RHS of Eq. (11) , in the referential configuration assumes the following expression:

$$\sum_{i=1}^{n} \int_{\Omega_i} \underset{\sim}{t} \delta \underset{\sim}{u} dA + \sum_{i=1}^{n} \int_{V_i} \underset{\sim}{f} \delta \underset{\sim}{u} dV = \sum_{i=1}^{n} \int_{\Omega_r i} \underset{\sim}{t} \delta \underset{\sim}{u} \det(\bar{J}) d\Omega_r + \sum_{i=1}^{n} \int_{V_{ri}} \underset{\sim}{f} \delta \underset{\sim}{u} \det(J) dV_r \tag{16}$$

where $\det(\bar{J})$ is the determinant of a scalar metric representing the ratio of differential areas and Ω_r is the loaded area in the referential configuration. Finally substituting Eqs. (12), (15), and (16) into Eq. (11) the following equation is obtained:

$$\sum_{i=1}^{n}\int_{V_{ri}} \underset{\sim}{C}\left(\underset{\sim}{\nabla}_{r}\,\underset{\sim}{u}\underset{\sim}{J}^{-1}\right)\delta\left(\underset{\sim}{\nabla}_{r}\,\underset{\sim}{u}\underset{\sim}{J}^{-1}\right)\det(J)\,dV_r+\sum_{i=1}^{n}\int_{V_{ri}}\rho\left[\underset{\sim}{u}''-2\underset{\sim}{\nabla}_{r}\,\underset{\sim}{u}'\,\underset{\sim}{J}^{-1}\underset{\sim}{X}'\right.$$
$$-\left(\underset{\sim}{\nabla}_{r}\,\underset{\sim}{u}\quad\underset{\sim}{J}^{-1}\right)\cdot\underset{\sim}{X}''+\underset{\sim}{\nabla}_{r}\left(\underset{\sim}{\nabla}_{r}\,\underset{\sim}{u}\,\underset{\sim}{J}^{-1}\right)\underset{\sim}{J}^{-1}\underset{\sim}{X}'\,\underset{\sim}{X}'$$
$$\left.+\,\underset{\sim}{\nabla}_{r}\,\underset{\sim}{u}\,\underset{\sim}{J}^{-1}\cdot\left(\underset{\sim}{\nabla}_{r}\,\underset{\sim}{X}'\underset{\sim}{J}^{-1}\right)\underset{\sim}{X}'\right]\delta\underset{\sim}{u}\det(J)\,dV==\sum_{i=1}^{n}\int_{\Omega_r i}\underset{\sim}{t}\delta\underset{\sim}{u}\det(\bar{J})d\Omega_r$$
$$+\sum_{i=1}^{n}\int_{V_{ri}}\underset{\sim}{f}\delta\underset{\sim}{u}\det(J)dV_r \tag{17}$$

5 Dynamic Energy Release Rate and Crack Growth Criterion

The dynamic ERR is evaluated by using the J-integral concept [23]. In the literature several expressions exist to evaluate dynamic energy release rate based on *J*-integral formulation. In particular, the ERR is defined as the rate of mechanical energy flow out of the body and into the crack tip per unit crack advance [4] and it should be recovered by the limit of the following balance equation:

$$J=\lim_{\varepsilon\to 0}\oint_{\Omega}\left[(W+K)\,n_1-\underset{\sim}{t}\frac{\partial\underset{\sim}{u}}{\partial\underset{\sim}{X}}\right]ds \tag{18}$$

where n_1 is the component with respect to X_1 of the normal along integration contour, $\underset{\sim}{u}$ is the displacement, $\underset{\sim}{t}$ the traction vector, $\underset{\sim}{f}$ is the body force, ρ is the volume density, *W* is the strain energy density and *K* is the kinetic energy (Fig. 4). However, since the analysis is developed in dynamic framework, the integral function in Eq. (18) loses its path independent property. Moreover, the evaluation of the limit procedure reported in Eq. (18), involves numerical complexities in the integral

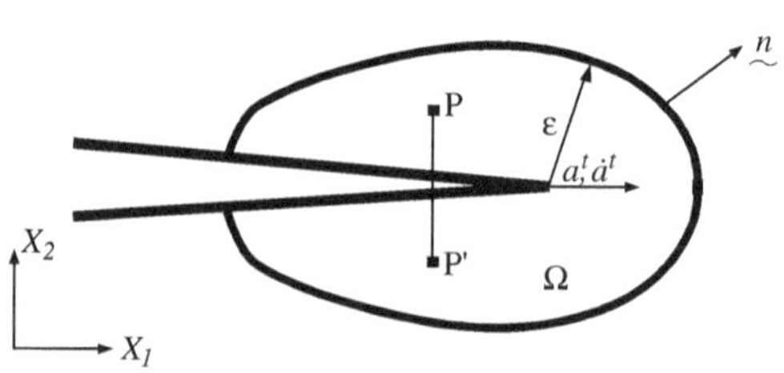

Fig. 4 Synoptic representation of the J-integral evaluation

calculation due to the high gradients of the crack tip fields. To overcome such difficulties, an alternative expression of the J-integral, which is path independent is utilized [23]:

$$J = \oint_{\partial\Omega} \left[(W + K) n_1 - \underset{\sim}{t} \frac{\partial \underset{\sim}{u}}{\partial \underset{\sim}{X}} \right] ds + \int_{\Omega} \left[\rho \left(\ddot{\underset{\sim}{u}} - \underset{\sim}{f} \right) \nabla \underset{\sim}{u} - \rho \dot{\underset{\sim}{u}} \nabla \dot{\underset{\sim}{u}} \right] dA \tag{19}$$

In those cases where mixed mode loading conditions are involved, the total value of the ERR and the corresponding modal components are evaluated by the use of the decomposition methodology of the J-integral expression [8, 22]. In particular, the J-integral is decoupled into an additive form, involving a direct decomposition of the ERR. It is worth noting that, the former formulation was proposed by Rigby and Aliabady [22] for static analysis and then it was generalized for a steady state crack growth by Greco and Lonetti [8]. In this framework, in order to take into account for the inertial contributions, new terms are introduced in the main equations, leading to similar expressions to those introduced in the previous formulations, but formally generalized to an unsteady crack growth, as:

$$\begin{aligned} J_{I,II} = G_{I,II} = & \oint_{\partial\Omega} \left[\left(W^{S,AS} + K^{S,AS} \right) n_1 - \sigma_{ij}^{S,AS} n_j \frac{\partial u^{S,AS}}{\partial x} \right] ds \\ & + \int_{\Omega} \left[\rho \left(\ddot{\underset{\sim}{u}}^{S,AS} - \underset{\sim}{f}^{S,AS} \right) \nabla \underset{\sim}{u}^{S,AS} - \rho \dot{\underset{\sim}{u}}^{S,AS} \nabla \dot{\underset{\sim}{u}}^{S,AS} \right] dA \end{aligned} \tag{20}$$

with $J = J_I + J_{II}$, (G_I, G_{II}) are the mode I and mode II ERR components, Ω is a closed path surrounding the crack tip front. It is worth noting that the superscripts $(^S)$ and $(^{AS})$ in Eq. (20), represent the symmetric and antisymmetric components with respect to a plane containing the crack tip. For the sake of brevity, the proof of Eq. (20) is not reported, because it is very similar to that involved in Rigby and Aliabadi [22] for the static case, except for the kinetic energy functionals and integral function over the crack tip region, which can be easily decoupled, according to the main methodology, as the sum of symmetrical and antisymmetrical fields, corresponding exactly to mode I and mode II contributions, respectively. However, expressions reported in Eq. (20) are validated numerically through comparisons arising from FE results based on VCCM.

In order to evaluate the crack growth phenomenon, a fracture criterion based on the crack tip variables should be introduced. Typically, dynamic crack criteria are commonly expressed by means of relationships involving stress intensity factor or ERR and crack tip speed. In the literature, a detailed discussion on the main problems regarding possible ways to analyze the crack tip behavior can be recovered in [3]. From an experimental point of view, several observations have shown that the crack growth is strictly dependent from the instantaneous crack tip speed [8, 24]. However, the literature dealing with the definition of a proper crack growth criterion is very limited. In the proposed modeling, the crack criterion is based on an asymptotic three parameters evolution law. In particular, it depends for low range of the crack tip speed on an initial value of the ERR, which is, typically, close to the initial crack toughness of the material. Moreover, as far as the ERR grows the crack tip speed reaches asymptotically the Rayleigh wave speed of the material, namely V_R. Therefore, the facture toughness criterion of the material is assumed to be governed by the following expression:

$$G_D = \frac{G_0}{1 - \left(\frac{c_t}{V_R}\right)^m} \tag{21}$$

where m is a material parameter predicting the evolution on the speed range, G_0 is the initial fracture toughness and c_t is the speed of the crack tip. It is worth noting that Eq. (21) should be useful from an engineering point of view, since it is based on well-known available data easily recoverable from the literature on the base of the material typology. However, other phenomena such as diffraction, scattering, attenuation, dispersion and local dynamic stress concentration are considered in this context to produce local negligible effects in comparison to those associated with larger scale displacements.

In the case of a mixed mode loading condition, the crack growth, typically, depends from the mixed mode ratio [3]. Therefore, a generalization of Eq. (21) is thus necessary to accurately reproduce the relationship between crack tip speed and ERR. In the proposed modeling it is assumed that, the dynamic fracture toughness of the ERR mode components follows a similar expression to Eq. (21):

$$G_{ID}(c_t) = \frac{G_{0I}}{1 - \left(\frac{c_t}{V_R}\right)^m}, \quad G_{IID}(c_t) = \frac{G_{0II}}{1 - \left(\frac{c_t}{V_R}\right)^m} \tag{22}$$

in which (G_{0I}, G_{0II}) correspond to the initial toughnesses of the ERR mode components. Therefore, the toughness criterion under mixed mode loading conditions is based on the following additive expression, which predicts the crack growth as far as it becomes positive:

$$g_f = \frac{G_I}{G_{ID}(c_t)} + \frac{G_{II}}{G_{IID}(c_t)} - 1 \leq 0 \tag{23}$$

where G_I, G_{II} are the actual mode components at a generic time.

6 Governing Equations for ALE Formulation

In the framework of ALE formulation the computational mesh may moves arbitrary with respect to the material body. In particular, the mesh movements should be addressed to reduce distortions of the mesh elements and to handle for changes of the geometry produced by the crack growth. It is worth noting that the equilibrium equations given by Eq. (17) depend on the position of the mesh points. Therefore, in order to account mesh movements of the domain, a rezoning mesh method is required. The aim of the mesh regularization technique is to provide an easy way to move the position of the crack tip front, keeping the computational mesh undistorted during the whole calculation. In the proposed modeling, crack tip motion is consistent to the crack growth criterion based on Eq. (23). Once the movement of the geometry is predicted by the crack criterion, a prescribed speed at the crack tip front is applied and thus the motion of the crack tip is activated. The mesh motion in the domain and in the boundary lines affected by the mesh motion should be regular to avoid distorted mesh and irregular elements, which strongly limit the accuracy of the main parameters predicting the advancing behavior.

In the proposed formulation, a smoothing variational method based on Winslow approach is utilized, in which the horizontal and vertical mesh displacements, namely $\Delta X_1 = X_1 - r_1$ and $\Delta X_2 = X_2 - r_2$, are evaluated by solving the following expressions [21]:

$$\nabla^2_{\underset{\sim}{X}} \Delta X_1 = 0, \qquad \nabla^2_{\underset{\sim}{X}} \Delta X_2 = 0. \tag{24}$$

Internal and external boundary conditions need to be introduced to reproduce the crack growth. Without loss of generality and for clearness of exposition, a single delamination model based on a double cantilever beam (DCB) scheme is considered. In particular, with reference to the geometric model reported in Fig. 5, the external boundary conditions must be written to reproduce a mesh motion, which is fixed in the vertical direction for all the contour lines and for the horizontal lines, the evolution of the mesh points should avoid any displacements along the longitudinal direction. Moreover, in order to simulate crack tip advance, mixed mode criterion given by Eq. (23) should be checked. Once the crack growth is predicted, the crack speed is determined by solving Eq. (23) and internal boundary conditions on the crack speed must be taken into account. Furthermore, in order to avoid high distortions of the mesh elements where the integration is performed to evaluate the ERR mode components, it is assumed that the mesh geometry in the region surrounding the crack tip is moved rigidly. Therefore, the mesh elements during the motion remain practically in the undistorted configuration, avoiding irregularities in the mesh description and inaccuracies in the integration procedure.

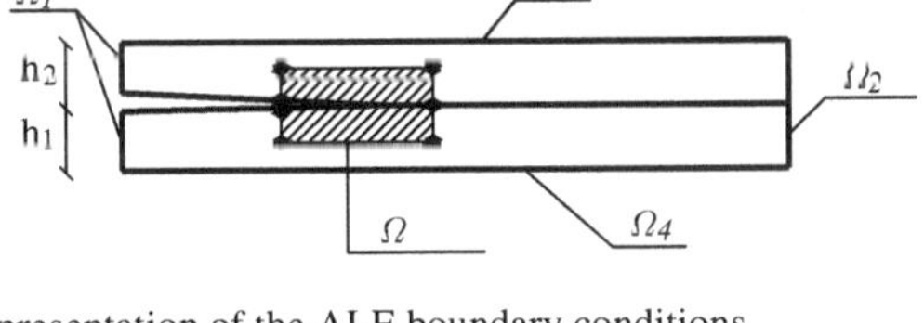

Fig. 5 Schematic representation of the ALE boundary conditions

With reference to model reported in Fig. 5 and assuming initial conditions to be homogeneous, the boundary conditions are defined by the following relationships:

$$\begin{array}{ll}
(\Delta X_1 = 0, \Delta X_2 = 0) & \text{on } \Omega_1 \cup \Omega_2, \\
\Delta X_2 = 0 & \text{on } \Omega_3 \cup \Omega_4 \\
\Delta X_1' = 0 \Leftrightarrow \mathit{if}\ g_f < 0 & \text{on } \Omega, \\
\Delta X_1' = c_t \Leftrightarrow \mathit{if}\ g_f \geq 0 & \text{on } \Omega, \\
\Delta X_2' = 0 & \text{on } \Omega \\
\multicolumn{2}{l}{\Delta X_1(0) = 0, \Delta X_2(0) = 0, \Delta \dot{X}_1(0) = 0, \Delta \dot{X}_2(0) = 0}
\end{array} \tag{25}$$

where g_f is the fracture criterion given by Eq. (21) or (23), G_D representing the dynamic fracture toughness and c_t is extracted by solving Eq. (21) or (23).

Weak forms of smoothing differential equations are derived, by multiplying Eq. (24) by a weight functions $w_1(X_1, X_2), w_2(X_1, X_2)$ and then integrating by part. Moreover, the boundary conditions regarding the prescribed crack tip speed, i.e. Eq. (25), is taken into account for by means of non-ideal weak constraint based on the Lagrangian multiplier method [25]. Therefore, the resulting equations regarding the ALE formulation are:

$$\begin{aligned}
&\int_V \nabla_{\underset{\sim}{X}} \Delta X_1 \nabla_{\underset{\sim}{X}} w_1 dA + \oint_\Omega \left[\delta\lambda \left(X_1' - c_t\right) + \lambda \delta \dot{X}_1\right] ds = 0, \\
&\int_V \nabla_{\underset{\sim}{X}} \Delta X_2 \nabla_{\underset{\sim}{X}} w_2 dA = 0,
\end{aligned} \tag{26}$$

where λ is the Lagrangian multiplier. Finally, introducing the vector $\Delta \underset{\sim}{X}^T = [\Delta X_1, \Delta X_2]$, Eq. (26) is written in the referential coordinate system by using transformation rules given by Eqs. (5) leading to the following relationship:

$$\int_{V_r} \left(\nabla_{\underset{\sim}{r}} \Delta \underset{\sim}{X} \underset{\sim}{J}^{-1}\right) \cdot \left(\nabla_{\underset{\sim}{r}} \underset{\sim}{w} \underset{\sim}{J}^{-1}\right) \det(J)\, dV_r + \oint_{\Omega_r} \left[\delta\lambda \left(\underset{\sim}{X}' - \underset{\sim}{c}_t\right) \underset{\sim}{i} + \lambda \delta \dot{\underset{\sim}{X}} \underset{\sim}{i}\right] \left(\bar{\bar{J}}\right) ds = 0, \tag{27}$$

with $\underset{\sim}{w}^T = [\underset{\sim}{w}_1, \underset{\sim}{w}_2]$ the weight function vector and $\underset{\sim}{i}^T = [1, 0]$ the propagation direction vector of the interfacial crack and $\underset{\sim}{c}_t^T = [c_t, 0]$ is the crack tip speed vector.

7 Finite Element Approximation

Governing equations given by Eqs. (17) and (27) introduce a non linear set of equations, which have been solved numerically, using a user customized finite element program, i.e. COMSOL Multiphysics TM version 3.5 [25]. Finite element expressions are written for 2D plane stress modeling, utilizing Lagrangian interpolation shape functions. The governing equations regarding the plane stress and ALE formulations are solved by using a finite element approach. In particular, isoparametric shape functions are used to represent plane stress and ALE variables:

$$\begin{aligned}
&\underset{\sim}{u}(r,t) = \sum_{i=1}^{n_d} \zeta_i \left(\underset{\sim}{r}\right) \underset{\sim i}{u}(t), \quad \underset{\sim}{\dot{u}}(r,t) = \sum_{i=1}^{n_d} \zeta_i \left(\underset{\sim}{r}\right) \underset{\sim i}{\dot{u}}(t), \quad \underset{\sim}{\ddot{u}}(r,t) = \sum_{i=1}^{n_d} \zeta_i \left(\underset{\sim}{r}\right) \underset{\sim i}{\ddot{u}}(t),\\
&\underset{\sim}{X}(r,t) = \sum_{i=1}^{n_d} \zeta_i (r) X_i, \quad \underset{\sim}{\dot{X}}(r,t) = \sum_{i=1}^{n_d} \zeta_i (r) \dot{X}_i, \quad \underset{\sim}{\ddot{X}}(r,t) = \sum_{i=1}^{n_d} \zeta_i (r) \ddot{X}_i\\
&\underset{\sim}{\lambda}(s) = \sum_{i=1}^{n_d} \zeta_i (r) \Lambda_i
\end{aligned} \tag{28}$$

where n_d represents the number of nodes of the master finite element. Substituting Eq. (28) in the governing equations, given by Eqs. (17) and (27), the following discrete equilibrium equations are obtained:

$$\begin{aligned}
&\sum_{i=1}^{n} \underset{\sim i}{M} \underset{\sim i}{U''} + \sum_{i=1}^{n} \underset{\sim i}{C} \underset{\sim i}{U'} + \sum_{i=1}^{n} \left(\underset{\sim i}{K} + \underset{\sim 0i}{K} + \underset{\sim 1i}{K} + \underset{\sim 2i}{K} \right) \underset{\sim i}{U} + \sum_{i=1}^{n} \underset{\sim i}{T} + \sum_{i=1}^{n} \underset{\sim i}{P} = 0\\
&\underset{\sim}{W} \cdot \Delta \underset{\sim}{X} + \underset{\sim}{Q} \cdot \Delta \underset{\sim}{X'} + \underset{\sim}{L} = 0,
\end{aligned} \tag{29}$$

where $\underset{\sim i}{M}$ is the consistent mass matrix, $\underset{\sim i}{K}$ is the stiffness matrix and $\left[\underset{\sim i}{C}, \underset{\sim 0i}{K}, \underset{\sim 1i}{K}, \underset{\sim 2i}{K} \right]$ are the damping and stiffness matrix associated to the ALE formulation, $\left(\underset{\sim i}{T}, \underset{\sim i}{P} \right)$ are the external load vectors, $\underset{\sim}{W}$ is the ALE discretization matrix, $\underset{\sim}{Q}$ and $\underset{\sim}{L}$ are the damping matrix and Lagrange multiplier vector concerning the ALE formulation. The algebraic equations are solved by using an implicit time integration scheme based on variable step-size backward differentiation formula (BDF). The equations are solved by means of a coupled approach in which no splitting operators are invoked. In particular, at a generic time step, the unknown variables are both the displacements vector and the position of the mesh points. It is worth noting that the coupled procedure with respect to the operator splitting technique offers a more detailed description of the advancing phenomena [21]. However, from a computational point of view, the operator splitting methods provide an easy way to implement and to compute the actual solution [26].

The integration procedure is able to evaluate the presence of non-symmetric convective terms given by the relative motion between material and referential systems. At a generic time, the non-linear equations are solved by using a Newton-Raphson iteration scheme and a linearization process of the above equations systems is performed in the step-by step solving procedure.

During the time integration, due to the fast speeds involved in the crack advance, a small time step size is utilized. In order to compute accurately the ERR with the aid of J-integral formula, a fine discretization mesh and a standard numerical integration method (quadrature) is adopted on the contour line and on the area surrounding the crack tip. At this aim, sensitivity analysis on the mesh size element is developed to chose a proper mesh discretization.

The time integration procedure at each iteration step checks the crack advance criterion; in the case it is satisfied, a prescribed crack speed is applied to the crack tip area by solving the toughness criterion given by Eq. (23) and thus moving boundary conditions are enforced to produce the crack tip motion. It is worth noting that during the crack tip motion, the discretization mesh is affected by high distortions in the crack region, leading to an inaccurate evaluation of the crack tip fields. However, in order to avoid such problems, a remeshing algorithm is applied. The remeshing procedure is performed by using COMSOL finite element program, by checking that the minimum value of the mesh quality parameter regarding the geometry of the element in the undistorted configuration should be grater then a fixed tolerance.

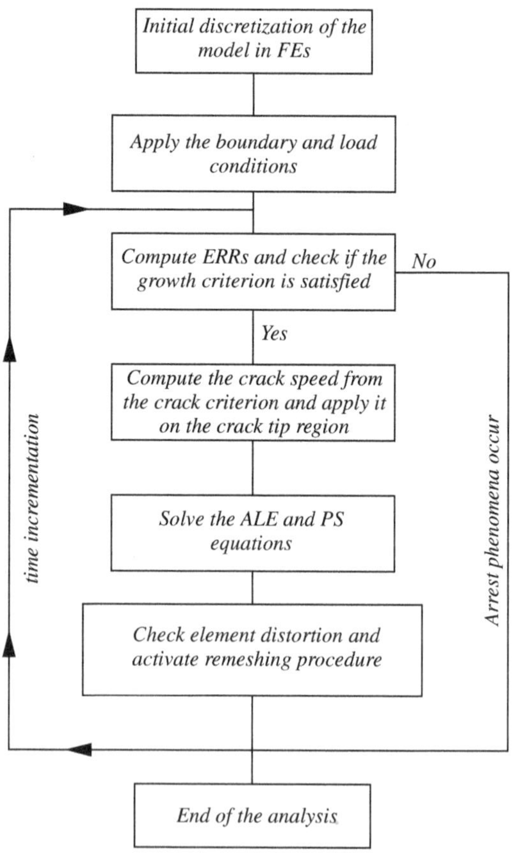

Fig. 6 Flow chart of the FEM integration algorithm

Once, this condition is not satisfied, the remeshing technique is performed on the basis of the deformed geometry in the actual reference system and a restart with an updating procedure from the previous converged time step is developed. More details on the remeshing algorithm can be recovered in Comsol [25]. A synoptic representation of the integration procedure is reported in Fig. 6.

8 Results

Comparisons with experimental results are performed to validate the proposed modeling for mode I, mode II loading conditions. Moreover, a sensitivity analysis is developed to investigate the dynamic behavior of interfacial cracks under mixed mode loading conditions.

A DCB specimen, under mode I loading condition is analyzed [27]. The material reported in the experimental tests is AS 3501-6 Graphite/Epoxy, whose properties are summarized in Table 1. The laminate is formed by unidirectional laminas and the specimen is 260 mm long, 20 mm wide and 3.7 mm thick. The loading scheme is based on a displacement controlled condition, in which the end points of the DCB model are opened at a loading rate equal to 0.1 mm/s. However, the crack propagation is produced introducing a strip of adhesive film at the crack tip, which enforces the crack to grow at high speeds. In order to reproduce correctly the experimental results, at first the laminate ends are displaced statically to an initial value of the ERR, leaving the crack tip fixed in the initial position. During this stage, the specimen stores strain energy, which is released once the crack tip is allowed to grow. It is worth noting that since the relationship between crack tip speed and ERR toughness is not provided in the experimental results, the material parameters (m, G_0), involved in the definition of the crack criterion, are taken as adjustable variables.

In Fig. 7 comparisons with experimental results, in terms of time history of the crack tip displacements are reported. The agreement between proposed and experimental results is noted. In Fig. 8, the evolution the strains and kinetic energies are also reported. The analysis shows that during the crack growth, the strain energy is transformed into kinetic energy. Moreover, the crack tip speed reaches its maximum value especially during the initiation phase with value comparable to the ones of the material wave characteristic. Once, the model is validated through comparisons with experimental data, sensitivity results are developed. In particular, the influence of the loading rate on the crack growth is analyzed, in Fig. 9, in which the time history of the crack tip speed for different loading rates is reported. Moreover, in Fig. 10 the relationships between the crack speed, the crack displacement and

Table 1 Mechanical properties of unidirectional fiber-reinforced AS4 graphite epoxy

Material	E_1(MPa)	$E_2 = E_3$(MPa)	$G_{12} = G_{23}$(MPa)	ρ (Kg/m^3)	$\nu_{12} = \nu_{13}$
AS4 graphite/epoxy	142E3	10.3E3	7.2E3	1,580	0.3

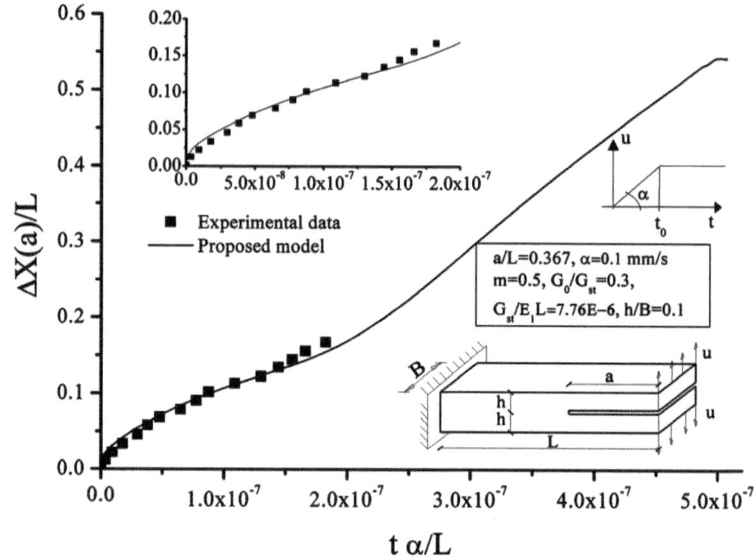

Fig. 7 Mode I dynamic crack growth in a DCB scheme. Comparison between experimental data [27] and proposed results: time history of the crack tip displacement ΔX(a)

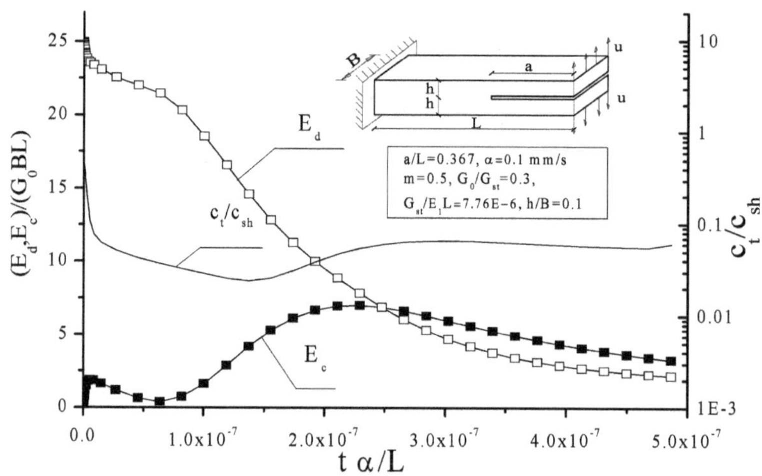

Fig. 8 Mode I dynamic crack growth in a DCB scheme. Time history of the strain energy (E_d), kinetic energy (E_c) and crack tip speed (c_t)

loading rates are reported. The analysis shows that, in the initiation phase, the rate effects produce the major amplifications of the crack tip fields. As a matter of this fact, the crack tip speed grows very rapidly reaching his maximum value during this stage. Subsequently the evolution of the crack tip speed tends to be regularized and a stable crack growth is observed.

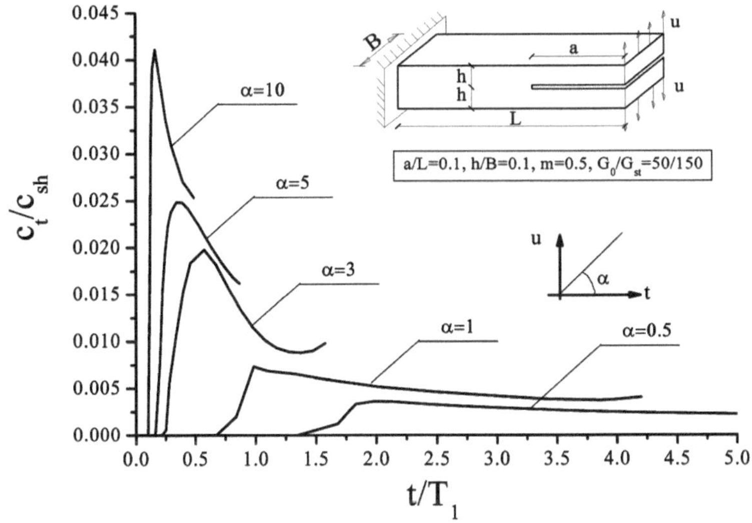

Fig. 9 Mode I dynamic crack growth in a DCB scheme. Time history of the crack tip speed (c_t) for different rates of the external loading

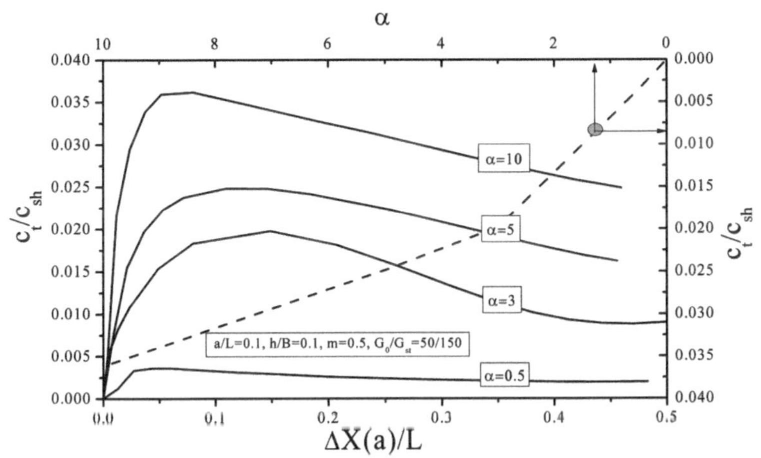

Fig. 10 Mode I dynamic crack growth in a DCB scheme. Relationship between crack tip speed (c_t), opening end displacement($\Delta X(a)$) and rate of the external loading (α)

Comparisons with experimental results available from the literature (Tsai et al. 2001) on a glass/epoxy S2/8553 unidirectional composite are proposed. The structural scheme refers to a modified end-notched flexural (ENF) scheme. The material properties are reported in Table 2. The specimen is 278 mm long, 12.7 mm wide and 11.6 mm thick, whereas the loading rate at the left end points is 0.025 mm/s. The experimental methodology utilized to analyze the structural behavior is similar

Table 2 Mechanical properties of unidirectional fiber-reinforced S2 glass epoxy

Material	E_1(MPa)	$E_2 = E_3$(MPa)	$G_{12} = G_{23}$ (MPa)	ρ (Kg/m^3)	$\nu_{12} = \nu_{13}$
S2/8553 glass/epoxy	43E3	12.7E3	4.46E3	2,100	0.3

to that reported for the previous case involving mode I loading condition. Moreover, since the experimental tests do not provide any data on the relationship between fracture toughness and crack speed, the material parameters (m, G_0) , involved in the definition of the crack criterion, are taken as adjustable variables. In Fig. 11, comparisons between proposed and experimental results in terms of crack tip displacements are reported. Moreover, the time histories of the crack speed and the dynamic amplification factors of the ERR are reported in Figs. 12 and 13, respectively. It is worth noting that the trend of the crack tip evolution predicted by the proposed model is in agreement with the experimental data. However, during the first part corresponding to the initiation phase of the growth phenomenon, the prediction of the crack motion is not fully accurate. This behavior can be justified by the experimental methodology, since especially in the initiation phase, in order to produce high speeds during the crack advance, a thin adhesive layer is utilized. Moreover, the experimental data show that, during the initial phase, non standard complex phenomena affect the crack growth (contact, scattering dissipation), which are not taken into account from the utilized toughness criterion.

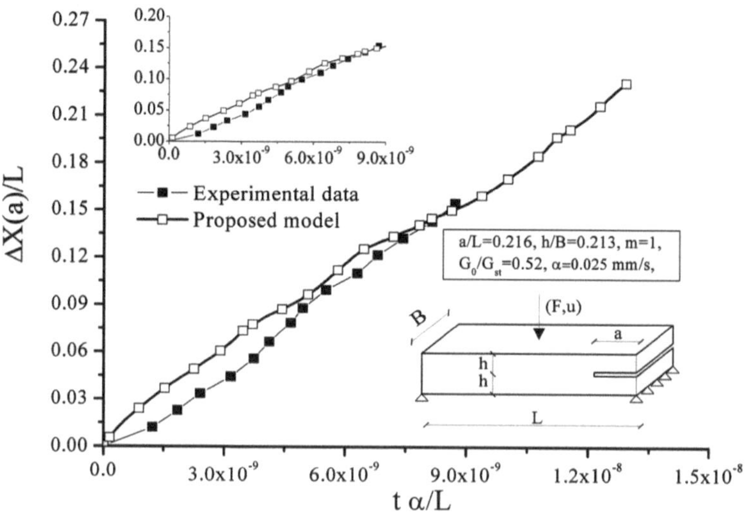

Fig. 11 Mode II dynamic crack growth in a ENF scheme. Comparison between experimental data [2] and proposed results: time history of the crack tip displacement $\Delta X(a)$ normalized on the total length of the laminate (L)

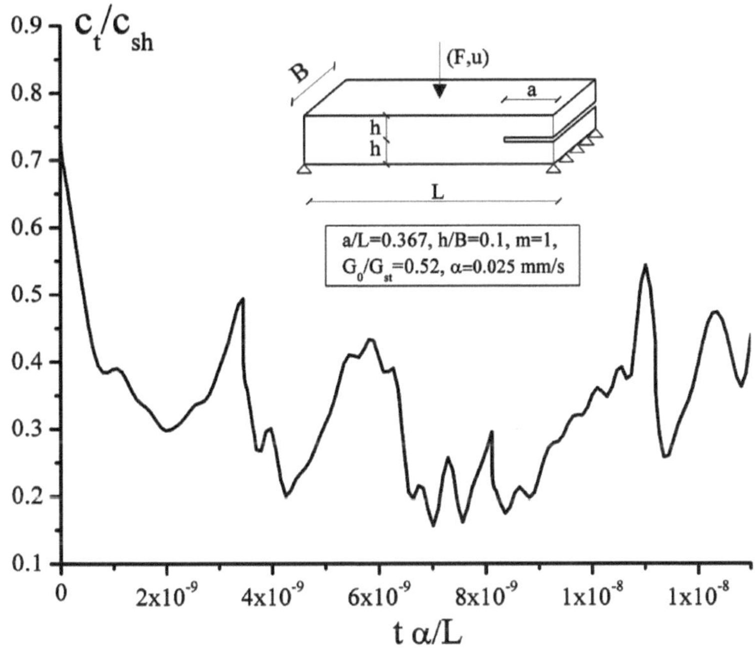

Fig. 12 Mode II dynamic crack growth in a ENF scheme. Time history of the crack tip speed (c_t) normalized on the shear wave speed of the material (c_{sh})

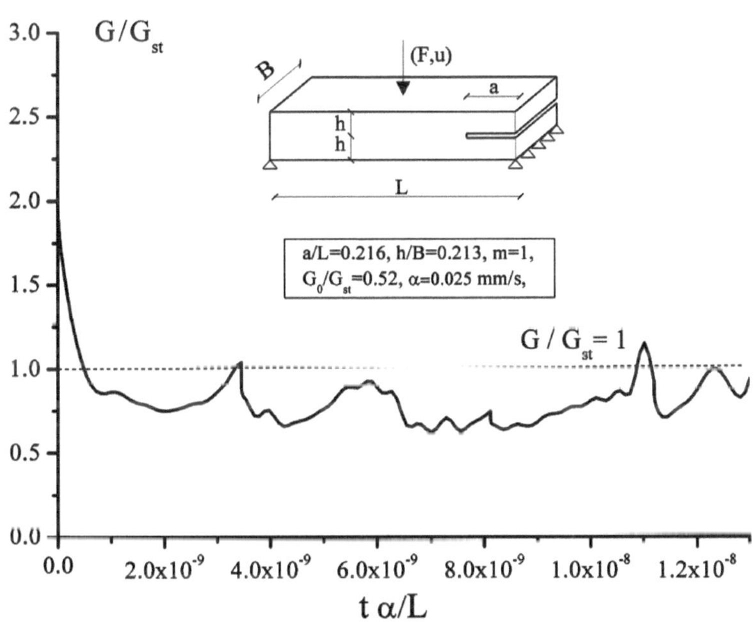

Fig. 13 Mode II dynamic crack growth in a ENF scheme. Time history of the ERR (G) normalized on the static value (G_{st})

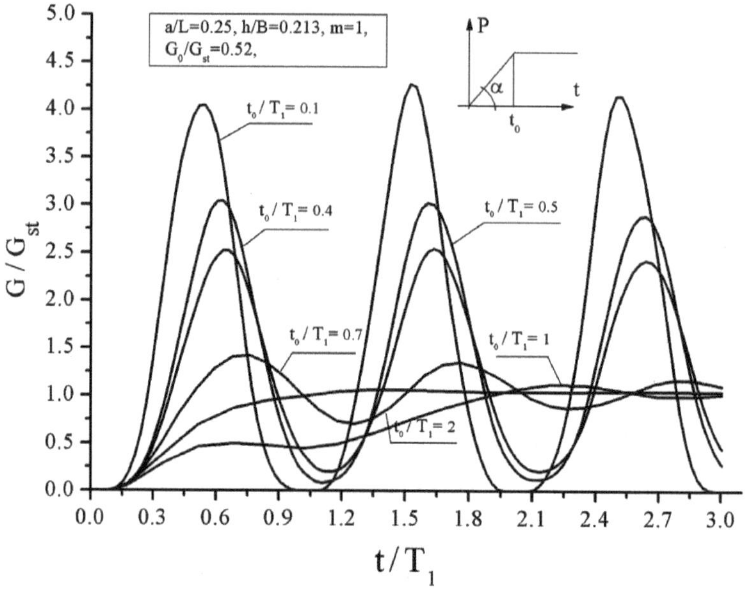

Fig. 14 Mode II dynamic crack growth in a ENF scheme. Time history of the dynamic amplification factors (G/G_{st}) for different rates of the external loading, with G_{st} representing the static value of the ERR

Results concerning dynamic amplification effect produced by different loading rates are reported in Fig. 14. In particular, the time history of the dynamic amplification factor of the ERR is evaluated in terms of the loading rate of the applied loads. The results show that when the loading rate is much greater than the first period of vibration, (T_1), the solution coincides with the static one, whereas as far as the loading rate grows the amplification effects are increased leading to high value of the DAFs.

Finally, an analysis on a loading case involving mixed mode is developed. In particular, the laminate structure refers to a DCB scheme with the same material and adjustable parameters previously reported for the case involving pure mode I loading condition. The laminate geometry presents the following data:

$$L = 268\,\text{mm}, B = 14\,\text{mm}, H = 6.8\,\text{mm}, \text{ with } a/L = 0.1, h_1/h_2 = 0.66$$

where H is the total thickness of the laminate, h_1 and h_2 are the thicknesses of the lower and upper sublaminates. The main aim of the proposed investigation is to evaluate on a typical laminated composite structure, the effects of the loading rate on the ERR computation and the crack growth. In particular, the analysis is performed on a laminate subjected to opening end displacements. The time evolution is based on a linear-constant function, whereas the amplitude of the maximum applied forces is chosen in such a way to produce crack growth at different intensities and loading

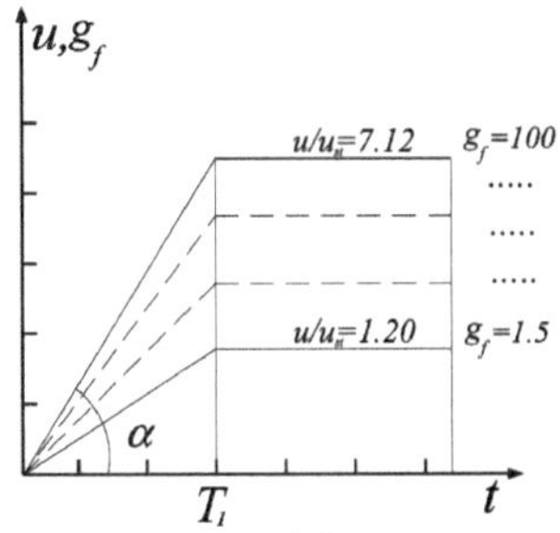

(a)

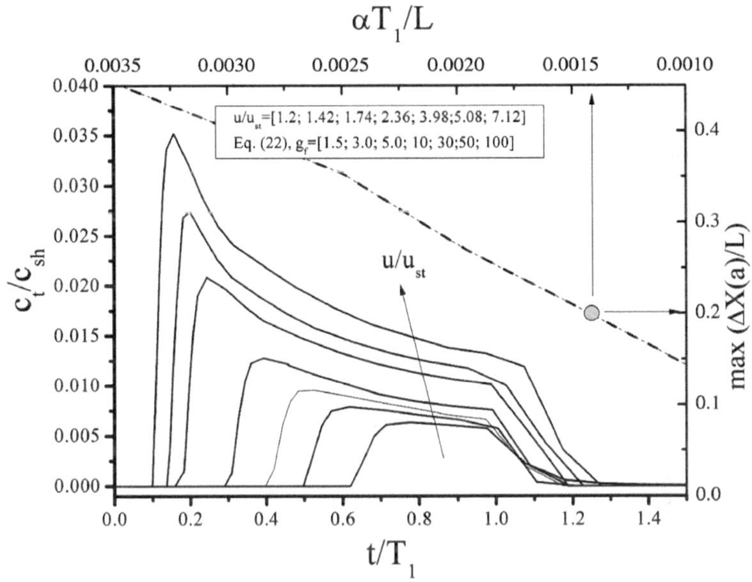

(b)

Fig. 15 Mixed mode dynamic crack growth in a DCB scheme. (**a**) Schematic representation of the loading curves. (**b**) Time history of the crack tip speed for different loading curves

rates (Fig. 15a). In Fig. 15b, the time history of the crack tip speed normalized on the first fundamental period of the laminate (T_1) is reported for different load curves. In Fig. 16, time history of the tip displacement relationship as a function of the loading rate is proposed. The results show that for larger values of the loading rate, both displacements and speeds at the crack tip tend to increase. In particular, during the initiation phase, i.e. when the crack growth is enforced by the loading curve, the displacements and the crack tip speeds grow very rapidly. In this phase, the kinetic energy of the system is increased, producing high speeds in the crack growth. Subsequently, as far as the constant value in the loading curve is reached, the speed of the crack is reduced and then arrest phenomena of the crack are observed.

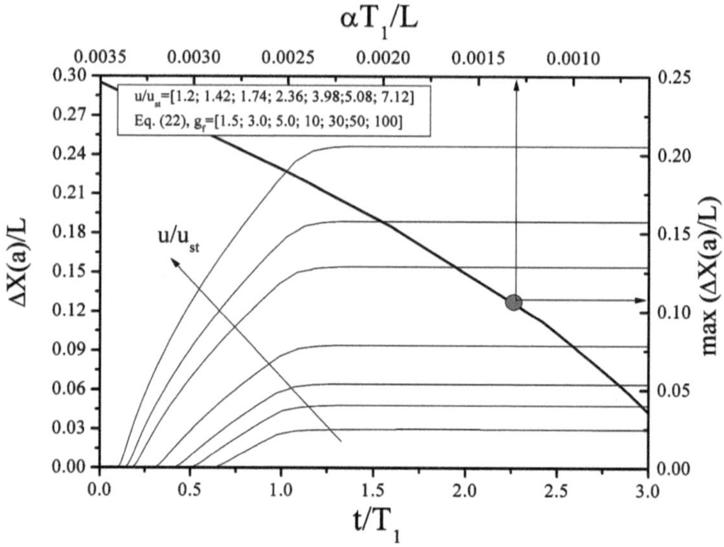

Fig. 16 Mixed mode dynamic crack growth in a DCB scheme. Time history of the of the crack tip displacement for different loading curves

9 Conclusion

A FE model to predict dynamic crack growth in composite structure is proposed. The moving mesh strategy combined with a Fracture Mechanics approach is able to predict properly the time dependent behavior of delamination phenomena. The proposed modeling is based on a generalized mixed mode dynamic fracture toughness, which depends on a limited number of adjustable variables. Comparisons with experimental results for loading conditions involving pure mode I and mode II cases at high speeds of the crack tip show the reliability of the proposed formulation. The parametric study shows the influences of the loading rates effects on the dynamic ERR, which determine high amplifications of the fracture variables characterizing the crack tip evolution.

References

1. E.J. Barbero, *Finite Element Analysis of Composite Materials*, CRC, Boca Raton, FL (2007)
2. J.L. Tsai, C. Guo, C.T. Sun, Dynamic delamination fracture toughness in unidirectional polymeric composites. Compos. Sci. Technol. **61**(1), 87–94 (2001)
3. K. Ravi-Chandar, *Dynamic Fracture* (Elsevier, Amsterdam/London, 2004)
4. L.B. Freund, *Dynamic Fracture Mechanics* (Cambridge University Press, Leyden, 1990)
5. K. Ravi-Chandar, W.G. Knauss, An experimental investigation into dynamic fracture: I. Crack initiation and arrest. Int. J. Fract. **25**, 247–262 (1984)

6. J.W. Hutchinson, Z. Suo, Mixed mode cracking in layered materials. In *Advances in Applied Mechanics*, vol. 29, ed. by J.W. Hutchinson, T.Y. Wu (Academic Press, New York, NY, 1992), pp. 63–191
7. D. Bruno, F. Greco, P. Lonetti, Dynamic mode i and mode ii crack propagation in fiber reinforced composites. Mech. Adv. Mater. Struct. **16**, 442–455 (2009)
8. F. Greco, P. Lonetti, Mixed mode dynamic delamination in fiber reinforced composites. Compos. B. Eng. **40**(5), 379–392 (2009)
9. J. Lambros, A.J. Rosakis, Dynamic decohesion of bimaterials: Experimental observations and failure criteria. Int. J. Solids. Struct. **32**, 2677–2702 (1995)
10. A. J. Rosakis, O. Samudrala, R.P. Singh, A. Shukla, Intersonic crack propagation in bimaterial systems. J. Mech. Phy. Solids. **46**, 1789–1814 (1998)
11. K. Ravi-Chandar, Dynamic fracture of nominally brittle materials. Int. J. Fract. **90**(1), 83–102 (1998)
12. G. Yagwa, Y. Sakai, Y. Ando, in *Fast Fracture and Crack Arrest*, vol. 627, ed. by G.T. Hahnand, M.F. Kanninen (ASTMSTP, Philadelphia, PA, 1977), pp. 109–122
13. P.N.R. Keegstra, J.L. Head, C.E. Turner, in *Numerical Methods in Fracture Mechanics*, ed. by A.R.L. Moore, D.R.J. Owen (University College, Swansea, 1978), pp. 634–647
14. M.F. Kanninen, C.H. Papelar, *Advanced Fracture Mechanics* (Oxford University Press, New York, NY, 1985)
15. D. Xie, S.B. Biggers, Calculation of transient strain energy release rates under impact loading based on the virtual crack closure technique. Int. J. Impact. Eng. **34**(6), 1047 1060 (2007)
16. A. Corigliano, S. Mariani, A. Pandolfi, Numerical analysis of rate-dependent dynamic composite delamination. Compos. Sci. Technol. **66**, 766–775 (2006)
17. L. Iannucci, Dynamic delamination modelling using interface elements. Comput. Struct. **84**, 1029–1048 (2006)
18. H.M. Koh, R.B. Haber, Formulation of the Eulerian-Lagrangian kinematic description. J. Appl. Mech. **53**, 839–845 (1986)
19. T. Nishoka, S.N. Atluri, *Computational Methods in the Mechanics of Fracture*, ed. by S.N. Atluri (Elsevier Science Publishers, NewYork, NY, 1986), pp. 335–383.
20. T. Nishioka, The state of the art in computational dynamic fracture mechanics. JSME Int. J. Ser. A **37**(4), 313–333 (1994)
21. J. Donea, A. Huerta, J.P.H. Ponthot, A. Rodrıguez-Ferran, In *Arbitrary Lagrangian-Eulerian Methods Encyclopedia of Computational Mechanics,* eds. E. Stein, R. de Borst, T. Hughes John (Wiley & Sons, New York, 2004)
22. R.H. Rigby, M.H. Aliabadi, Decomposition of the mixed-mode J-integral-revisited. Int. J. Solids. Struct. **35**, 2073–2099 (1998)
23. T. Nishioka, H. Tokudome, M. Kinoshita, Dynamic fracture-path prediction in impact fracture phenomena using moving finite element method based on Delaunay automatic mesh generation. Int. J. Solids. Struct. **38**, 5273–5301 (2001)
24. A.R. Shahani, M.R. Amini Fasakhodi, Finite element analysis of dynamic crack propagation using remeshing technique. Mater. Des. **30**, 1032–1041 (2008)
25. Comsol, *Reference Manual* (Comsol AB, Stockholm, 2007)
26. J.P. Ponthot, T. Belytschko, Arbitrary Lagrangian-Eulerian formulation for element-free Galerkin method. Comput. Methods Appl. Mech. Eng. **152**, 19–46 (2001)
27. C. Guo, C.T. Sun, Dynamic mode-I crack-propagation in a carbon/epoxy composite. Compos. Sci. Technol. **58**, 1405–1410 (1998)

Computational Analysis of Loading–Unloading and Non-homogeneity Effects in Metallic Hollow Sphere Structures

Branca F. Oliveira, Luiz A.B. da Cunda, Andreas Öchsner, and Guillermo J. Creus

1 Introduction

Metallic hollow sphere structures (MHSS) differ from traditional open or closed-cell metal foams [1], which are produced by expansion of gases among other methods. MHSS are characterised by more consistent geometrical and mechanical properties [2, 3].

Fiedler and Öchsner [3] and Fiedler [4] address the mechanical properties of bonded MHSS to determine Young's modulus and initial yield stress dependence on loading direction and geometrical arrangement, and include experimental load-displacement results under compressive load, both for partially glued (Fig. 1) and syntactic foams.

A significant issue is the way in which the behaviour of the metal foam is related to the behaviour of a single cell [5]. For the case of MHSS a very interesting analysis of the elasto-plastic behaviour is given in Lim et al. [6], considering single and multiple spheres and comparing numerical and experimental results. This work indicates dispersion of the experimental results, as well as significant differences between experimental and numerical data. Some of these differences may be attributed to the presence of neighbouring spheres, an effect analysed in a previous paper [7]. The influence of defects on dispersion of results has been numerically studied also by Kepets et al. [8] and may depend on several factors. At any rate, the statistical dispersion of experimental results on foams (elastic and plastic properties) is much larger than for the parent metal. For example, Ramamurty and Paul [9] found that to obtain a variability of the order of 4%, at least seven tests with foams have to be performed, while for the parent material, a variability of 2% is obtained with 3 tests. As it is well known [10] that the statistical deviation of results on damage and fracture is much larger than those related to plastic behaviour.

B.F. Oliveira (✉)
Virtual Design Group, Universidade Federal do Rio Grande do Sul, 90035-190
Porto Alegre, RS, Brazil
e-mail: branca@ufrgs.br

A. Öchsner et al. (eds.), *Materials with Complex Behaviour*, Advanced Structured Materials 6, DOI 10.1007/978-3-642-12667-3_6, © Springer-Verlag Berlin Heidelberg 2010

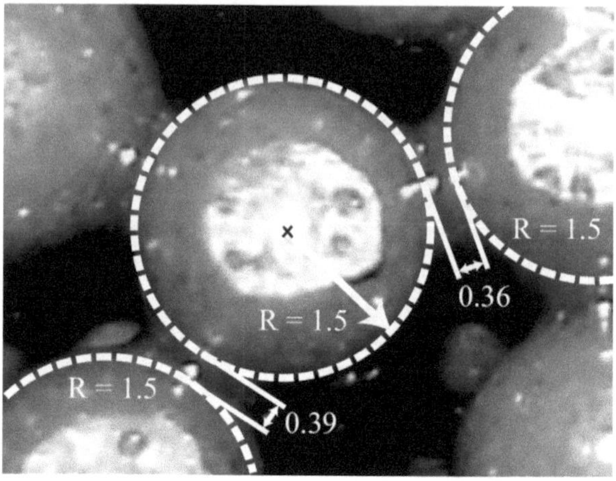

Fig. 1 Partially glued metallic hollow sphere structure [4]

The present work focuses on the computational analysis of two particular forms of MHSS [3]. Results are compared with the experimental results of Fiedler [4]. The behaviour of single isolated spheres is also numerically analysed, focusing on loading-unloading behaviour.

The content of the chapter is as follows. Section 2 gives a short account of the Gurson damage formulation to be used in the analyses. Section 3 studies the behaviour of a single sphere, considering loading-unloading situations. Section 4 presents a numerical study for a MHSS material, comparing the results obtained with the experimental results of Fiedler [4].

2 Gurson Damage Model

The Gurson damage model was developed to describe the mechanical effect of high plastic deformations in ductile metals. The loss of resistance is governed by the porosity level. The (isotropic) damage variable employed is the volumetric void fraction, represented by f and defined by $f = V_v/V$, where V_v is the volume of voids in a representative small volume V, corrected for effects such as stress concentration; f is defined at each point of the continuum. The presence of voids alters the elasto-plastic constitutive relations. The equations usually employed in computational damage analyses, the Gurson–Tvergaard model [11–13], consider a yield surface defined by

$$\Phi = \sqrt{\frac{3}{2}s_{ij}s_{ij}} - \bar{\omega}\sigma_y = 0, \tag{1}$$

where

$$\bar{\omega} = \left[1 - 2\alpha_1 f \cosh\left(-\frac{\alpha_2 3p}{2\sigma_y}\right) + \alpha_3 f^2\right]^{1/2}, \tag{2}$$

$$s_{ij} = \sigma_{ij} + p\delta_{ij}, \tag{3}$$

where σ_{ij} are the Cauchy stresses and σ_y is the yield stress in simple tension. The so called hydrostatic pressure p, or simply pressure, is defined as $p = -1/3\ \sigma_{ii}$. The adopted values [14, 15] to the yield surface material parameters in Eq. (2) are $\alpha_1 = 1/f_U = 1.5$, $\alpha_2 = 1.0$ and $\alpha_3 = (\alpha_1)^2$. The parameter $f_U = 1/\alpha_1$ is the maximum volumetric void fraction admissible before rupture in the absence of pressure. Another possible interpretation for the α_1 and α_2 parameters is that they work as multipliers acting on porosity f and pressure p, respectively.

In Fig. 2, yield surfaces for different levels of void content are shown, in a plot of normalised deviatoric stress versus normalised pressure.

It can be seen that the elastic domain depends on the hydrostatic pressure p. When the volumetric void fraction f decreases, the influence of pressure decreases, leading to a larger elastic domain. For $f = 0$, the model reduces to the von Mises model, which is independent of hydrostatic pressure. It should be noted here that in the absence of hydrostatic pressure, the coefficient $\bar{\omega}$ reduces to

$$\bar{\omega} = 1 - \alpha_1 f = 1 - \frac{f}{f_U}. \tag{4}$$

The plastic strain rate tensor is given by

$$D_{ij}^p = \dot{\lambda}\frac{\partial \Phi}{\partial \sigma_{ij}}, \tag{5}$$

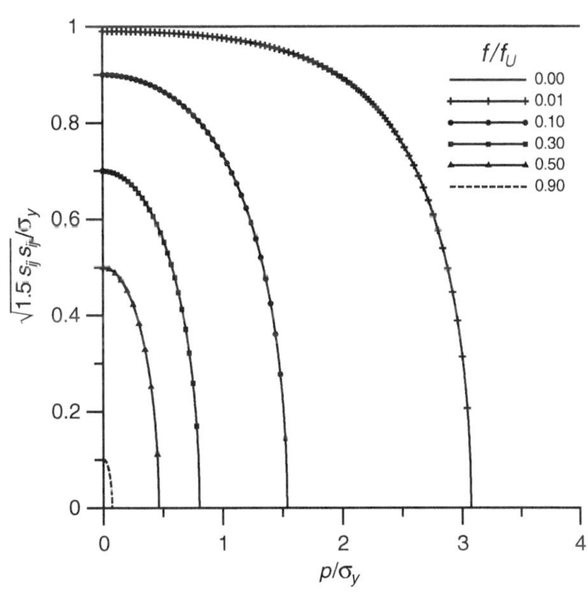

Fig. 2 Gurson yield surface for different values of volumetric void fraction content

where $\dot{\lambda}$ is the plastic multiplier. The equivalent plastic strain rate is defined by

$$\dot{\varepsilon}^p = \sqrt{2/3D^p_{ij}D^p_{ij}} \tag{6}$$

The basic mechanisms of damage evolution are nucleation, growth and coalescence of voids. Nucleation occurs mainly due to material defects, in the presence of tension. Growth occurs when the voids (pre-existent or nucleated) change their size according to the volume change in the continuum. Coalescence is related to the fast rupture process that occurs after that the volumetric void fraction reaches a limit, indicated by f_C. Coalescence consists in the union of neighbour voids due to the rupture of a ligament.

The equations that govern damage evolution are modelled in a simplified form as follows. First, it is assumed that the total void rate is given by

$$\dot{f} = \begin{cases} \dot{f}_n + \dot{f}_g & f \leq f_C \\ \dot{f}_c & f > f_C \end{cases}, \tag{7}$$

where $\dot{f}_n$ is the void nucleation rate, $\dot{f}_g$ is the void growth rate and $\dot{f}_c$ is the void coalescence rate. Thus, as long as f is smaller than a characteristic value f_C, only nucleation and growth develop. Above f_C, only coalescence takes place.

The nucleation rate is proportional to the rate of equivalent plastic strain

$$\dot{f}_n = A(\varepsilon^p)\dot{\varepsilon}^p. \tag{8}$$

For $A(\varepsilon^p)$, Chu and Needleman (1980) propose the statistical distribution

$$A(\varepsilon^p) = \frac{f_N}{s_N\sqrt{2\pi}} \exp\left[-\frac{1}{2}\left(\frac{\varepsilon^p - \varepsilon_N}{s_N}\right)^2\right], \tag{9}$$

where f_N is the nucleation void volumetric fraction, ε_N is the plastic strain value for nucleation and s_N is the standard deviation for the distribution. Sometimes it is assumed that nucleation takes place only with negative pressure (i.e. tension) [17, 18], which implies that $A(\varepsilon^p) = 0$ if pressure is positive.

Voids increase or decrease their volume according to the volume variation in the continuum. The growth rate of voids is controlled by mass conservation through the expression

$$\dot{f}_g = (1 - f)D^p_{ii}. \tag{10}$$

Coalescence is usually described [19] by the relation

$$\dot{f}_c = \frac{f_U - f_C}{\Delta\varepsilon}\dot{\varepsilon}^p, \tag{11}$$

where $\Delta\varepsilon$ is a material parameter.

An alternative way of taking coalescence into account [14, 17] is to change the volumetric void fraction f in the Gurson yield surface, Eqs. (1) and (2), by a corrected volumetric void fraction f^* given by the Eq. (12), with the material parameter f_F representing the volumetric void fraction that leads to rupture. In this case, only nucleation and growth are considered in Eq. (7). This is the formulation adopted in the present analysis.

$$f^* = \begin{cases} f & f < fc \\ fc + \frac{(1.0-fc)}{(f_F-fc)}(f - fc) & f > fc \end{cases}. \tag{12}$$

3 Loading–Unloading Analysis of a Single Sphere

In this section the effect of reversal loading on the behaviour of a single sphere is studied (Fig. 3). A similar sphere, compressed between two parallel planes, was studied experimentally and numerically by Lim et al. [6] considering elastoplastic behaviour, without damage considerations. Oliveira et al. [7] studied a sphere with damage and simplified neighbouring cells considerations. The difference between the spheres studied by Lim et al. [6] and Oliveira et al. [7] is the hardening adopted, kinematic in the first study and isotropic in the second.

In the present study, employing isotropic hardening, a single sphere is first compressed in vertical direction to a final height of 10% of its initial radius. In a second stage, after the vertical compression is removed, the sphere is compressed in the horizontal radial direction by a cylindrical surrounding surface until an external radius of 0.62 mm is reached. The first stage develops from time 0 to 0.9 and the second stage develops from 0.9 to 1.5. Figure 3 shows the mesh and boundary conditions employed. The sphere is modeled as an axisymmetric body with 375 linear quadrilateral elements. Because of symmetry considerations, only a quarter of the sphere

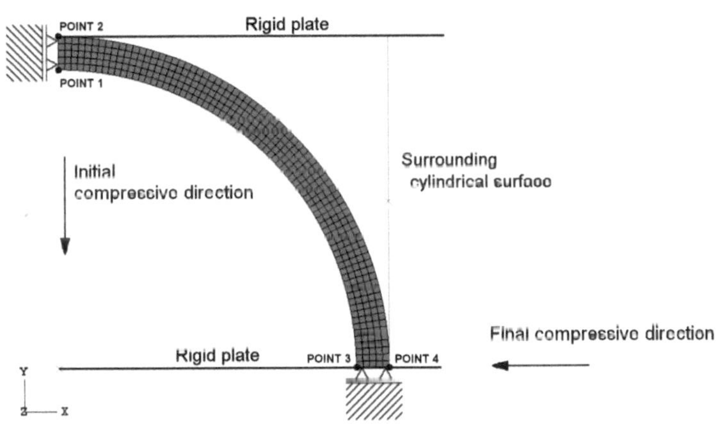

Fig. 3 Mesh and boundary conditions employed in the sphere compression

is modeled. The plate of the test machine is modeled as a rigid plane with prescribed displacements and the contact algorithm is activated. The sphere analysed had an external radius of 1.0 mm and wall thickness of 0.1 mm. The material constants used [6] were: elastic modulus $E = 200$ GPa, initial yield stress 200 MPa and Poisson's ratio $\nu = 0.3$. The Gurson model with isotropic hardening was employed, with a hardening modulus of 250 MPa.

To define porosity, one approach must be adopted: strain governed nucleation of voids or an initial fraction of voids. Both approaches were considered in this work. When strain governed nucleation, Eqs. (8) and (9), was adopted, the material constants employed were $f_N = 5\%$, $\varepsilon_N = 0.3$ and $s_N = 0.1$. When an initial fraction of voids was considered, $f_0 = 5\%$ was employed. Figures 4 and 5 show subsequent stages of vertical and lateral compression respectively.

Figures 6, 7 and 8 present the evolution of pressure, equivalent plastic strain and porosity at points 3 and 4 indicated in Fig. 3, considering strain governed nucleation model. In Fig. 9, the evolution of porosity using the initial porosity formulation is presented.

Looking at the results for point 3, it can be noticed that, if the strain governed nucleation model is employed, although this is the point with the highest equivalent plastic strain, it develops no damage during the deformation cycle (Fig. 8), because the equivalent plastic strain nucleation ε_N is reached in compression. It is a characteristic of the strain governed nucleation approach in the Gurson model that as

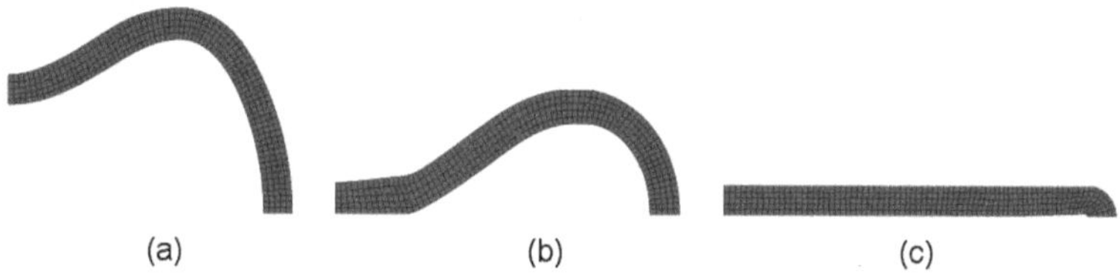

Fig. 4 Deformed meshes at different levels of vertical compression – first stage of compression: (**a**) 33%; (**b**) 66% and (**c**) 100%

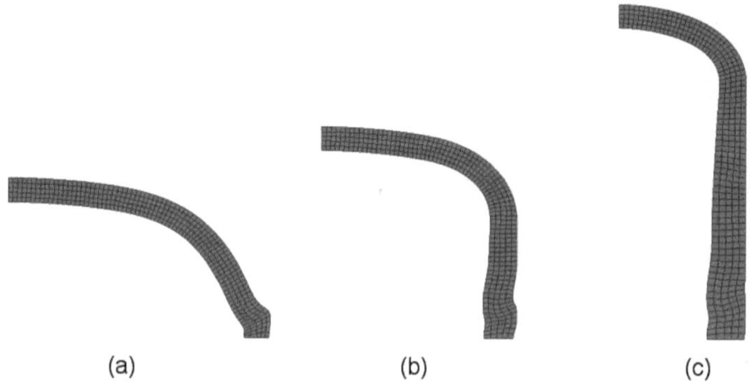

Fig. 5 Deformed meshes at different levels of lateral compression – second stage of compression: (**a**) 33%; (**b**) 66% and (**c**) 100%

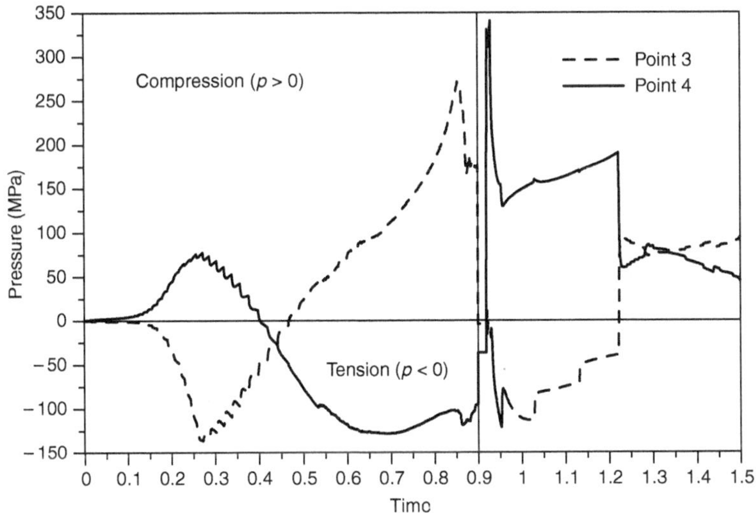

Fig. 6 Evolution of pressure in points 3 and 4 (Fig. 3), considering strain governed nucleation model

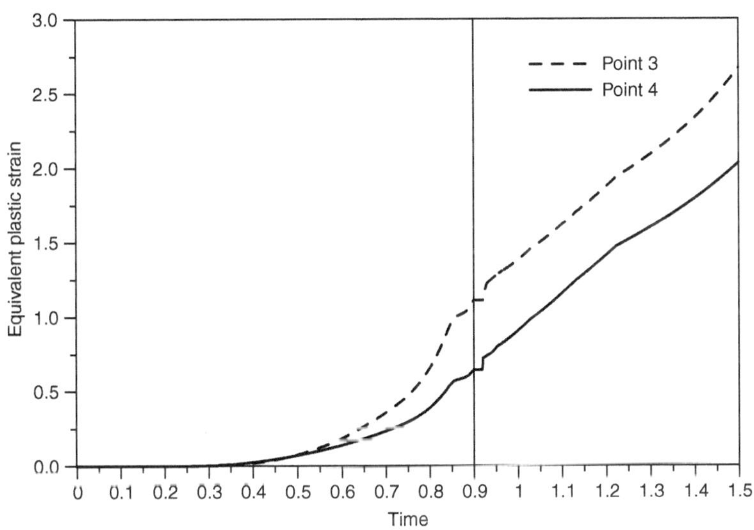

Fig. 7 Evolution of equivalent plastic strain in points 3 and 4 (Fig. 3), considering strain governed nucleation model

the material reaches the mean nucleation plastic strain in compression, no nucleation occurs under a subsequent tension. As this behaviour does not agree with experimental observations, alternative strain governed nucleation models have been proposed [18]. This problem does not arise when the initial porosity approach is employed.

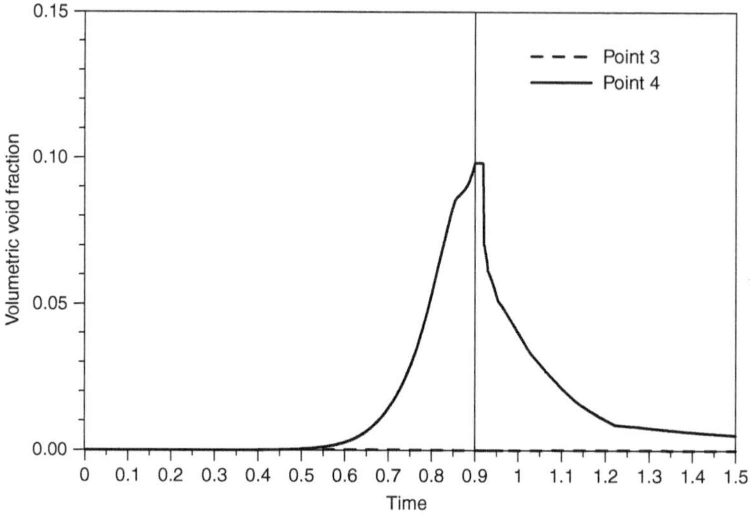

Fig. 8 Evolution of volumetric void fraction in points 3 and 4 (Fig. 3), considering strain governed nucleation model

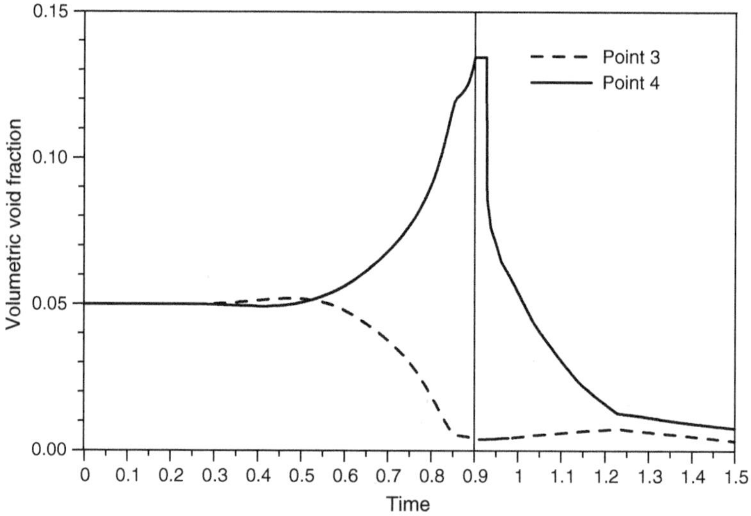

Fig. 9 Evolution of volumetric void fraction in points 3 and 4 (Fig. 3), initial porosity approach

4 MHSS Analysis

In this section the finite element results for cellular materials are presented in comparison with the experimental results of Fiedler [4]. Figure 10 shows the experimental set-up and the foam specimen before and after the test.

In the numerical simulation of the MHSS behaviour, two geometries (Fig. 11) were considered. The first one, known as partial geometry, in which the spheres are glued at their contact point and the second one, known as syntactic geometry, in which the spheres are totally embedded in the resin matrix. The models used in the analyses were made to fit global densities for the sets resin-metal of 1.2 g/cm^3 (syntactic) and 0.6 g/cm^3 (partial). The metal spheres had an external radius of 1.5 mm and the resin thickness between spheres was 0.36 mm, according the observation of the specimen of MHSS shown in Fig. 1.

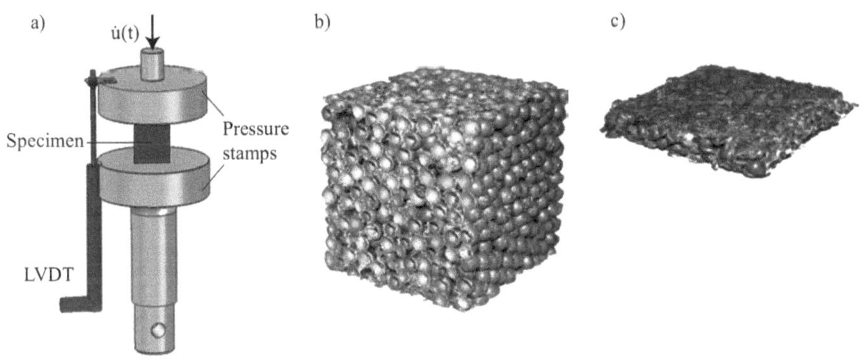

Fig. 10 (**a**) Experimental setup; (**b**) specimen before test; (**c**) specimen after test [4]

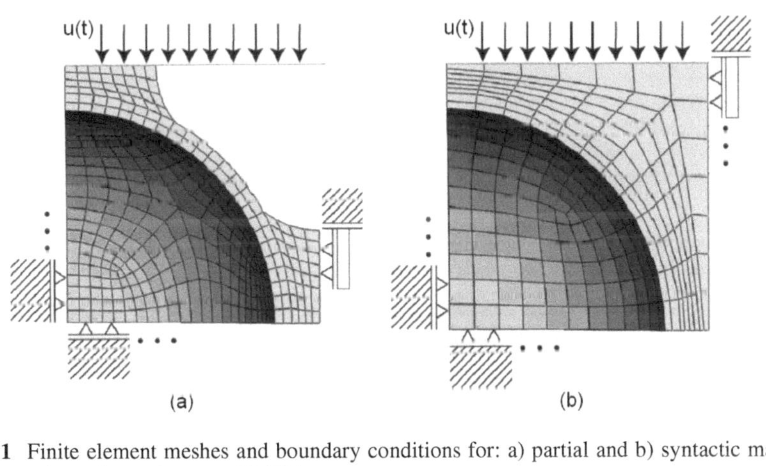

Fig. 11 Finite element meshes and boundary conditions for: a) partial and b) syntactic material representative volume elements (RVE's)

Material constants used were elastic modulus $E = 110$ GPa, $\nu = 0.30$, initial yield stress $\sigma_y{}^0 = 300$ MPa and $\rho = 6.95$ g/cm^3 for the metal of the sphere [3] and $E = 2.46$ GPa, $\nu = 0.34$, compression yield stress $\sigma_y{}^{0,c} = 113$ MPa and $\rho = 1.13$ g/cm^3 for the resin. Both metal sphere and matrix were modeled as elastoplastic. Damage was considered only for the metallic spheres. The boundary conditions employed in the vertical faces reproduce symmetry conditions on the left face and periodic boundary condition on the right face, maintaining the right face plane and vertical, but allowing horizontal displacements.

4.1 Partial Geometry: Single and Multi-cell Models

In this section some results obtained studying the behaviour of the partial geometry are presented. The models employed for the partial case are the corresponding RVE (representative volume element) model (Fig. 11a) and a model composed by one layer with five cells (RVE's) in each direction, as shown in Fig. 12.

Figure 13 shows experimental [4] and numerical macroscopic stress versus normalised displacement plots for the case of the partial geometry. Numerical results were obtained considering a RVE as shown in Fig. 11a (confined RVE), the same RVE without restrictions on the lateral right face (free RVE) and the 5×5 cells model shown in Fig. 12. It can be observed in Fig. 13 that the boundary conditions in the RVE model have little influence on the results. This is because in the partial geometry the spheres are quite free except at the small region in which they are glued.

Three significant changes in the stiffness of the numerical results can be identified in Fig. 13, indicated by points 1, 2 and 3. This localised change in the stiffness corresponds to new contact surfaces activated between RVE and the contact matrices [7]. In the experimental results, this stiffness change is not so localised due to the non-uniformity of the geometry.

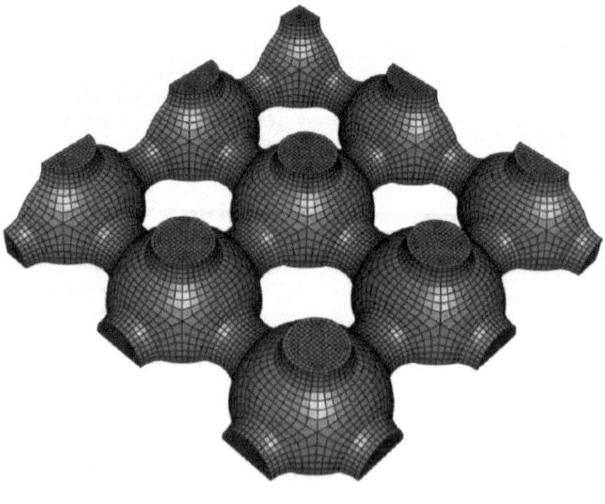

Fig. 12 Mesh for the 5×5 cells model in the undeformed configuration

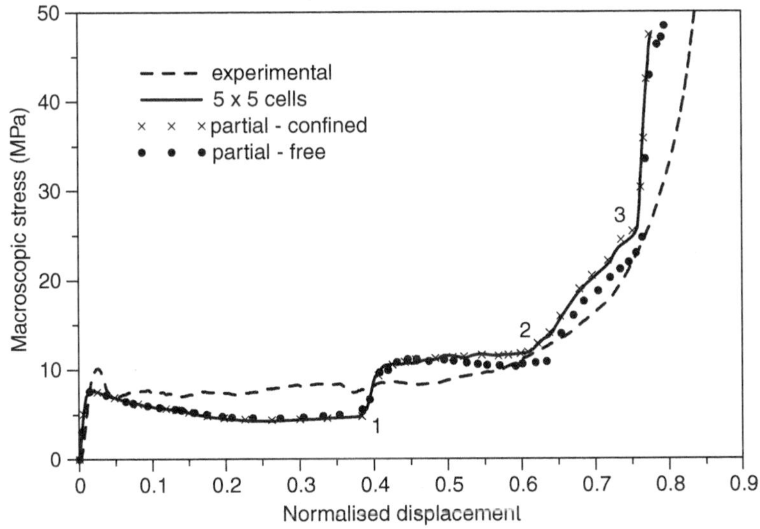

Fig. 13 Macroscopic stress versus normalised displacement plots considering a confined RVE, a free RVE and the 5×5 cells model (partial geometry)

4.2 Spheres with Non-homogeneous Properties

Some numerical analyses were performed to determine the influence of non-homogeneity on the load-displacement results. To consider this effect, one layer model, (5×5×1 model) as shown in Fig. 12, in which random variation of the yield stress (± 10%) of the individual spheres were considered. Besides, models with five layers (5×5×5) as shown in Fig. 14, were also studied. However, the analyses could not be completed because of strong instabilities in the numerical results. As far as consistent results were obtained, no major influence of non-homogeneity was

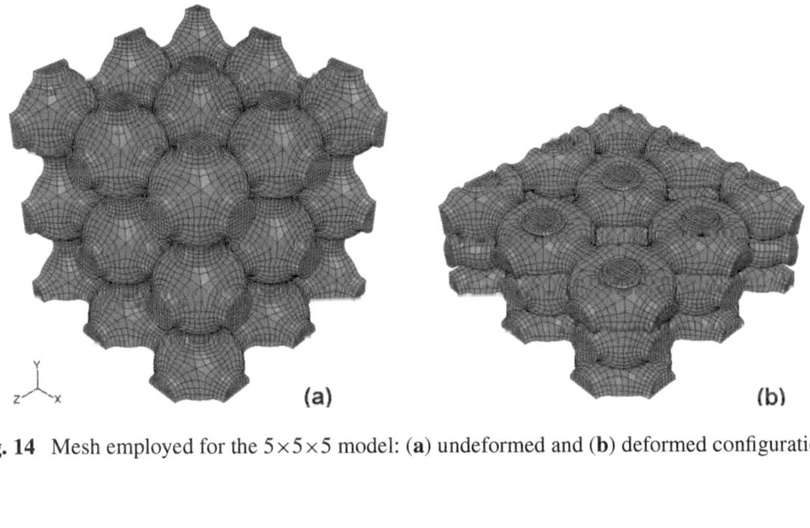

Fig. 14 Mesh employed for the 5×5×5 model: (**a**) undeformed and (**b**) deformed configuration

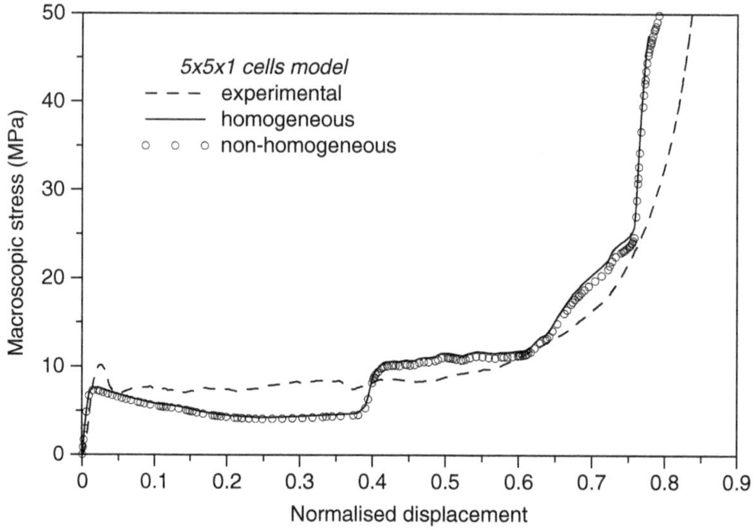

Fig. 15 Macroscopic stress for homogeneous and non-homogeneous models with 5×5×1 cells

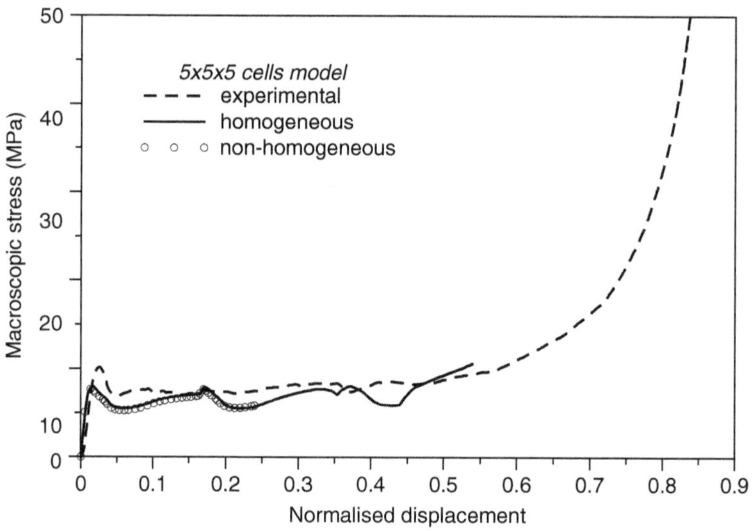

Fig. 16 Macroscopic stress for homogeneous and non-homogeneous models with 5×5×5 cells; results with non-homogeneous cells were limited to smaller loads due to numerical instability

detected, as shown in Fig. 15 for the one layer model and in the Fig. 16 for the five layers model. It can observed that employing a greater number of layers the changes in stiffness (points 1, 2 and 3 in Fig. 13) are smoothed up and the numerical results become closer to the experimental ones. The study of these situations has to be pursued.

4.3 Syntactic Material: Use of Drucker–Prager Model for the Resin

Considering the analysis of the syntactic RVE (Fig. 11b), it was observed that the resin is subjected not only to compressive but also, for larger strains, to tensile stresses. Fiedler and Öchsner [3] report the resin properties with two yield stresses, one value in tension (61.5 MPa) and another value in compression (113 MPa). Thus, use of the Drucker-Prager model, which considers different yield stresses in tension and compression seems appropriate. Three situations were studied: von Mises with yield stress 61.5 MPa, von Mises with yield stress 113 MPa and Drucker-Prager with tension yield stress 61.5 MPa and compression yield stress 113 MPa. Figure 17 shows macroscopic stress versus normalised displacement plots for the studied cases.

In the experimental results, an accentuated initial peak can be observed, that is not present in the numerical results. This effect is probably related to resin characteristics not properly represented in the numerical simulation, such as debonding between resin and steel or cracks and failure of the resin.

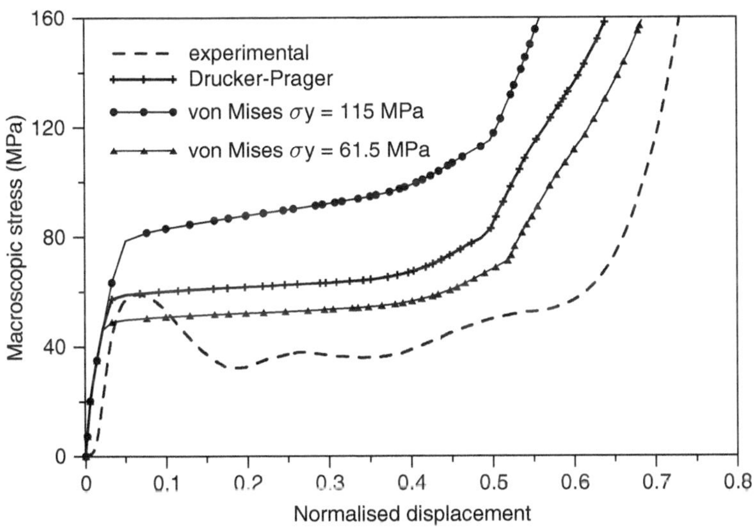

Fig. 17 Macroscopic stresses for the syntactic foam: comparison of experimental data with numerical results that use von Mises and Drucker–Prager constitutive models

5 Conclusions and Final Remarks

The analysis of a single sphere indicates that damage has important local effects but smaller influence on the global load-displacement plot [7], which depends on boundary conditions. However, the local state of damage is important because it can give rise to local rupture and eventually originate the final collapse of the structure. In Sect. 3, it is shown that loading-unloading can affect damage local state.

In the case of partial foams, analytical and experimental curves are reasonably close. The jumps in the numerical curve are not observed in the experiments, probably due to the presence of physical and geometrical random properties that smoothes up such details. Tests with non-homogeneous models have shown some trends but presented numerical problems, and will be pursued.

In the case of syntactic foams, a large difference is observed in the load for a given displacement after first buckling. This is probably due to the consideration of the matrix as a ductile elastoplastic material, which is not adequate to represent the resin behaviour. The use of the Drucker-Prager model for the matrix seems to offer a better representation of the maximum load but still not fully adequate.

The application of the Gurson model needs the choice of seven or more parameters (as compared to one or two in the case of von Mises plasticity) in addition to the election of nucleation mechanism and thus requires strong interaction between experimental and numerical research.

Acknowledgments Moises H. Krützmann helped in the computational modeling stage. The financial support of CAPES, CNPq (Projects 572851/2008-1, 307787/2009-5, 480237/2007-7, 301068/2006-2) and PROPESQ/UFRGS is also acknowledged.

References

1. M. Ashby, A. Evans, N. Fleck, J. Hutchinson, H. Wadley, *Metal Foams: A Design Guide* (Butterworth-Heinemann, Oxford, UK, 2000)
2. T. Fiedler, B. Sturm, A. Öchsner, J. Gracio, G. Kuhn, Modelling of the mechanical behaviour of adhesively bonded and sintered hollow-sphere structures. Mech. Comp. Mater. **42**, 559–570 (2006)
3. T. Fiedler, A. Öchsner, On the anisotropy of adhesively bonded metallic hollow sphere structures. Scr. Mater. **58**, 695–698 (2008)
4. T. Fiedler, Numerical and experimental investigation of hollow sphere structures in sandwich panels. Doctoral Thesis, Universidade de Aveiro, Portugal, 2007
5. B.F Oliveira, L.A.B. Cunda, A. Öchsner, G.J. Creus, Comparison Between RVE and full mesh approaches for the simulation of compression tests on cellular metals. Materialwiss Werkst. **39**, 133–138 (2008)
6. T.J. Lim, B. Smith, D.L. McDowell, Behaviour of a random hollow sphere metal foam. Acta. Materialia. **50**, 2867–2879 (2002)
7. B.F. Oliveira, L.A.B. Cunda, A. Öchsner, G.J. Creus Hollow sphere structures: a study of mechanical behaviour using numerical simulation. Materialwiss Werkst **40**, 144–153 (2009)
8. M. Kepets, T.J. Lu, A.P. Dowling, Modeling of the role of defects in sintered FeCrAlY foams. Acta. Mechanica. Sinica. **23**, 511–529 (2007)
9. U. Ramamurty, A. Paul, Variability in mechanical properties of a metal foam. Acta. Materialia. **52**, 869–876 (2004)
10. J. Lemaitre, J.L. Chaboche, *Mécanique des matériaux solides* (Dunod, Paris, 1988)
11. A.L. Gurson, Continuum theory of ductile rupture by void nucleation and growth: Part I-yield criteria and flow rules for porous ductile media. ASME. J. Eng. Mater. Technol. **99**, 2–15 (1977)
12. V Tvergaard, Influence of voids on shear band instabilities under plane strain conditions. Int. J. Fract. **17**, 389–407 (1981)

13. V. Tvergaard, On localization in ductile materials containing spherical voids. Int. J. Fract. **18**, 237–252 (1982)
14. V. Tvergaard, A. Needleman, Analysis of the cup-cone fracture in a round tensile bar. Acta Metallurgica **32**, 157–169 (1984)
15. J. Koplik, A. Needleman, (1988) Void growth and coalescence in porous plastic solids. Int. J. Solids. Struct. **24**, 835–853 (1988)
16. ABAQUS *Theory Manual, Version 5.2* (Hibbitt, Karlsson and Sorensen, Inc., Pawtucket, RI, 1992)
17. L.A.B. Cunda, G.J. Creus, A note on damage analyses in processe with nonmonotonic loading. Comput. Model. Simul. Eng. **4**, 300–303 (1999)
18. V. Tvergaard, (1982) Material failure by void coalescence in localized shear bands. Int. J. Solids. Struct. **18**, 659–672 (1982)
19. C.C. Chu, A. Needleman, (1980) Void nucleation effects in biaxially stretched sheets. ASME J. Eng. Mater. Technol. **102**, 249–256 (1980)

Dielectric Spectra Analysis: Reliable Parameter Estimation Using Interval Analysis

Adrien Brochier, Maëlenn Aufray, and Wulff Possart

1 Introduction

Dielectric spectroscopy (DES) is widely applied to polymers, monomers and other insulating materials because it is an extremely effective method for characterizing the molecular dynamics over many orders of magnitude of time or frequency, respectively. In the measurement, the complex dielectric function

$$\varepsilon^*(\omega) = \varepsilon'(\omega) - i\varepsilon''(\omega) \tag{1}$$

with the angular frequency ω, is measured at constant temperature. This function is called the dielectric spectrum. Commonly, dielectric spectra are modelled by a sum of relaxation processes, but the choice of a reasonable physical model for the relaxator is critical. Most of the usual models, reviewed briefly in the introductory section, result from phenomenological considerations providing limited physical foundation. Moreover, the fitting algorithm turns out to be crucial in terms of reliability and unambiguity of the dielectric model function determined. As discussed in Sect. 2, common softwares use least square approximation fitting algorithms which need initial values for the fit parameters. This could imply some predestination of the fit results. In this work, a parameter estimation algorithm which is free of these limitations will be developed (See Sect. 3). The new algorithm S.A.D.E. not only provides the chosen dielectric model function by a confidence interval for each model parameter like the frequency position and the intensity of all relaxations: it also indicates the number of relaxations that are necessary to model the measured spectrum.

M. Aufray (✉)
Chair Adhesion and Interphases in Polymers, University of Saarland,
66041 Saarbrücken, Germany
e-mail: maelenn.aufray@ensiacet.fr

A. Öchsner et al. (eds.), *Materials with Complex Behaviour*, Advanced Structured Materials 6, DOI 10.1007/978-3-642-12667-3_7, © Springer-Verlag Berlin Heidelberg 2010

1.1 The Dielectric Spectroscopy and Its Models

1.1.1 Relaxations

The Debye relaxators [1] describe the dielectric relaxation response of an ideal, non-interacting population of freely rotating dipoles to an alternating external electric field

$$\varepsilon^* = \varepsilon_\infty + \frac{\varepsilon_S - \varepsilon_\infty}{1 + i\omega\tau_0} = \varepsilon_\infty + \frac{\Delta\varepsilon}{1 + i\omega\tau_0} \tag{2}$$

Where ε_S is the static permittivity ($\varepsilon_S = \lim_{\omega\to 0} \varepsilon'(\omega)$), ε_∞ is the optical dielectric constant ($\varepsilon_\infty = \lim_{\omega\to\infty} \varepsilon'(\omega)$) and τ_0 is the characteristic relaxation time of the medium. Let us note that the Debye model refers to a well-defined physical situation. All other relaxator models reported in the literature imply phenomenological modifications of the Debye relaxator without well–defined physical background. For example, the Havriliak–Negami (HN [2]) equation

$$\varepsilon^* = \varepsilon_\infty + \frac{\varepsilon_S - \varepsilon_\infty}{(1 + (i\omega\tau_0)^\alpha)^\beta} \tag{3}$$

is a mixture of the Cole–Cole [3,4] and the Cole–Davidson [5] equations, accounting for the asymmetry and broadness of the measured dielectric dispersion curve by the additional phenomenological parameters α and β. Developed to describe the dielectric relaxation of some polymers, the HN function is now one of the most popular models for dielectric relaxation although no exact physical meaning can still be given to the coefficients α and β.

Then, for dielectric spectra containing several relaxations, it is possible to sum a number of relaxation processes, according to Eq. (2) or (3), i.e. irrespective of the model used. Here is the example, for n relaxations represented by the Debye model,

$$\varepsilon^* = \varepsilon_\infty + \sum_{j=0}^{n} \frac{\Delta\varepsilon_j}{1 + i\omega\tau_j} \tag{4}$$

and the Havriliak–Negami model.

$$\varepsilon^* = \varepsilon_\infty + \sum_{j=0}^{n} \frac{\Delta\varepsilon_j}{(1 + (i\omega\tau_0)^{\alpha_j})^{\beta_j}} \tag{5}$$

1.1.2 DC-Conductivity

At high temperatures and $\omega\to 0$, a contribution of a DC-conductivity (σ_{DC}) can be observed in the dielectric spectra of real polymer samples. It contributes only to the imaginary part of the measured complex dielectric permittivity, as long as the imaginary part of the generalized complex conductivity can be neglected in the low frequency region (i.e. $\sigma'' \ll \varepsilon'$). With this presumption, the following equation links the measured quantities $\tilde{\varepsilon}'$, $\tilde{\varepsilon}''$ to the true dielectric material quantities ε', ε'' and the DC electric conductivity (σ_{DC}):

$$\tilde{\varepsilon}'(\omega) = \varepsilon'(\omega) \text{ and } \tilde{\varepsilon}''(\omega) = \varepsilon''(\omega) + \frac{\sigma_{DC}}{\omega\varepsilon_0} \tag{6}$$

At high frequencies, the contribution from the DC-conductivity becomes negligible and hence

$$\lim_{\omega\to\infty} \tilde{\varepsilon}''(\omega) = \varepsilon'' \tag{7}$$

At low frequencies, the ε'' term is negligible compared to σ_{DC}, leading to the equation

$$\lim_{\omega\to 0} \log_{10}(\tilde{\varepsilon}''(\omega)) = \log_{10}\left(\frac{\sigma_{DC}}{\varepsilon_0}\right) - \log_{10}(\omega) \tag{8}$$

And hence, as a first approximation, the equation

$$\tilde{\varepsilon}''(\omega) \cdot \omega \cdot \varepsilon_0 = \sigma_{DC} \tag{9}$$

gives a good fit of the DC-conductivity [6–9]. However, some relaxations can take place even at these low frequencies. They are very difficult to fit using classical approaches (i.e. common algorithms) because they can be masked by the DC-conductivity.

As a consequence, the fitting process has to lead to a set of parameters which makes the model to fit both the real and the imaginary part, and has to detect hidden relaxations.

1.1.3 Polarization at Electrodes and Phase Boundaries

Electrode polarization is a parasitic effect in dielectric experiments which can mask the pure dielectric response of the sample material [10]. Moreover, as described by Maxwell, Wagner and Sillars (MWS polarization), phases in heterogeneous media are to be treated as macroscopic volume elements with different ε^* and conductivities σ^* [11–13]. The most basic geometrical situation was considered by Maxwell [11]. This consists of a plate capacitor filled with n dielectric sheets of non-complex dielectric properties and DC conductivities ε_1, σ_1, ε_2, σ_2, ..., ε_n, σ_n. This resulted in differential equations relating the field across the dielectric as a function of the current though the strata. Maxwell showed his model to be capable of explaining the observed data for dielectric relaxation in such systems. By considering small spheres with material properties ε_2, σ_2 dispersed so as to preclude electrostatic interaction with one another through a medium with properties ε_1, σ_1 Wagner [12] was able to develop Maxwell's analysis further. This analysis was developed by Sillars [13] for a disperse 2-phase system by assuming that the matrix material behaved as a perfect dielectric (i.e. $\sigma_1 = 0$). The inclusions are spheroids with axis a in the field direction, and with b and c equal to one another. The geometry is wholly described by two variables, q, the volume fraction of dielectric 2, and the axial ratio a/b. The conducting inclusions behave as point dipoles in the dielectric matrix, and a full analysis yields relations similar to the Debye equations:

$$\varepsilon' = \varepsilon_\infty + \frac{\varepsilon_1}{1+\omega^2\tau^2} \tag{10}$$

$$\varepsilon'' = \frac{\varepsilon_1\omega\tau}{1+\omega^2\tau^2} \tag{11}$$

Where λ is a particular function of a/b, and ε_∞, N and τ are given by:

$$\tau = \frac{\varepsilon_1(\lambda-1)+\varepsilon_2}{\sigma_2}\cdot\varepsilon_0 \tag{12}$$

$$N = \frac{\lambda^2}{\varepsilon_1(\lambda-1)+\varepsilon_2} \tag{13}$$

$$\varepsilon_\infty = \varepsilon_1\left(1+\frac{q\cdot\lambda(\varepsilon_2-\varepsilon_1)}{\varepsilon_1(\lambda-1)+\varepsilon_2}\right) \tag{14}$$

Sillars includes the dimensionless quantity, q, which is a function of the ratio a/b. The non-linear variation of this quantity implies that little can be deduced about the dielectric properties of a heterogeneous material unless the shapes of the inclusions are known. Now, applying an electric voltage to the heterogeneous material, the mobile charges in phase 2 can be blocked and piled up at the phase boundaries. This picture also applies to the electrodes on the dielectric sample in the similar case of polarization at a blocking electrode. The experimental example presented in the last part of this work reveals that the MWS polarization causes a strong rise both in the real part (where electrode polarization is more visible) and in the imaginary part of the permittivity (where the electrode polarization superimposes to the DC-conductivity) with decreasing frequency. In this work, the electrode polarization will be modeled by one strong Debye relaxator (recall that the MWS equation is very similar to the Debye equation). The only way to separate all the phenomena (electrode polarization, DC-conductivity, and maybe low-frequency dielectric material relaxations) is a simultaneous fit of the real and imaginary part of the permittivity using a formula taking all these phenomena into account.

1.2 Modeling Problems: Simultaneous Fit and Choice of the Model

The main problem is that different physical processes can occur at the same frequency and temperature: the example of some small relaxations hidden by the DC-conductivity is very clear. Therefore, the only way to separate the phenomena is a simultaneous fit of the real and imaginary part of the permittivity. So the simple fit of the imaginary part of the permittivity (used by most of researchers, except Axelrod et al. [8]) does not give a good solution as relaxations or other phenomena can be missed. The fit of the complex function of the permittivity will be done by our algorithm S.A.D.E. presented in the next paragraphs: the relaxations and the electrode polarization are fitted by the Debye model and the DC-conductivity by its specific function. Of course, we cannot determine *ex ante* the number of Debye relaxations needed to fit our curves, but the program S.A.D.E. will try to fit the curve using from *0* to *n* relaxations and one term due to DC-conductivity.

$$\varepsilon^* = \varepsilon_\infty + \sum_{j=0}^{n} \frac{\Delta\varepsilon_j}{1 + i\omega\tau_j} - i\frac{\sigma_{DC}}{\omega\varepsilon_0} \tag{15}$$

The goal of the data fitting is to find the parameter values of the applied physical model that match the data most closely. The models to which the data are fitted depend on adjustable parameters. Therefore, the fitting process requires both the choice of a physical model and the choice of a suitable computing algorithm. Most of the scientific softwares fit experimental data by using some variants of the least squares approximation method but the success of the fit is not guaranteed. The next section will summarize the least squares approximation method and its disadvantages. A different algorithm will be presented which makes the fit by the Debye model of experimental data with several relaxations possible in an efficient way, even if some of them are hidden.

2 Data Fit Methods: From Least Square Approximation to the Interval Analysis

2.1 Introduction

Let $(x_i, y_i)_{1 \leq i \leq n}$ be a set of experimental data, and $f(x, p)$ be a physical model depending on a vector of parameters $p = (p_1, \ldots, p_k)$. If the measurement accuracy is known, it is natural to assume that a given vector of parameters p^* leads to a "good" fit if it makes the model consistent with the measurement error e_i on each data point, that is:

$$\forall 1 \leq i \leq n, y_i - e_i < f(x_i, p^*) \leq y_i + e_i$$

A graphical visualization of this criterion is shown in Fig. 1.

Interval analysis makes possible the computation with intervals rather than with real numbers, and thus it leads to a very similar criterion for checking whether a "vector of intervals" contains good parameters. Thus, by cutting the multidimensional space in which good parameters are searched into finitely many small pieces

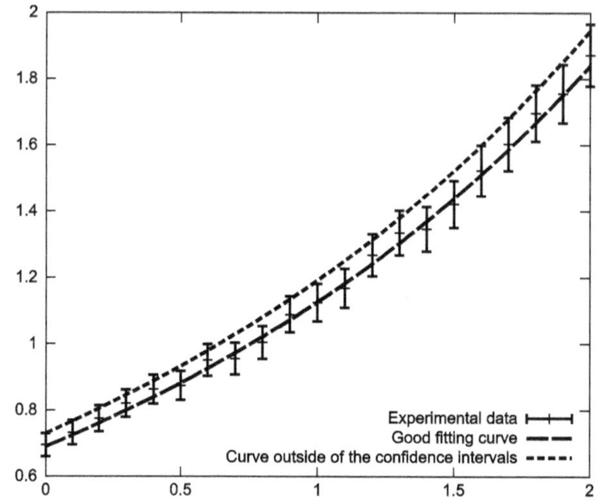

Fig. 1 Fit and confidence intervals

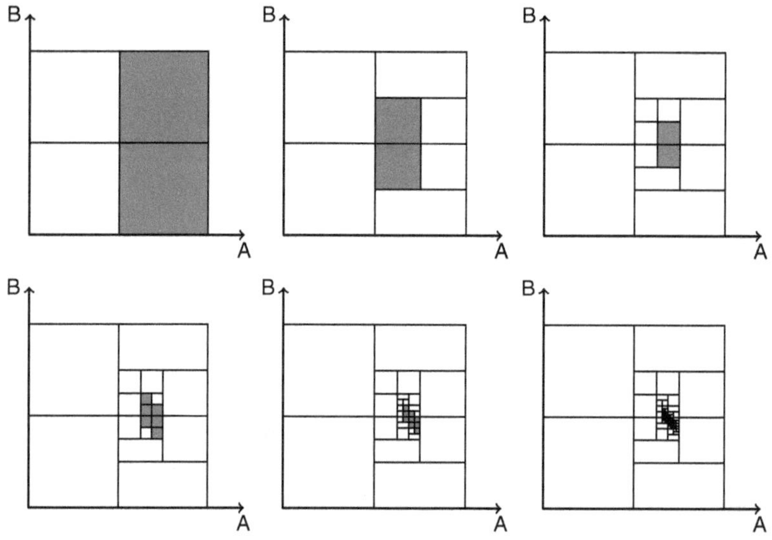

Fig. 2 Modelization of artificial data using a model depending on two parameters A and B. At every step of the algorithm, the set of feasible parameters is guaranteed to be contained is the area covered by gray squares. White squares doesn't contain any feasible parameters. Thus, at each step the approximation of the set of feasible parameters becomes more accurate

in a suitable way, it is possible to compute the best approximation of the set of good parameters. As shown in Fig. 2, this algorithm is *global*, as at each step the area covered by gray squares is guaranteed to contain the true set of good parameters.

2.2 The Least Square Approximation

The main goal of data fitting is to find the parameters that make the model best describing the data. The commonly used method, namely the Least Square Approximation, tries to minimize the distance between the measured and the calculated points, i.e. to find:

$$\min_{p \in \mathbb{R}^k} \left(\sum_{i=0}^{n} (f(x_i, p) - y_i)^2 \right) \tag{16}$$

If the model is linear this can be done analytically but if not, some classical minimization methods (generally based on gradient descent algorithms) are used. Such an algorithm takes a set of some initial parameter values as input and generates a sequence of model parameter vectors which are supposed to converge to one particular parameter vector producing the minimum according to Eq. (16). This method has some disadvantages however:

- The choice of initial values: The key for success of this algorithm is the proper choice of initial values but there is no general method to do that. In some cases, each single fit needs a patient observation of the data and a large amount of unsuccessful testings before finding good initial values. Moreover, when using HN or Debye models, the number of relaxations has to be known in advance. If the number of relaxations is higher than 1 and the relaxations are superimposed, it is very often impossible to find correct initial values for each relaxation.
- The convergence speed: As the algorithm works as an iterative process, it "jumps" at each iteration step from one value to another which is closer to a solution. One critical choice is the size of such a jump. Although this choice is partially done by the algorithm using the gradient and often some additional methods, for example the Levenberg–Marquardt algorithm [14,15], the step must be adjusted according to the nature and the order of magnitude of the parameters.
- The local minimum problem: The algorithm starts from a given vector of parameters and tries to "follow the slope" to find a minimum of the function. Thus, the solution provided by the algorithm could correspond to a local minimum which seems to be a good fit, whereas only the global minimum is physically meaningful.
- The complexity of the model: The model has to be well conditioned in order to make the algorithm work well. If the model is unstable (i.e. if a small variation of the parameters leads to a big variation of the corresponding computed value) or if the number of parameters is too large, convergence is not guaranteed. In particular, if there are some symmetries (i.e. if some permutation of the parameters does not change the value of the function) the algorithm could "hesitate" between different correct possibilities and not converge at all. This problem appears in all cases where we describe the measured dielectric function by a sum of individual physical model functions of the same mathematical type. There are two possibilities to solve this problem:

1. assuming a condition on the parameters which breaks the symmetry, for example that

$$\tau_1 < \tau_2 < \ldots < \tau_r \tag{17}$$

but this is not possible using the least squares approximation, or
2. choose some perfect initial values which clearly distinguish the different relaxations, which, as it was already noticed, is often a complicated problem.

- The measurement accuracy: Although it is possible to attribute a weight factor to each measured data point according to the accuracy, the quality of the fit result is not directly linked to the measurement accuracy and there is no guarantee that the result will be consistent with it. In particular, a result is always provided, even if no parameter set is consistent with the measurement accuracy (for example if the chosen number of relaxations is too small to describe the data correctly). Moreover, the method supposes that the errors are randomly distributed, which is not always true.
- The bound on the parameters: In general, there is no way to impose some specific constraints for the parameters (for example that they must not be negative).

2.3 The Interval Analysis

Interval analysis was first introduced in order to have a true representation of real numbers for numerical computing: for example, assuming that $\pi = 3.1415$ generates some numerical error which propagates or amplifies during the computing process decreasing the quality of the result. By using an interval instead of a float precision number and assuming that $\pi = [3.1415; 3.1416]$ leads to an interval as the result of the calculation and this interval is guaranteed to contain the true result. Therefore, intervals could also be used in order to manipulate a large range of real numbers simultaneously, and thus to make an approximation of a complex set which is easily handled in computing.

Let $\mathbb{R}$ be the set of real numbers. An interval denoted with $[x]$ is a closed connected subset of $\mathbb{R}$. The lower and upper bound of $[x]$ are denoted by x^- and x^+, respectively. Let $\mathbb{IR}$ be the set of all real intervals, then elementary real operations are extended to intervals according to the following formula:

$$[x] \circ [y] = \{x \circ y \mid x \in [x],\ y \in [y]\} \text{ for } \circ \in \{+, -, *, /\} \tag{18}$$

leading to, for example:

$$[1; 3] + [2; 4] = [3; 7] \tag{19}$$

A vector of intervals is called a box, and $\mathbb{IR}^n$ denotes the set of all n–dimensional boxes. Arithmetic operations with boxes are defined componentwise. The size of an interval $[x] = [x^-; x^+]$ is defined by:

$$\text{Size}([x]) = x^+ - x^- \tag{20}$$

The size of a box is the size of its greatest component. A bisection procedure will also be used, which cuts a box into two parts and returns the two parts. For a given box $[p] = ([p_1], [p_2], \ldots, [p_k])$, the procedure finds the index i of the greatest component, and returns 2 boxes:

$$([p_1], [p_2], \ldots, [p_i^-; p_i^- + (p_i^+ - p_i^-)/2], \ldots, [p_k]) \tag{21}$$

and

$$([p_1], [p_2], \ldots, [p_i^- + (p_i^+ - p_i^-)/2; p_i^+], \ldots, [p_k]) \tag{22}$$

Let us keep in mind that we do not always get an interval when we calculate the image of an interval by a real function. For example,

$$\sqrt{[4;9]} = [-3; \quad 2] \cup [2;3] \tag{23}$$

provides not one but two intervals. Thus, for more complicated operations, some approximations have to be applied in order to keep a consistent representation for computing. In this case for example, the only possible choice is to take the "interval square root" of [4, 9] to be $[-3, 3]$.

More generally, let

$$f : \mathbb{R} \longrightarrow \mathbb{R} \tag{24}$$

be a real function. A so-called "inclusion function" for f is an interval function

$$[f] : \mathbb{IR} \longrightarrow \mathbb{IR} \tag{25}$$

which verifies

$$\forall [x] \in \mathbb{IR}, \ f([x]) \subset [f]([x]) \tag{26}$$

In other word, the image of an interval (or a box) $[x]$ by an inclusion function for f is always still an interval (or a box) which contains the true image of $[x]$. Of course, there are infinitely many inclusion functions for a given real function. One of them is minimal but could be difficult to find. Then, the so-called natural inclusion function will be used. The natural inclusion function is simply obtained by replacing each operator in the associated real function by its interval equivalent, and each usual function (sin, cos, exp, $\sqrt{\ }$) by a suitable interval counterpart. It is important to note that the natural inclusion function depends on how the real function is written and especially on the repetition of the same variable. For example, the two real functions

$$f_1(x) = x \text{ and } f_2(x) = x + x - x \tag{27}$$

are obviously equal, but their associated natural inclusion function are not. Indeed:

$$\begin{cases} [f_1]([1;2]) = [1;2] \\ [f_2]([1;2]) = [1;2] + [1;2] - [1;2] = [0;3] \end{cases} \tag{28}$$

To be sure that the inclusion function is an acceptable approximation of the real one, the inclusion function has to meet two conditions: It has to respect the inclusion

$$\forall [x], [y] \in \mathbb{IR}, \ \ [x] \subset [y] \Rightarrow [f]([x]) \subset [f]([y]) \tag{29}$$

and, given a sequence of intervals (or boxes) with a size converging to *0*, the size of the image of this sequence by the inclusion function has to converge to 0 too.

$$\forall ([x_n])_{n \in \mathbb{N}} \in \mathbb{IR}^{\mathbb{N}}, \ \ (\text{Size}([x_n]) \to 0) \Rightarrow (\text{Size}([f]([x_n])) \to 0) \tag{30}$$

2.4 Data Fit and Set Inversion

As outlined in the introducting Sect. 2.1, modelization problems can be reformulated in the language of interval analysis. With $(x_i, y_i)_{1 \leq i \leq n}$ being a set of measured data, $(e_i)_{1 \leq i \leq n}$ being the corresponding measurement accuracy and $f(x,p)$ being a model depending on several parameters $p = (p_1, p_2, \ldots, p_k)$, each measured value is associated with an interval according to the measurement accuracy:

$$[y_i] = [y_i - e_i, y_i + e_i] \tag{31}$$

Thus, a vector of parameters p is called feasible if

$$\forall \ 1 \leq i \leq n, \ f(x_i, p) \in [y_i] \tag{32}$$

and is called unfeasible otherwise. This definition has to be extended to the intervals. Obviously, a box of parameters $[p]$ is feasible if and only if it contains only feasible parameters, and is unfeasible if it contains only unfeasible parameters. But there is one more possibility: a box can contain both feasible and unfeasible parameters (see Fig. 3). Such a box is called indeterminate. More formally, a box $[p]$ is called

- feasible if

$$\forall 1 \leq i \leq n, \ \ [f](x_i, [p]) \subset [y_i] \tag{33}$$

- unfeasible if

$$\exists i, \ \ [f](x_i, [p]) \cap [y_i] = \emptyset \tag{34}$$

- indeterminate otherwise

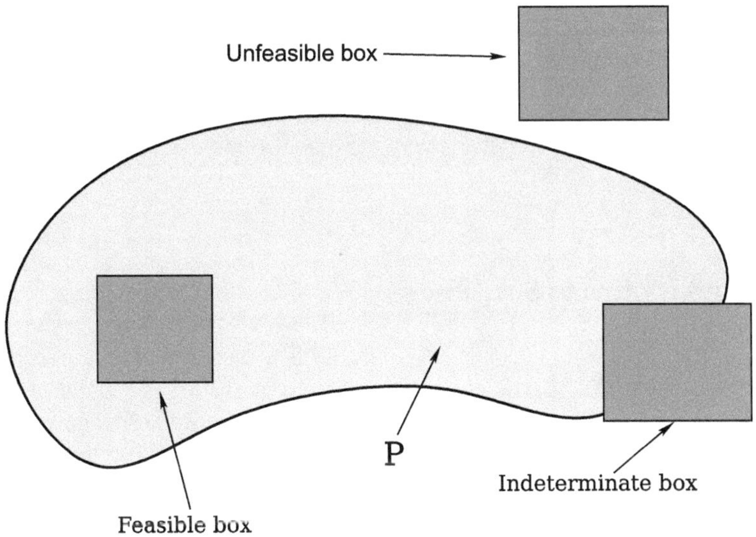

Fig. 3 Feasible, unfeasible and indeterminate boxes; P denotes the complete set of feasible parameters

Thus, the set P of all feasible parameters can be defined as the inverse image of the $[y_i]$ by a specific function. Thus, let

$$F : \mathbb{R}^k \to \mathbb{R}^n \tag{35}$$

$$p \to \begin{pmatrix} f(x_1, p) \\ f(x_2, p) \\ \vdots \\ f(x_n, p) \end{pmatrix} \tag{36}$$

be the function taking a vector of parameter P and returning all the corresponding calculated values and

$$\lfloor y \rfloor = \begin{pmatrix} [y_1] \\ [y_2] \\ \vdots \\ [y_n] \end{pmatrix} \tag{37}$$

the boxes of all measured intervals as defined in Eq. (31). So the set P of all feasible parameters is exactly:

$$P = \{p \mid F(p) \in [y]\} = F^{-1}([y]) = \bigcap_{i=1}^{n} f_i^{-1}([y_i]) \tag{38}$$

where $f_i(p) = f(x_i, p)$. Thus, the set of feasible parameters can also be considered as the set of parameters which satisfies a system of constraints.

Theoretically, it is very easy to test if a box of parameters $[p]$ is feasible or not but in practice it requires the ability to calculate $f(x_i, [p])$ which is not always an interval or easy to calculate. That is why the (natural) inclusion function $[f]$ will be used instead.

3 From S.I.V.I.A. to S.A.D.E.

S.I.V.I.A. (Set Inversion Via Interval Analysis) is a set inversion algorithm introduced by Jaulin [16, 17]. The algorithm is able to solve a large range of various problems by finding a list of boxes approximating the inverse image of a given set of real numbers or vectors (i.e. a subset of $\mathbb{R}^k$) by a given function. S.I.V.I.A. is a "branch and bound" algorithm: it first tests if a given box is feasible or unfeasible. If the box is indeterminate, the box is cut into 2 parts and each part is tested recursively. As S.I.V.I.A. has exponential complexity, a subroutine called contractor, which increases the speed of the main algorithm, is first presented.

3.1 Contractor

Given a box $[p]$, a contractor (C) is a subroutine which decreases the size of $[p]$ by removing some unfeasible parameters (as illustrated in the Fig. 4). So a function C is a contractor if and only if:

$$\forall [p] \in \mathbb{IR}^k,\ C([p]) \subseteq [p] \text{ and } C([p]) \cap P = [p] \cap P \tag{39}$$

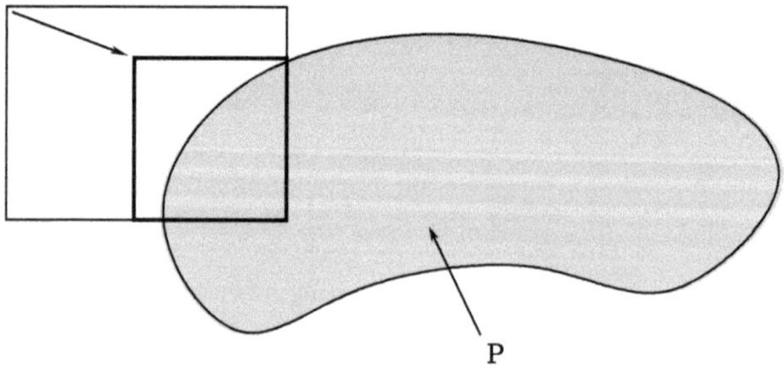

Fig. 4 Action of a contractor: it reduces the size of the current box without removing any feasible parameter. Thus, the resulting box fits the set P of feasible parameters more accurately

There are several ways to implement a contractor. One of them uses a "forward-backward propagation". Let $[p_0]$ be a box of parameters and

$$\begin{cases} f(x_1,p) \in [y_1] \\ f(x_2,p) \in [y_2] \\ \vdots \\ f(x_n,p) \in [y_n] \end{cases} \tag{40}$$

be the system of constraints corresponding to a fit problem. For each single constraint, the value of $f(x_i,[p_0])$ is calculated (forward propagation) and compared to the interval $[y_i]$. Then, the difference is propagated back to the parameter box $[p_0]$. Finally, the found box is contracted using the following constraint, and so on. This process is repeated as long as the contraction has a significant effect. For example, let

$$[p_1] \times [p_2] = [2;3] \times [0;1] \tag{41}$$

be a box of parameters,

$$[y] = [6;10] \tag{42}$$

be an interval,

$$f(p_1,p_2) = p_1 \times \exp(p_2) \tag{43}$$

be a real function and

$$f(p_1,p_2) \in [y] \tag{44}$$

be a single constraint. The aim of the contractor is to remove some values from $[p_1]$ and $[p_2]$ which do not satisfy the constraint:

- Forward propagation: The expression is simply calculated but in order to make the backward propagation possible, each partial calculation is stored as some variable.

$$\begin{cases} [z] = \exp([p_2]) = [\exp(0);\exp(1)] = [1;2.7183] \\ [y_p] = [p_1] \times [z] = [2;3] \times [1;2.7183] = [2;8.1549] \end{cases} \tag{45}$$

- the values from $[y_p]$ which do not satisfy the constraint are removed

$$[y^*] = [y_p] \cap [y] = [2;8.1549] \cap [6;10] = [6;8.1549] \tag{46}$$

- backward propagation: the result is propagated back to remove any inconsistent value from the initial parameters:

$$\begin{cases} [p_1^*] = ([y^*]/[z]) \cap [p_1] = [2.2072; 3] \\ [z^*] = ([y^*]/[p_1^*]) \cap [z] = [2; 2.7183] \\ [p_2^*] = (\log([z^*])) \cap [p_2] = [0.6931; 1] \end{cases} \tag{47}$$

Thus, this process returns a new box $[2.2072; 3] \times [0.6931; 1]$ which is smaller than the initial one, but contains exactly the same feasible parameters.

3.2 Set Inversion Via Interval Analysis (SIVIA)

It is now possible to describe the S.I.V.I.A. algorithm. An initial box $[P_0]$ which represents the initial search space is taken as input: The initial box is chosen large enough to be sure that the expected solution set P of parameters satisifes the condition $P \subset [P_0]$. The second input value is a small positive real number η which determines the precision of the algorithm. If a box is indeterminate but smaller than η, this box is still accepted. This guarantees that the algorithm terminates. The fact that the inclusion function satisfies the conditions (29) and (30) guarantees also the correctness of the algorithm, i.e. that by choosing the value of η small enough, it is possible to make the approximation of P as precise as desired. The original version of S.I.V.I.A. returns an inner and an outer approximation of the set P of all feasible parameters, but for the sake of simplicity, the algorithm presented in this article only returns an outer approximation. Some basic structures used in the algorithm are defined here:

- A List is a structure which can store a list of elements. An element is added to the list with the function Push (List, element).
- A Stack is a LIFO (Last In, First Out) structure. The function Push (Stack, element) adds an element to the top of the Stack, and the function Pop (Stack) returns the element which is on the top of the stack and removes it from the stack.

The precise description of the algorithm is given Fig. 5. S.I.V.I.A. (Set Inversion Via Interval Analysis) has some advantages compared to LSA (Least Squares Approximation):

- This is a global algorithm. If there are some parameters which satisfy the constraints, they will all be found.
- Conversely, if there is no parameter which satisfies the constraints, an empty set will be returned. Therefore, S.I.V.I.A. also gives a strong criterion to estimate the number of relaxations for a given dielectric spectrum.
- All the returned parameters belong to the initial search space. So there is no risk to find non-sense parameters, such as negative values.

- It is possible to add some arbitrary constraints that the parameters have to satisfy. It is also possible to use in only one step, several systems of constraints coming from different sources (for example to fit simultaneously the real and imaginary part of a set of data) or to use several sets of data coming from repeated measurements (data accumulation) in order to increase the quality of the result.
- The returned list of boxes can be reinterpreted to find an interval for each parameter, which is guaranteed to contain the true parameter value, and which size is directly linked to the measurement accuracy.

Fig. 5 S.I.V.I.A. (Set Inversion Via Interval Analysis)

```
List L
Stack S
Push (S, [P0])
While Not ( is Empty (S) )
      [pc] ← Pop (S)
      Contract ( [pc] )
      If [pc] is feasible
            Push ( L, [pc] )
      End If
      If [pc] is indeterminate
            If Size ( [pc] ) ≤ η
                  Push ( L, [pc] )
            Else
                  { [p1], [p2]} = Bisection ( [pc] )
                  Push ( S, [p1] )
                  Push ( S, [p2] )
            End If
      End If
End While
Return L
```

Figure 6 shows a symbolic set of parameters approximated by a list of boxes which entirely cover the parameter set with an accuracy determined by the chosen value for the algorithm parameter η (the maximal size of indeterminate boxes which are still accepted). The list of boxes has to be post-processed in order to

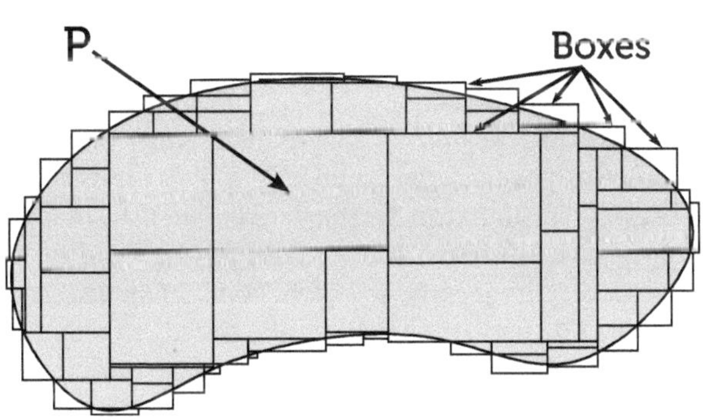

Fig. 6 The complicated set P is approximated by many boxes which are easier to handle

obtain an interval for each parameter of the applied physical model. Thus, S.I.V.I.A. is really efficient for a broad range of hard problems, such as robots localization, parameter estimation or stability analysis, and especially for very badly conditioned problems. For such problems, S.I.V.I.A. provides a detailed description of the part of the parameter space which solves the problem. However, in the specific case of experimental data fitting using a physical model with a lot of parameters, this feature of S.I.V.I.A. results in some redundancy of the parameters. This is caused by the symmetry of the model, i.e. the set P is not a connected set but it contains several connected components that correspond to all possible permutations of the parameters which let the model invariant. Then, the post-processing of the list could be time-consuming. So it is necessary to find a way to select a single connected component.

3.3 *How to Modify S.I.V.I.A. for Dielectric Spectroscopy*

In dielectric spectroscopy, the dielectric function ε^* measured as a function of frequency provides the experimental data set which consists of the two sub-sets for the real and the imaginary parts ε' and ε'', respectively. For dielectric data fit, the results given by S.I.V.I.A. are too precise: the list S.I.V.I.A. returns is not easy to process and a perfect approximation of P is not necessary as only an interval for each parameter is needed. Thus, it is possible to modify this algorithm to make it applicable to most complex cases.

3.3.1 Returned Values

As only an interval is needed for each parameter, the first idea is to approximate the bounding box of P, instead of P itself. The bounding box of P is simply the smallest box $[P]$ containing P. Each time the algorithm finds a feasible box of parameters, it will store the convex union of the current result and the newly found box instead of pushing this new box into a List. The convex union of 2 intervals is defined by:

$$\left[[x] \cup [y]\right] = [Min(x^-, y^-); Max(x^+, y^+)] \tag{48}$$

The convex union of 2 boxes is defined componentwise. In other words, the convex union of 2 boxes is the bounding box of the classical union of the 2 boxes. Thus, the returned value is not a List anymore, but a single box. The occupied memory space depends only on the size of the Stack, which is generally negligible.

Another positive consequence is, as the convex union of 2 boxes is bigger (in the sense of inclusion) than the standard union, that some parameters can already belong to the currently found box without having been processed before. This situation appears very frequently, so before contracting and testing a box, the algorithm will first check if the box which will be tested, already belongs to the currently found box. If yes, then the algorithm has nothing to do and can move along to the following box in the Stack saving a lot of computing time (see Fig. 7).

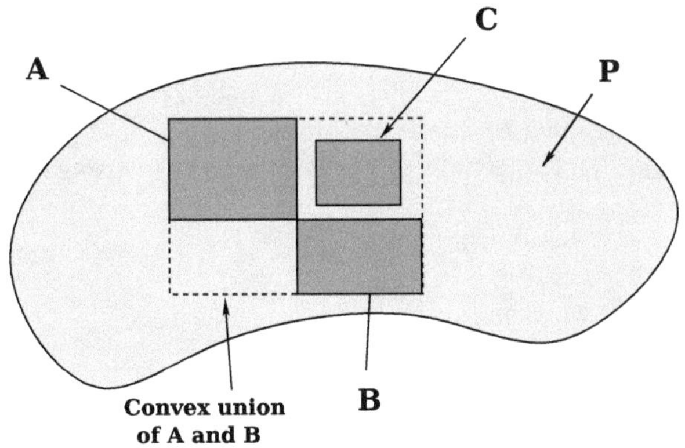

Fig. 7 A and B are feasible boxes. C is accepted without having been processed as it already belongs to the convex union of A and B

The computing time decreases a lot by using this simple technique. However, this technique is not justified if P is a non-connected parameter set. Indeed, the bounding box of a non-connected set includes all the connected components of the set which are generally far from each other. Thus, having a method to select one single connected component is now not only a question of saving computing time, but also a question of relevance of the result (as illustrated by Fig. 8).

3.3.2 Dealing with Symmetry

It is now necessary to find a way to select one single connected component from the set of feasible parameters. The easiest way is to assume that, for example (in the case of r relaxation times τ_i associated with r dielectric relaxations),

$$\forall i \in \{1 \ldots r-1\}, \quad \tau_i < \tau_{i+1} \tag{49}$$

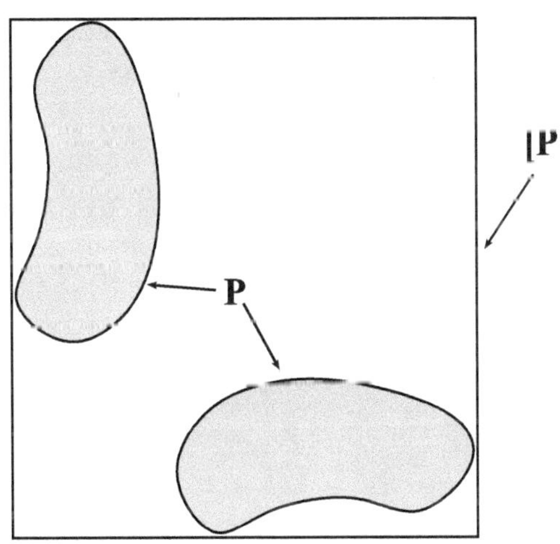

Fig. 8 Bounding box of a non-connected set. Because of the symmetries, the set of feasible parameters is divided into two components which are far one from the other. Thus, the bounding box of the whole set is not relevant at all, whereas the bounding box of one of the components is

But, as the algorithm deals with intervals, this constraint cannot be applied directly. However, it is obvious that this relation implies that the upper bound of $[\tau_i]$ cannot be greater than the upper bound of $[\tau_{i+1}]$, and conversely the lower bound of $[\tau_{i+1}]$ cannot be smaller than the lower bound of $[\tau_i]$. So the algorithm will remove some (and potentially all) values from $[\tau_i]$ by using:

$$\text{For } i = 1 \text{ To } r-1, \quad [\tau_i] \leftarrow \left[\tau_i^-; \min(\tau_i^+, \tau_{i+1}^+)\right] \tag{50}$$

and

$$\text{For } i = 1 \text{ To } r-1, \quad [\tau_{i+1}] \leftarrow \left[\max(\tau_i^-, \tau_{i+1}^-); \tau_{i+1}^+\right] \tag{51}$$

Again, there is a positive consequence of that. The size of $[\tau_i]$ could decrease by this process, or $[\tau_i]$ could become empty. Obviously, there are much more parameters which do not satisfy this constraint than parameters which do, so a lot of boxes will be simply removed during this process (namely, when $\tau_{i+1}^+ < \tau_i^-$, or when $\tau_i^- > \tau_{i+1}^+$ for some i) even if they are mathematically feasible, decreasing considerably the computing time. Moreover, the proportion of boxes which do not satisfy this constraint increases when the number of parameters increases. Then, the increase of computing time due to the increase of the number of parameters is partially compensated by the decrease of computing time due to this process.

3.3.3 Bisection

As the parameter τ_i (relaxation times) and the corresponding interval $[\tau_i]$ could cover many orders of magnitude it makes no sense to cut this interval in the middle. Their size will often be greater than the size of the measured parameter $[\Delta\varepsilon_i]$ [see Eq. (15)]. Thus, plenty of computing time would be wasted by making a lot of useless bisections. Instead, it will be useful to define a specific size and a specific bisection for τ_i. So, by rewriting

$$[\tau_i] = [10^{T_i^-}; 10^{T_i^+}] \tag{52}$$

the size of τ_i is defined by:

$$\text{Size}([\tau_i]) = T_i^+ - T_i^- \tag{53}$$

and the bisection by:

$$\text{Bisection}([\tau_i]) = [10^{T_i^-}; 10^{T_i^- + (T_i^+ - T_i^-)/2}], [10^{T_i^- + (T_i^+ - T_i^-)/2}; 10^{T_i^+}] \tag{54}$$

So, exactly for the same reason that it is reasonable to use a logarithmic axis for the frequency to plot a function, the algorithm will use a "logarithmic bisection" for $[\tau_i]$.

3.4 S.A.D.E. Algorithm

By using these modifications, it is possible to give a new algorithm, S.A.D.E. (as S.I.V.I.A. Applied to DiElectric spectroscopy) which will also find the optimum number of relaxations. This algorithm is described in Fig. 9. The algorithm will loop until it finds the minimum number of relaxations which make the result being non-empty. Note that if it is possible to fit the data with r relaxations, trying to fit it with more than r relaxations will give too many degrees of freedom and will lead to a very bad result. In fact, if there are more than r relaxations, the additional relaxations are masked by the measurement errors. Then, the numbers of relaxations provided by S.A.D.E. should be considered as optimum.

4 S.A.D.E. Examples

4.1 A First Test by Using Home-Made Data

First, the widely used Havriliak-Negami function was tested. Using fixed parameters for only one relaxation ($\varepsilon_\infty = 4$, $\Delta\varepsilon = 2$, $\tau = 1s$, $\alpha = 0.5$ and $\beta = 0.5$), a perfect data set was created using the Havriliak–Negami model [according to Eq. (3)]. Of course, as it was a "home-made" data set, the experimental error on data points would be zero. Hence, this data set was tested as a real measured data set by adding

```
Box [P] ← ø
Integer relaxation Number ← 1
Stack S
While [P] = ø
    Push ( S, [P0] )
    While Not ( is Empty (S) )
        [pc] ← Pop (S)
        If Not ( [pc] ⊂ [P] )
            Break Symetries ( [pc] )
            Contract ( [pc] )
            If [pc] is feasible,
                [P] ← [[P] ∪ [pc]]
            End If
            If [pc] is indeterminate
                If Size ( [pc] ) < η
                    [P] ← [[P] ∪ [pc]]
                Else
                    {[p1],[p2]}= Bisection ([pc])
                    Push (S, [p1])
                    Push (S, [p2])
                End If
            End If
        End If
    End While
    relaxation Number ← relaxation Number + 1
End While
Return [P]
```

Fig. 9 S.A.D.E. (as Set inversion via interval analysis Applied to DiElectric spectroscopy)

Table 1 Calculated parameters, from a "home made" Havriliak–Negami relaxation

Parameter	Interval	
ε_∞	3.96	4.04
$\Delta\varepsilon$	1.9853	2.0538
τ	0.983181 s	1.013406 s
α	0.4847	0.5112
β	0.4836	0.5059

an accuracy of 1%. Then, with a 0.01 box size (η, described in Sect. 3.2) and within less than 1 second computing time on a common desktop computer, the resulting parameter intervals were obtained by S.A.D.E. and presented in Table 1.

This result is very promising since the boundaries of the intervals deviate by less than 4% from the exact parameter values and hence it validates the program S.A.D.E..

Secondly, the Debye function was tested. This function is the only one having a physical meaning and will be used for our experiments in the next part. Using fixed parameters for only one relaxation ($\varepsilon_\infty = 2$, $\Delta\varepsilon = 3$, $\tau = 10^{-6}$s), a perfect data set for the Debye model (see equation (2)) was created. Of course, as it was a "home-made" data set, the experimental error on data points would be zero. Again, this data set was tested as a real measured data set by adding an arbitrary error of 1%. With a 0.1 box size and with less than 1 second computing time on a common desktop computer, the resulting parameter intervals were calculated and presented in Table 2.

As for the previous example, this result is very promising since the boundaries of the intervals deviate by less than 2% from the exact parameter values. Moreover, as this model was simpler than the HN model (less parameters), S.A.D.E. was able to calculate the parameters with a smaller error although the box size was ten times bigger.

Table 2 Calculated parameters, from a "home made" Debye relaxation

Parameter	Interval	
ε_∞	1.97963	2.02038
$\Delta\varepsilon$	2.93991	3.05322
τ	9.89954×10^{-7} s	1.01024×10^{-6} s

4.2 Test with Real Experimental Curves

4.2.1 Experimental Details

The pure diglycidylether of bisphenol A (DGEBA) DER 332 from DOW Chemical was studied by dielectric broadband measurements in a frequency range of *0.1* Hz to 10^6 Hz using a Novocontrol High Resolution Dielectric Alpha Analyser with automatic temperature control by a Quatro cryosystem. For the examples presented here, 200 data points were measured in the frequency range at −90 and 60°C.

The viscous DGEBA was placed between stainless steel electrodes, with a teflon spacer in order to have a well-defined geometry.

As the number of relaxations is higher for real data points than for Home-made curves, the common desktop computer was changed to a Dual Opteron machine containing two 2.4 GHz/64 Bits CPUs and at least 4 GB of RAM. The operating system was Suse Linux 10.0. A parallelized version of the algorithm was implemented (in the C++ programming language) in order to take advantage of the two CPUs.

4.2.2 Application of S.A.D.E. to Multiple Relaxations Data

Figures 10 and 11 show ε' and ε'' as measured at −90°C for pure DGEBA. By eye, only two relaxation processes would be recognized for the curves. However, the shape of the spectra is complex. We choose a sum of Debye relaxators to describe it [see Eq. (15)]. Then, S.A.D.E. identifies the five relaxations represented in Figs. 10 and 11. The five relaxation curves were drawn using the middle point of each computed interval (in the logarithmic sense for the parameters τ_i). With a 0.1 box size and with 382 s computing time on the cluster, the resulting parameter intervals were calculated and presented in Table 3.

This result is very promising since the boundaries of each of the intervals are not far apart. Of course, the width of the intervals can be reduced further as they depend on the error of the experimental data points and on the given box. But the smaller the box η is made, the higher the calculation time will be. Let us note that S.A.D.E. tries first to fit these data points by only one Debye relaxator. As the fit failed, it tries with two to five relaxations (using five Debye relaxators lead to the determination of eleven parameter intervals!). If we try to fit the data points with more than 5

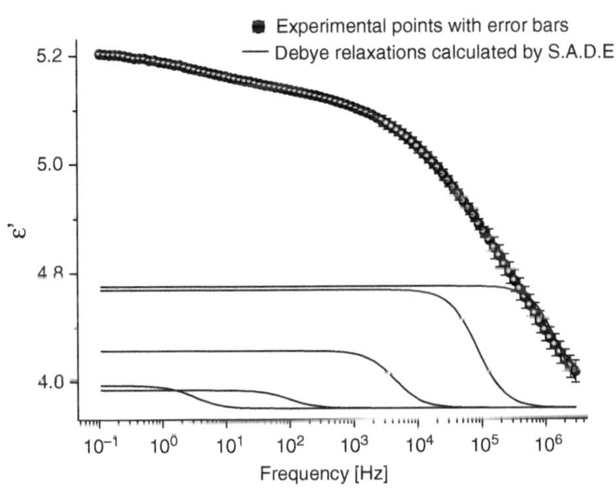

Fig. 10 Dielectric spectroscopy real permittivity of the DGEBA DER 332 prepolymer at −90°C and the corresponding five Debye relaxations as calculated by S.A.D.E.

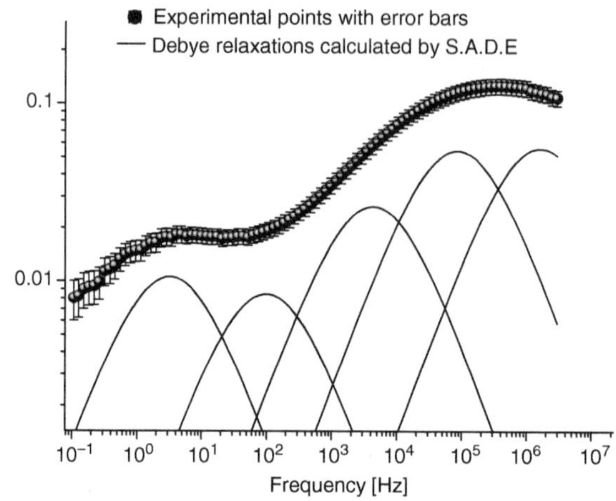

Fig. 11 Dielectric spectroscopy imaginary permittivity of the DGEBA DER 332 prepolymer at −90°C and the corresponding five Debye relaxations as calculated by S.A.D.E.

Table 3 Calculated parameters, using the equation (4), from experimental data set at −90°C presented in the Figs. 10 and 11

Parameter	Interval	
ε_∞	4.50144	4.60168
$\Delta\varepsilon_1$	0.176763	0.270755
τ_1	7.23×10^{-8} s	1.24×10^{-7} s
$\Delta\varepsilon_2$	0.164731	0.269318
τ_2	1.54×10^{-6} s	2.21×10^{-6} s
$\Delta\varepsilon_3$	0.0771269	0.133832
τ_3	2.74×10^{-5} s	4.70×10^{-5} s
$\Delta\varepsilon_4$	0.0239589	0.0437055
τ_4	1.19709×10^{-3} s	2.05353×10^{-3} s
$\Delta\varepsilon_5$	0.0321849	0.0524569
τ_5	3.69994×10^{-2} s	6.26434×10^{-2} s

relaxations, the additional relaxations will be masked by the measurement errors. Then, the numbers of relaxations provided by S.A.D.E. is considered as optimum. It is worth noting that these parameters were not found by classical least square approximation fitting routines.

4.2.3 Application of S.A.D.E. to the Global Problem: Data with Relaxations, Conductivity and Electrode Polarization

Figures 12 and 13 show ε' and ε'' measured at 60°C for the pure DGEBA DER 332. At first glance, only conductivity, electrode polarization and the beginning of a relaxation process (at high frequencies) can be identified for that measured dielectric spectrum. But two relaxations are found by S.A.D.E., and the DC-conductivity is well defined. As for the previous figures, the relaxation curves were drawn using the

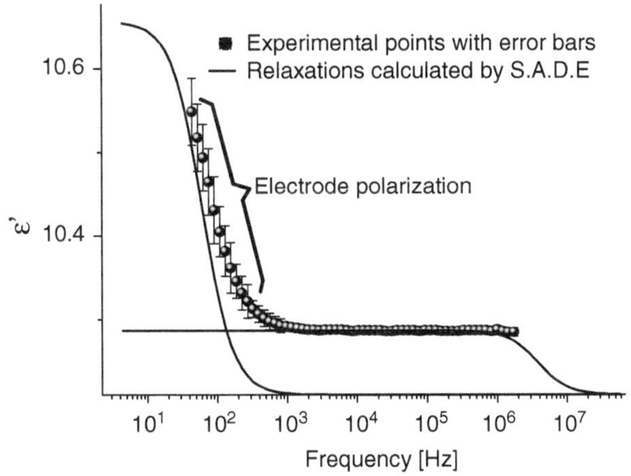

Fig. 12 Dielectric spectroscopy real permittivity of the DGEBA DER 332 prepolymer at 60°C and the corresponding two Debye relaxations as calculated by S.A.D.E.

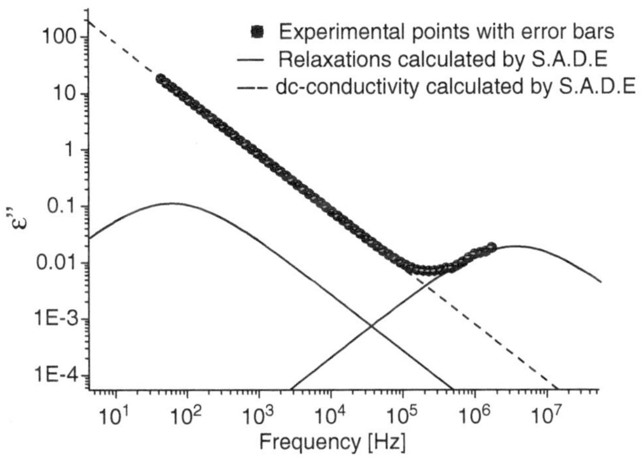

Fig. 13 Dielectric spectroscopy imaginary permittivity of the DGEBA DER 332 prepolymer at 60°C and the corresponding two Debye relaxations as calculated by S.A.D.E.

middle point of each computed interval (in the logarithmic sense for the parameters τ_i). With a 0.1 box size and with 4336 s computing time on the cluster, the resulting parameter intervals were calculated and presented in Table 4.

The first relaxation (at high frequencies, with the smaller relaxation time τ_1) corresponds to a dipole relaxation of the material. By just looking at the curve, it is not possible to say if some relaxators take place as only conductivity and electrode polarization could be identified, but S.A.D.E. clearly identified this material relaxation. In addition, the second relaxation (at low frequencies, with the higher

Table 4 Calculated parameters, using the Eq. (15), from experimental data set at 60°C presented in the Figs. 12 and 13

Parameter	Interval	
ε_∞	10.169	10.2691
$\Delta\varepsilon_1$	0.0308334	0.123296
τ_1	1.56023×10^{-8} s	6.97831×10^{-8} s
$\Delta\varepsilon_2$	0.222722	0.671
τ_2	1.00487×10^{-3} s	4.17685×10^{-3} s
σ_{DC}	$3.95878 \times 10^{-8}\text{S} \times \text{m}^{-1}$	$4.9343 \times 10^{-8}\text{S} \times \text{m}^{-1}$

relaxation time τ_2) corresponds to the electrode polarization. As mentioned in the paragraph "Polarization at electrode and phase boundaries" (see Sect. 1.1.3), the electrode polarization is well modeled as a strong Debye relaxator. In fact, close to the electrodes the mobile charges can be blocked and piled up at the phase boundaries, as described by Maxwell, Wagner, and Sillars [11–13], whose relation is similar to the Debye equation (see Eqs. (2) and (10)). Electrode polarization is reasonable since the DGEBA contains traces of ions from synthesis. Let us note that these relaxations were not visible in the data plot (Figs. 12 and 13) and not possible to fit by classical ways. Finally, the DC-conductivity was calculated simultaneously with the Debye relaxations: it perfectly fits the rise observed in the imaginary part of the material permittivity (Fig. 13).

In conclusion, as S.A.D.E. found some Debye-like relaxations we can be sure that a Debye relaxator also describes electrode polarization reasonably well.

5 Conclusion

Dielectric spectroscopy is an extremely versatile method for characterizing the molecular dynamics over a large range of time scales. Unfortunately, the extraction of model parameters by data fitting is still a crucial problem which is now solved by our program S.A.D.E.

S.A.D.E. is based on the algorithm S.I.V.I.A. which was proposed and implemented by Jaulin [16,17] in order to solve constraint satisfaction problems. The problem of dielectric data analysis is reduced to a problem of choosing the appropriate physical model. In this article, Debye relaxations were used and validated to fit the relaxations of a DGEBA prepolymer and the polarization of the spectrometer electrodes. The conductivity was evaluated too.

Acknowledgement The authors acknowledge Professor Springborg who granted the access to the computing cluster and Michael Bauer for his help in using it. We also thank Luc Jaulin for his most useful advices.

Software Availability

S.A.D.E. is freely available on the website of M. Aufray (http://maelenn.aufray.free.fr). S.A.D.E. is protected by copyright (c) 2006 Brochier, and is distributed under the terms of the GNU general public license.

References

1. P. Debye, *Polar Molecules* (Chemical Catalog, Reprinted by Dover Publications, New York, 1929)
2. Havriliak, S. Negami, Polymer **8**, 161 (1967), http://www.sciencedirect.com/science/article/B6TXW-48FBVVM-4T/2/ffb6daa2b39f773a1666f9d2cb8d4ed8
3. K.S. Cole, R.H. Cole, J. Chem. Phys. **9**, 341 (1941)
4. K.S. Cole, R.H. Cole, J. Chem. Phys. **10**, 98 (1942)
5. D. Davidson, R.H. Cole, J. Chem. Phys. **18**, 1417 (1950)
6. M. Mangion, M. Wang, G. Johari, J. Polym. Sci. B Polym. Phys. **30**(5), 433 (1992), http://www3.interscience.wiley.com/cgi-bin/fulltext/104051384/PDFSTART
7. Corezzi, S. Capaccioli, G. Gallone, A. Livi, P. Rolla, J. Phys. Condens. Matter 9, 6199 (1997), http://www.fisica.unipg.it/~corezzi/
8. N. Axelrod, E. Axelrod, A. Gutina, A. Puzenko, P. Ben Ishai, Y. Feldman, Meas. Sci. Technol. **15**, 1 (2004)
9. F. Stickel, E. Fischer, R. Richert, J. Chem. Phys. **104**(5), 2043 (1996)
10. R.M. Fuoss, J.G. Kirkwood, J. Am. Chem. Soc. **63**(2), 385 (1941)
11. J. Maxwell, *A Treatise on Electricity and Magnetism,* vol 1. (Dover Publications, New York, 1954)
12. K.W. Wagner, Archiv für Elektrotechnik **2**(9), 371 (1914). URL http://www.springerlink.com/content/xr0617448810/?p=a72f93d829804c93afa018301302c470&pi=624
13. R. Sillars, Proc. Inst. Electr. Eng. **80**, 378 (1937)
14. D.W. Marquardt, J. Soc. Ind. Appl. Math. **11**(2), 431 (1963)
15. K. Levenberg, Q. J. Appl. Math. **2**, 164 (1944)
16. Jaulin, E. Walter, Automatica **29**(4), 1053 (1993), http://www.sciencedirect.com/science/article/B6V21-47WVRTR-5T/2/35ab6e2405d16f0886be1fcfc21d687c
17. Jaulin, E. Walter, Math. Comput. Simul. **35**(2), 123 (1993), http://www.sciencedirect.com/science/article/B6V0T-45DHW9D-2/2/f8024c6b00508f6f3ce7087b39f78f59

Part II
Modeling and Simulation of Non-classical Materials and Structures

Numerical Modeling of Complex Structures: Shells and Biological Cells

J.N. Reddy, R.A. Arciniega, G.U. Unnikrishnan, and V.U. Unnikrishnan

1 A Refined Shell Finite Element

1.1 Introduction

In solid and structural mechanics the development of efficient computational models for the nonlinear analysis of shells has been one of the most important research activities. This is partially motivated by the need to analyze new materials such as composites and functionally graded shells. Shells made of laminated composites continue to be of great interest in many engineering applications [1]. In some applications these structures can experience large elastic deformations and finite rotations. Consequently, geometric nonlinearity plays an essential role in the behavior of the shell. Therefore, the choice of an appropriate mathematical model together with a consistent and robust computational procedure that can accurately represent nonlinear deformations and stresses in shell structures are of vital importance.

The most commons finite elements used for the analysis of shell structures are the continuum-based shell element (or degenerated solid element) and a 2D element based on a shell theory [2]. Finite elements based on shell theories describe, in a natural way, the behavior of the shell since they are written in terms of curvilinear coordinates. For this case, two different approaches can be identified depending whether or not there is an approximation of the geometry of the mid-surface (i.e., finite element domain in the parametric space of the mid-surface). Formulations in which the mid-surface is represented by a chart and interpolate the covariant components of the field variables are called tensor-based finite element models. It is often argued that this kind of interpolation automatically causes difficulties with the rigid body modes of curved structures because they cannot be properly represented [3]. We do not share this view point. It is demonstrated in this paper (from a heuristic perspective) that with a suitable formulation of the shell kinematics and choice of

J.N. Reddy (✉)
Department of Mechanical Engineering, Texas A&M University, College Station, TX 77843-3123, USA
e-mail: jnreddy@tamu.edu

A. Öchsner et al. (eds.), *Materials with Complex Behaviour*, Advanced Structured Materials 6, DOI 10.1007/978-3-642-12667-3_8, © Springer-Verlag Berlin Heidelberg 2010

interpolation of the field variables over the finite element, these problems can be avoided.

The dominant trend to overcome locking in shells over the last decades has been the use of low-order finite elements with mixed formulations. A mixed finite element model is based on Hu-Washizu type mixed variational principle [4, 5]. Alternatively, high-order elements have been proposed for the analysis of shells. The claim of this approach is to use finite elements of sufficiently high degree to recover the convergence property in an optimal order. This is called p-version finite element technology (where p is the degree of the interpolation polynomial). In this case, there is no need to use mixed formulations, and displacement-based finite element formulations can be applied. Finite elements with high-order approximations have been utilized by Pitkäranta [6, 7] for linear analysis of shells (also, see [8]). The use of tensor-based finite element formulations together with high-order elements for the analysis of shell structures leads to an efficient computational approach which is straightforward to implement. Such model can be applied to linear and nonlinear analyses of shells made of isotropic, laminated composite and functionally graded materials (FGMs). To the knowledge of the authors, no applications of high-order elements to nonlinear shell analysis are found in the literature until the authors work.

The objective of this study is to develop a refined shell element using a tensor-based formulation. The approach is based on a first-order shell theory with seven independent parameters [9, 10] under the Lagrangian framework. This theory circumvents the use of a rotation tensor and preserves the additive structure of the configuration update of the shell. The element is capable of simulating finite displacements and rotations with no presence of membrane or shear locking.

The mathematical shell model is consistently derived using absolute tensor notation and the finite element model is developed in a straightforward way. Moreover, the simplicity of this approach makes it attractive for applications to contact mechanics and damage propagation. Previous works for linear analysis of laminated shells using tensor-based finite elements can be found in Arciniega and Reddy [11] and Reddy and Arciniega [12].

1.2 Theoretical Formulation

Let $\{\theta^i\}$ be a set of convected curvilinear coordinates, and $i = 1, 2, 3$. The mid-surface of the shell is defined by the coordinates $\{\theta^\alpha\}$. The displacement vector is assumed to be of the form

$$\mathbf{V}(\theta^i) = \mathbf{u}(\theta^\alpha) + \theta^3 \boldsymbol{\varphi}(\theta^\alpha) + (\theta^3)^2 \underline{\boldsymbol{\psi}}(\theta^\alpha) \quad (1)$$

where $\mathbf{u}$ denotes displacement vector of a point on the mid-surface, φ the rotation tensor, and $\underline{\psi}$ is the thickness stretch tensor with only component along the transverse normal

$$\mathbf{u}\left(\theta^{\alpha}\right) = u_i \mathbf{a}^i, \quad \boldsymbol{\varphi}\left(\theta^{\alpha}\right) = \varphi_i \mathbf{a}^i, \quad \underline{\boldsymbol{\psi}}\left(\theta^{\alpha}\right) = \psi_3 \mathbf{a}^3 \tag{2}$$

Equation (1) contains 7 independent variables. The quadratic term $\underline{\boldsymbol{\psi}}$ is included to avoid the Poisson or thickness locking; therefore, no enhanced methods are needed. Since the drilling degree of freedom is included in the model, full 3D constitutive equations are utilized. The use of a rotation tensor is avoided and the additive structure of the configuration update is preserved. For the given displacement field, we obtain the following Green strain tensor

$$\hat{\mathbf{E}} = \boldsymbol{\varepsilon}^0 + (\theta^3)\,\boldsymbol{\varepsilon}^1 + \mathrm{HOT} \tag{3}$$

where $\hat{\mathrm{E}}$ is the pull-back of the covariant Green strain tensor $\mathbf{E}$, and $(\boldsymbol{\varepsilon}^0, \boldsymbol{\varepsilon}^1)$ are the tensors associated with membrane and bending strains. We denote with $\mathbf{M}^n$ the stress resultant tensor, which is symmetric because of the symmetry of the second Piola-Kirchhoff stress tensor $\mathbf{S}$ (see [13]):

$$\mathbf{M}^n = \int_{-h/2}^{h/2} (\theta^3)^n \hat{\mathbf{S}}\, \mu\, d\theta^3 \tag{4}$$

where $\hat{\mathrm{S}}$ is a pull-back operator of a contravariant of the second Piola-Kirchhoff stress tensor $\mathbf{S}$, $n = 1{,}2$ and μ is the determinant of the shifter tensor $\boldsymbol{\mu}$. The weak form can be easily constructed from the principle of virtual work (see [1, 2]). The configuration of the shell is defined by the triplet (or seven parameters) $\Phi \equiv \left(\mathbf{u}, \boldsymbol{\varphi}, \underline{\boldsymbol{\psi}}\right) = (u_i, \varphi_i, \psi_3)$. We have

$$\begin{aligned}\mathcal{J}(\Phi, \delta\Phi) &= \mathbb{G}_{\text{int}}(\Phi, \delta\Phi) + \mathbb{G}_{\text{ext}}(\Phi, \delta\Phi) = 0 \\ &= \int_{M_R} \left(\mathbf{M}^0 \cdot \delta\boldsymbol{\varepsilon}^0 + \mathbf{M}^1 \cdot \delta\boldsymbol{\varepsilon}^1\right) d\Omega + \mathbb{G}_{\text{ext}}(\Phi, \delta\Phi)\end{aligned} \tag{5}$$

where $\delta\Phi := \left(\delta\mathbf{u}, \delta\underline{\boldsymbol{\varphi}}, \delta\underline{\boldsymbol{\psi}}\right) \in \mathcal{V}$ and $\mathcal{V}$ is the space of admissible variations.

Let $\mathscr{A}$ be the domain of the midsurface parametrization. Notice that the location of the midsurface is given by the mapping $\mathbf{r}(\cdot) : \mathscr{A} \subset \mathbb{R}^2 \to \mathscr{M}_R \subset \mathbb{R}^3$. The finite element domain is given in the parametric space $\mathscr{A}$. In that case the parameters of the midsurface are computed exactly at each Gauss point. The finite element equations are obtained by discretizing the covariant components of the displacements and rotations of the midsurface written in terms of the surface base vectors (tensor-based formulation). Namely

$$\mathbf{u}^{hp}(\underline{\boldsymbol{\theta}}) = \left(\sum_{j=1}^{m} u_k^{(j)} N^{(j)}(\xi,\eta)\right)\mathbf{a}^k$$
$$\underline{\boldsymbol{\varphi}}^{hp}(\underline{\boldsymbol{\theta}}) = \left(\sum_{j=1}^{m} \varphi_\beta^{(j)} N^{(j)}(\xi,\eta)\right)\mathbf{a}^\beta \tag{6}$$
$$\underline{\boldsymbol{\psi}}^{hp}(\underline{\boldsymbol{\theta}}) = \left(\sum_{j=1}^{m} \psi_3^{(j)} N^{(j)}(\xi,\eta)\right)\mathbf{a}^3$$

where $\underline{\boldsymbol{\theta}}(\xi,\eta) : \hat{\Omega}^e \in \mathbb{R}^2 \to \mathrm{A}^e \subset \mathrm{A}$. The elements are interpolated by using higher-order Lagrange polynomials which, in contrast to low-order finite element formulation, do not exhibit any locking problems. Finally, the linearization of the discrete form of Eq. (5) is carried out, leading to a symmetric system of equations. The resulting set of nonlinear algebraic equations is solved by the incremental procedure based on the Newton–Raphson method and the cylindrical arc-length method.

1.3 Numerical Results

In this section, numerical results obtained using the shell element described herein are presented for several bench mark problems. Regular meshes of Q25, Q49 and Q81 high-order elements with seven degrees of freedom per node are utilized in the finite element analysis. These elements are based on $p = 4$, 6, and 8, respectively. By increasing the p level or refining the finite element mesh, we mitigate locking problems. Full Gauss integration rule is employed in all examples.

We consider the nonlinear behavior of functionally graded shells. The first example consists of a shallow FGM shell panel made of ceramic and metal and subjected to a point load (Fig. 1). The geometric parameters used are $a = 508$ mm, $R = 2540$ mm, $h = 25.4, 12.7, 6.35$ mm, $\alpha = 0.1$rad. The material properties of the constituents are $E_c = 151$ GPa, $\nu_c = 0.3$, $E_m = 70$ GPa, $\nu_m = 0.3$. A typical material property (with the exception of Poisson's ratio, which is kept constant at 0.3) is assumed to vary continuously through the thickness of the shell as

$$P\left(\theta^3\right) = P_c f_c + P_m f_m, \quad f_c = \left(\frac{\theta^3}{h} + \frac{1}{2}\right)^n, \quad f_m = 1 - f_c$$

where the subscript c refers to ceramic and m refers to metal. The FGM shell is still isotropic with constant material properties in each surface. Figure 1 contains central deflection versus point load curves of the present approach for different volume fractions n (from fully metal to ceramic). The results show standard limit points and complex equilibrium curves with snap-through and snap-back behavior.

Next, we consider two examples of plates: the bending of a clamped plate under distributed moment and an annular plate subjected to a shear force. These are classical benchmark problems for large deformation analysis and they have been

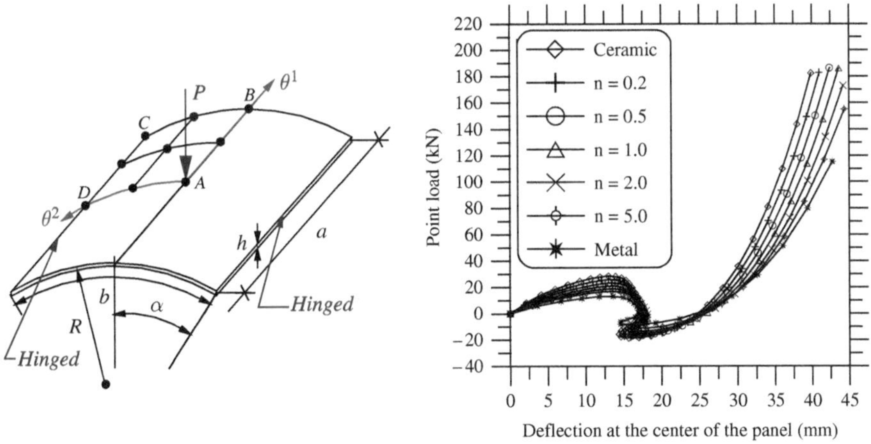

Fig. 1 Deflection at the center of the cylindrical panel under point load ($h = 6.35$ mm)

considered by many authors. In fact, these problems are good to test the capability of the finite element model to simulate finite rotations on shells. Figure 2 shows the deformed configurations of the strip ($E = 2.0 \times 10^{11}$, $\nu = 0.3$, $L = 0.5$, $b = 0.075$, $h = 0.0045$) and annular plate ($E = 21.0 \times 10^6$, $\nu = 0.0$, $R_i = 6$,

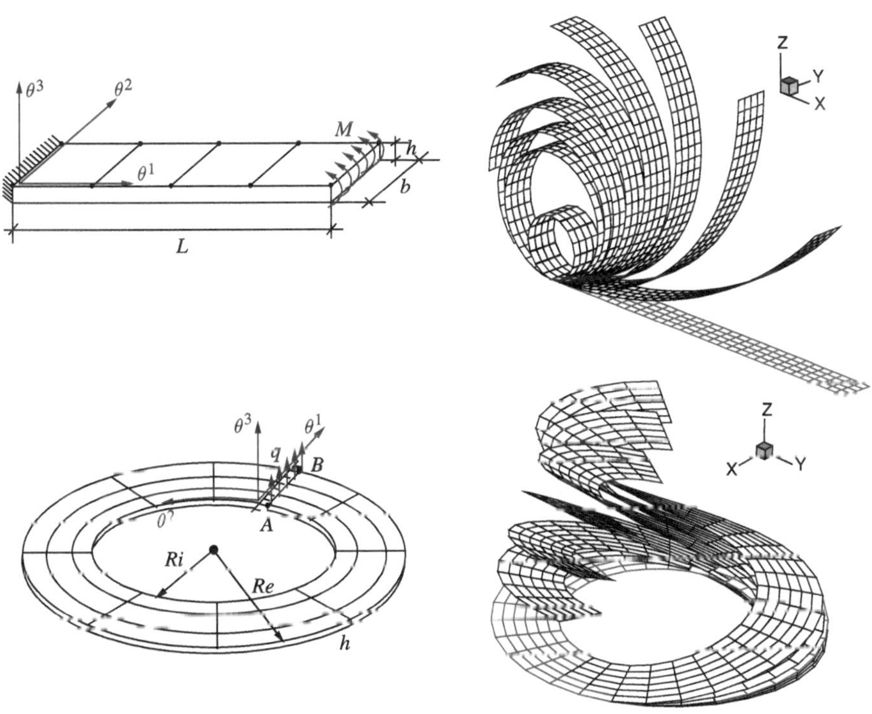

Fig. 2 Roll-up of a clamped plate and annular plate under end shear loading

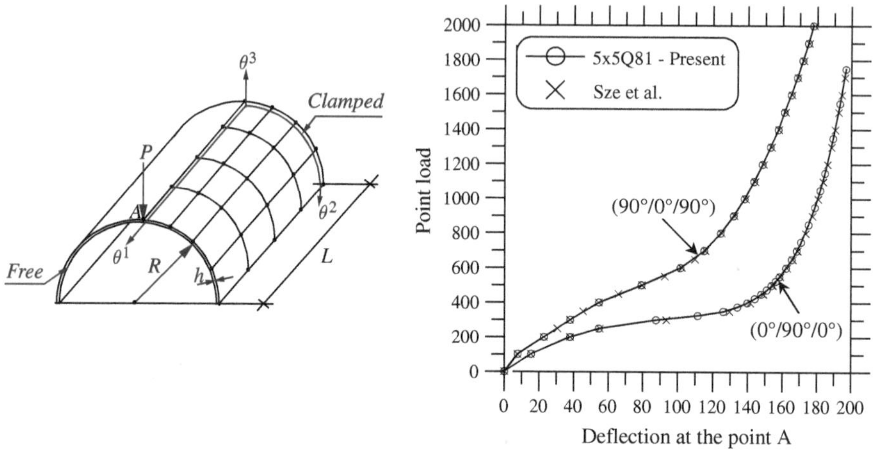

Fig. 3 Clamped semi-cylindrical shell under point load

$R_e = 10, h = 0.03$) for various load stages. The last deformed configurations demonstrate the ability of the present approach to handle extreme rotations of the plate (up to 720°) without any singularities.

Another well-known benchmark problem for finite deformation is the semi-cylindrical shell ($L = 304.8$, $R = 101.6$, $h = 3.0$). under point load. Figure 3 shows the deflection at point A versus the point load for composite shells ($E_1 = 2068.5, E_2 = E_3 = 517.125, G_{12} = G_{13} = 795.6, G_{23} = 198.89 \nu_{23} = 0.3$, $\nu_{12} = \nu_{13} = 0.3$). To validate the our results, the present results are compared with those of Sze et al. [14], who used Abaqus program to generate the results.

Finally, we just show the deformed configurations of a full hemispherical shell and a hyperboloid shell under two inward and two outward point loads. These challenging problems demonstrate the robustness of the present finite

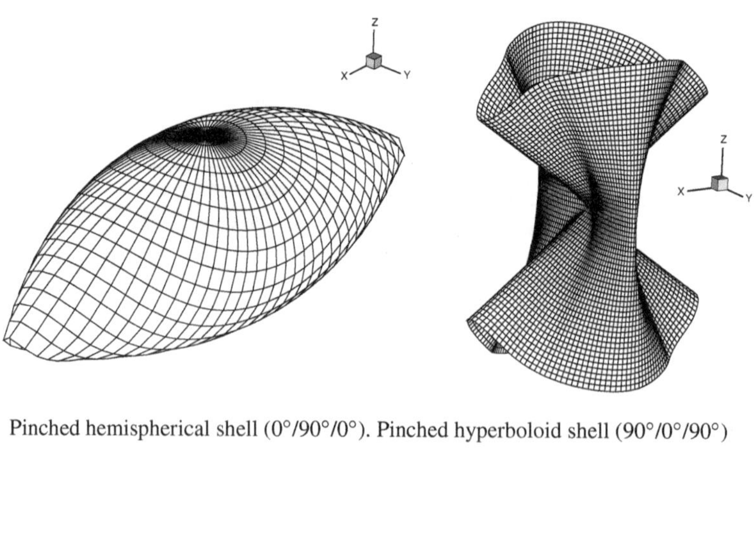

Fig. 4 Pinched hemispherical shell (0°/90°/0°). Pinched hyperboloid shell (90°/0°/90°)

element model and its applicability to arbitrary shell geometries with very strong geometric nonlinearity. The analysis is carried out for cross-ply composite laminates (90°/0°/90°) and (0°/90°/0°). For the hemispherical shell we utilized a mesh of 2×2 Q81 elements (in a quarter of the shell) while for the hyperboloid shell we used 5×5 Q81 elements (in an octant of the shell). The final configurations of both shells are shown in Fig. 4. For the details of geometry and material properties (see [15]).

2 Constitutive Modeling of Biological Cell

2.1 Introduction

Cell is the fundamental unit of every living organism. Various computational models of cells have been developed in literature to analyze its behavior, and can be broadly defined as the discrete cell models and continuum cell models. Most of the earlier works based on continuum hypothesis homogenize the entire cell and do not explicitly consider the effect of inhomogeneity of the cell, with some exceptions [16–18]. These models, even though they reduce the number of mechanical parameters, fail to capture the properties caused by the structural inhomogeneity of cytoplasm, like actin network layer, stress fibers etc (see Fig. 5). This becomes a crucial factor in the study of mechanical behavior of cells in-vivo as well as in the determination of mechanical parameters using experimental techniques like atomic force microscopy and micropipette suction.

Scanned images and experimental procedures have shown that there exist regions in cytoplasm having distinct physical properties. The contribution of the cytoskeletal filaments, especially the actin stress fibers, in influencing the mechanical properties of cells is well established in the literature [18, 19]. A precise representation of the anisotropic, nonlinear behavior of the cytoskeletal architecture is required for any computational analysis of a living cell. The homogenous material definition of the cell is far from being accurate, especially for an adherent cell in which stress fiber introduces significant inhomogeneity. In this work, a mechanical model

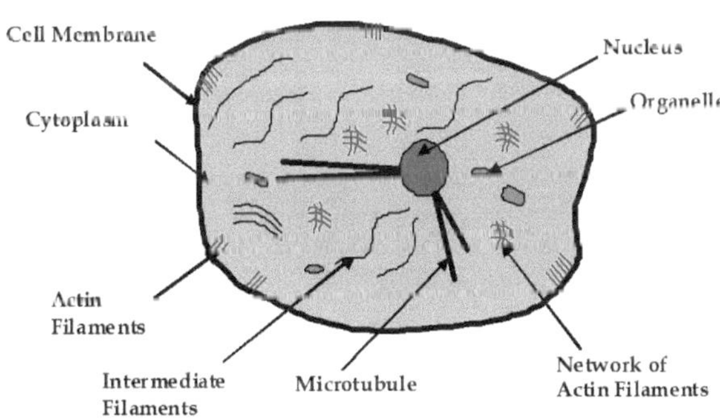

Fig. 5 A schematic representation of a generalized cell

of an adherent cell based on continuum micromechanics considering the structural inhomogeneity of the cytoplasm is developed. The homogenized cytoplasm is considered to be a matrix reinforced with stress fibers; the periplasm or the actin cortex as a layer of semi-flexible polymer networks and finally the nucleus.

2.2 Homogenization of Cell Structure

In the cortical region, actin filaments behaving as biopolymers forms a complex filamentous network whose properties are dependent on the individual filament characteristics. To capture the behavior of the cortical region, suitable material models are derived based on the actin filament and network characteristics. Cytoplasm is composed of a highly organized network of cytoskeletal filaments of actin, intermediate filaments (IF), and microtubules (MT). The distribution of the cytoskeletal filaments differs according to the type and environment of the cell, thus changing their material properties. In a tissue, the material inhomogeneity is introduced in the cytoplasm through the formation of actin fibers, and their directions are influenced by the external collagen alignment. The actin fibers, having a few microns in diameter, are formed from the bundling of actin filaments in the presence of actin binding proteins. A constitutive model incorporating the effect of actin bundles are developed in this study, by assuming the cytoplasm to be a "fiber-reinforced composite" satisfying the continuum hypothesis [20].

Using such an idealization, the effective property of the cytoplasm reinforced with stress fibers could be obtained by borrowing ideas from the widely accepted homogenization theories in composite material. The property of this homogeneous continuum is based on a statistically homogeneous volume element, called the representative volume element (RVE). The RVE is a representation of the material at the microscale, and is small in comparison to the macrostructure (e.g. whole cell, tissue), has negligible influence on macroscopic property, but large in comparison to the microstructure (e.g. protein molecules) for a meaningful sampling. Actin cortex can be considered as a semi-dilute polymer solution of actin filaments, crosslinked with actin binding proteins. The nonlinear stress-strain behavior of the actin network is captured by introducing a neo-Hookean material model using the calculated shear modulus with the cortex treated as an incompressible material, whose strain energy density is taken as

$$W(I_1) = \frac{\mu_0}{2}\,(I_1 - 3) \tag{7}$$

where μ_0 is the initial shear modulus, and I_1 is the first strain invariant. The cytoplasm, as observed in various experimental studies, has randomly distributed stress fibers in a matrix of microtubules and intermediate filaments. The matrix is assumed to be a hyperelastic material and nearly incompressible with a small strain shear modulus of 100 Pa. The linear effective modulus of the composite having a random distribution of fiber in a uniform matrix is given as

$$K_e = K_1 + v_0 \left\{ \frac{1}{K_0 - K_1} + \frac{3v_1}{3K_1 + 3\mu_0 + \mu_1} \right\}^{-1}$$
$$\mu_e = \mu_1 + v_0 \left\{ \frac{1}{\mu_0 - \mu_1} + \frac{2v_1}{5} \left[\frac{1}{\mu_0 + \mu_1} + \frac{3 - 4\upsilon_0}{\mu_0 + \mu_1(3 - 4\upsilon_0)} + \frac{1}{2\,(3K_1 + 3\mu_0 + \mu_1)} \right] \right\}^{-1} \tag{8}$$

where, μ_e is the shear modulus, K_e is the bulk modulus, υ Poisson's ratio, v_0 is the volume fraction of the matrix. The subscript number indicates 0 = matrix, 1 = fiber, e = effective matrix.

The nonlinear behavior of the composite system is modeled in this work by a simplified incremental approach. The stress-strain curve for the material after homogenization for different volume fractions of the fiber is shown in Fig. 6 and it can be seen that as the volume fraction of the fiber increases the cell becomes stiffer. The same effect is observed in many experimental procedures, which reported a decrease in the measured elastic modulus of cell, acted upon by actin disrupting chemicals. For sufficiently large stress fiber volume fractions, the Poisson's ratio decreases to the fiber Poisson's ratio, as shown in Fig. 6 (0.35 for a stress fiber volume fraction of 2.5%). This is a significant relationship as it partly explains the wide differences in the Poisson's ratio values reported from experimental studies, and ranging from a nearly incompressible value to the range of 0.25 + 0.05.

The effectiveness of the developed model in accurately interpreting the experimental results is illustrated through the numerical simulation of the Atomic Force Microscopy (AFM) indentation. In general, the properties derived from the interpretation of data from the AFM experiments are based on certain assumptions. These assumptions become invalid in the case of biological cell due to: large deformation compared to the cell size; inherent inhomogeneity of the cytoplasm, and so on. Thus, in the interpretation of the results obtained from the atomic force microscopy, a more detailed numerical approach like the finite element method is required. The implementation of the developed constitutive model of the cytoplasm in the numerical study of AFM is described below.

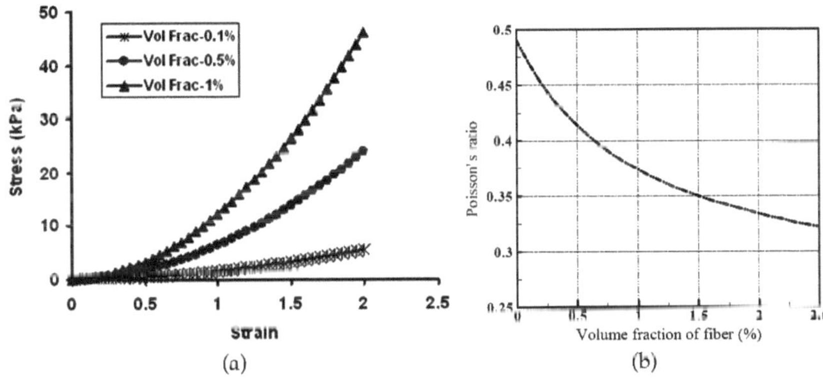

Fig. 6 (**a**) Stress-Strain curve of the effective composite (**b**) variation of Poisson's ratio with volume fraction of stress fiber

Finite element analysis [2] of an AFM indentation using spherical indenter of diameter 0.4 microns acting on the cell surface above the nucleus is carried out. The cell geometry considered is of 3.5 microns in half width, 3.0 microns in height, with a nucleus of 0.9 microns diameter at a height of 0.75 microns from the base, and the cortical region is assumed to be 0.2 microns thick. The cell is assumed to be axisymmetric with a rigid spherical indenter acting on the cell surface above the nucleus. Displacement boundary conditions are applied on the indenter and also at the base of the cell, while symmetric boundary conditions are taken along the axis of symmetry. The indenter is given a vertical displacement and the cell base is constrained in all directions to assume a perfect contact with the substrate. The symmetric half cell model is discretized using an axisymmetric finite element with a finer mesh towards the top of the cell and the finite element analysis is performed using commercial software, ABAQUS [21]. The strain distribution of the cell with stress fiber volume fraction of 0.1% subjected to an indentation of 0.5 microns is shown in Fig. 7. The actin cortical layer, which is in direct contact with the indenter, sustains the maximum deformation. The inner cytoplasm near the region of indentation also experiences very high strains and the intensity decreases away from the center.

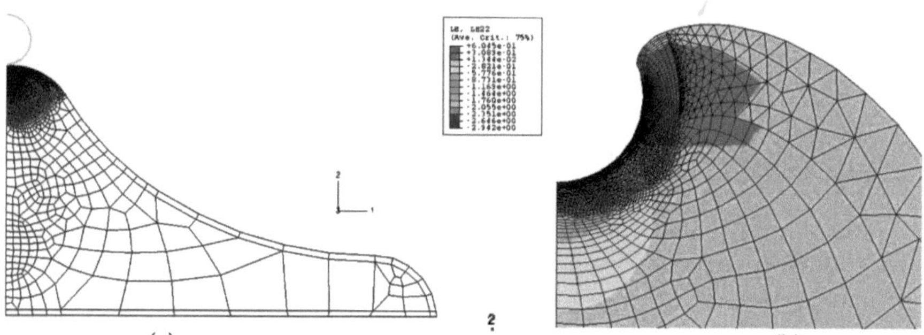

(a) (b)

Fig. 7 (**a**) Half cell axisymmetric finite element model of the cell (**b**) strain distribution obtained from the finite element analysis of cell

3 Conclusions

In the first part of the paper, a refined higher-order finite element for the nonlinear analysis of shells is discussed. A first-order shell theory with seven parameters including a thickness-stretch and accounting for full 3D constitutive equations is used. The family of high-order Lagrangian approximations are utilized to eliminate membrane and shear locking. I number of benchmark problems are analyzed using the element developed herein. Numerical examples for plates and cylindrical shells illustrate the validity of the present approach and accuracy of the developed shell element. In the second part of the paper, a parametric study of the material properties of actin, cytoplasm and nucleus, to determine the material properties of normal biological cells is carried out. This study was able to correlate the concentration of actin filament with the cellular material property, and ultimately to the experimentally determined force-deflection curves from AFM studies.

Acknowledgments The authors gratefully acknowledge the support of this research through Oscar S. Wyatt Endowed Chair funds at Texas A&M University.

References

1. J.N. Reddy, *Mechanics of Laminated Plates and Shells. Theory and Analysis,* 2nd ed. (CRC Press, Boca Raton, FL, 2004)
2. J.N. Reddy, *An Introduction to Nonlinear Finite Element Analysis* (Oxford University Press, New York, NY, 2004)
3. P.S. Lee, K.J. Bathe, Insight into finite element shell discretizations by use of the "basic shell mathematical model". Comput. Struct. **83**, 69–90 (2005)
4. E. Dvorkin, K.J. Bathe, A continuum mechanics based four-node shell element for general nonlinear analysis. Eng. Comput. **1**, 77–88 (1984)
5. J.C. Simo, M.S. Rifai, A class of mixed assumed strain methods and the method of incompatible modes. Int. J. Numer. Meth. Eng. **29**, 1595–1638 (1990)
6. H. Hakula, Y. Leino, J. Pitkäranta, Scale resolution, locking, and high-order finite element modeling shells. Comput. Methods Appl. Mech. Eng. **133**, 157–182 (1996)
7. Y. Leino, J. Pitkäranta, On the membrane locking of h-p finite elements in a cylindrical shell problem. Int. J. Numer. Methods Eng. **37**, 1053–1070 (1994)
8. J.P. Pontaza, J.N. Reddy, Least-square finite element formulation for shear deformable shells. Comput. Methods Appl. Mech. Eng. **194**, 2464–2493 (2005)
9. C. Sansour, A theory and finite element formulation of shells at finite deformations involving thickness change: Circumventing the use of a rotation tensor. Arch. Appl. Mech. **65**, 194–216 (1995)
10. M. Bischoff, E. Ramm, Shear deformable shell elements for large strains and rotations. Int. J. Numer. Methods. **40**, 4427–4449
11. R.A. Arciniega, J.N. Reddy, Consistent third-order shell theory with application to composite circular cylinders. AIAA J. **43**(9), 2024–2038 (2005)
12. J.N. Reddy, R.A. Arciniega, Shear deformation plate and shell theories: From Stavsky to present. Mech. Adv. Mater. Struct. **11**, 535–582 (2004)
13. J.N. Reddy, *An Introduction to Continuum Mechanics with Applications* (Cambridge University Press, New York, NY, 2008)
14. K.Y. Sze, X.H. Liu, S.H. Lo, Popular benchmark problems for geometric nonlinear analysis of shells. Fin. Elem. Analysis. Des. **40**, 1151–1569 (2004)
15. R.A. Arciniega, J.N. Reddy, Tensor-based finite element formulation for geometrically nonlinear analysis of shell structures. CMAME **196**, 1048–1073 (2007)
16. J. Ohayon, P. Tracqui, Computation of adherent cell elasticity for critical cell-bead geometry in magnetic twisting experiments. Ann. Biomed. Eng. **33**, 131–141 (2005)
17. H. Karcher, J. Lammerding, H. Huang, R.T. Lee, R.D. Kamm, M.R. Kaazempur Mofrad, A three-dimensional viscoelastic model for cell deformation with experimental verification. Biophys. J. **85**, 3336–3349 (2003)
18. J.D. Humphrey, On mechanical modeling of dynamic changes in structure and properties in adherent cells. Math. Mech. Solids **7**, 521–539 (2002)
19. C. Rotsch, M. Radmacher, Drug-induced changes of cytoskeletal structure and mechanics in fibroblasts: An atomic force microscopy study. Biophys. J. **78**, 520–535 (2000)
20. G.U. Unnikrishnan, V.U. Unnikrishnan, J.N. Reddy, Constitutive material modeling of cell: A micromechanics approach. J. Biomech. Eng.-T. ASME **129**, 315–323 (2007)
21. Hibbit, Karlsson and Sorensen Inc, HKS, *ABAQUS Standard*, Version 6.3-2 (HKS, Providence, RI, 2002)

Free Vibration Characteristics of Thermally Loaded Cylindrical Shell

Byung-Hee Jeon, Hui-Won Kang, and Young-Shin Lee

1 Introduction

The effect of temperature on the mechanics of solid bodies has highly increased because of rapid developments in space technology, high speed atmospheric flights and nuclear energy applications. Thermally induced vibration of plates and shells are of great interest in aircraft and machine designs and also in chemical, nuclear and astronautical engineering. The reason for this is that during the heating-up period of structures exposed to high intensity heat fluxes, the material properties undergo significant changes; consequently, the thermal effect on the modulus of elasticity of the material can affect the vibratory characteristics.

Many researchers have studied thermally induced vibration during the past several decades. Boley [1] and Boley and Barber [2] studied the first analytical investigation of thermally induced response of beam including inertia forces. They pointed out that this is a coupling problem of transient heat conduction and dynamic process of structures. Recently Avsec and Oblak [3] shows how the temperature field in beam has impact on vibrations of beams. There have developed the mathematical model where fundamental thermomechanical properties of state are functions of temperature. Mead [4] presented the modes and frequencies of a free rectangular Kirchoff plate subjected to in-plane stresses generated by prescribed non-uniform surface temperature distributions which are doubly symmetrical about the plate central axes. Pradeep and Ganesan [5] deal with the thermal buckling and vibration behavior of multi-layer rectangular viscoelastic sandwich plates. A decoupled thermo-mechanical analysis is made by using finite element method. Ganesan and Pradeep [6] analyzed the buckling and vibration behavior using a semi analytical finite element method on a 316L stainless-steel cylinder filled with hot liquid.

The documented experimental demonstration of thermally induced vibration was performed by Beam [7] at NASA Ames. Michael and Vician [8] investigated experimental modal data which show the effect of heat on the modal characteristics of

Y.-S. Lee (✉)
BK21 Mechatronics Group, Chungnam National University, Daejeon 305-764, Korea
e-mail: leeys@cnu.ac.kr

A. Öchsner et al. (eds.), *Materials with Complex Behaviour*, Advanced Structured Materials 6, DOI 10.1007/978-3-642-12667-3_9,

2,024 aluminum, A110-AT titanium and fiberglass plate and analytical vibration results which show the effect of modeling the structural material property changes and the thermal stresses on the accuracy of the results for non-uniform and transient heating profiles. Blandino and Thornton [9] worked on experiments and analyses for the first detailed study of thermally induced vibrations of internally heated beams which belong to the class of vibrations called self-sustaining oscillations. Huber and Bowman [10] investigated on the effect of longitudinal vibration on the capillary limit of a copper/water heat pipe with a tightly wrapped screen wick. Lee et al. [11] are investigated by finite element analysis the thermal stress and vibration characteristics of ATJ graphite disk under high temperature condition. The experiment of thermal heat is conducted using a CO_2 laser. Murphy et al. [12] show combined theoretical and experimental results for rectangular plates which was fully clamped (out-of-plane) and uniformly heated. Amabili and Carra [13] studied on the nonlinear forced vibrations and post buckling of isotropic rectangular plates subjected to thermal variations. Laboratory experiments have been performed on two plates of different thickness. In the other papers, thermally induced vibration characteristics of beams and plates which was conducted by analysis and laboratory experiment was showed.

This paper describes the first study of thermally induced vibration characteristics of a cylindrical shell made by different materials. The dimensions of the cylindrical shell were 5 m length, 0.4 m. diameter, 0.002 m thickness and its material was combined by stainless-steel 304 and aluminum 6,061. This configuration was used for the thermally induced vibration experiment in order to evaluate the thermally effect on the vibration characteristics. Finite element simulations were conducted and compared with experiments.

2 Experimental Procedure

2.1 Test Specimen

The specimen used in the experiment was a cylindrical shell consisting of skins and plates used to connect each skin. Materials of skins and plate were stainless-steel 304 and aluminum 6,061 materials, respectively. The cylindrical shell has a 0.4 m diameter (D), a 0.002 m thickness and a total length of 5 m. Figure 1 is the test specimen shape. The cylindrical shell was comprised of five ions. First and fifth

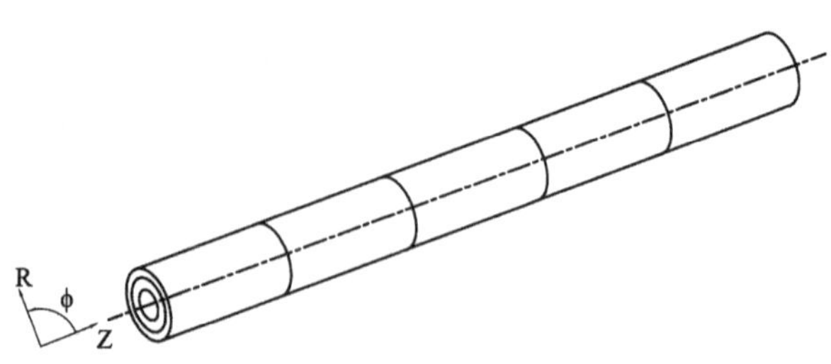

Fig. 1 Test specimen shape

section was mode of aluminum 6,061 and the others were mode of stainless-steel 304. Plates connected each shell by using 24 screws in circumference direction. The test specimen weight was approximately 240 kg.

2.2 *Test Configuration*

Test configuration for this experiment consisted of three major subsystem as shown Fig. 2. These included the heating system, the data acquisition system and the vibration control system. This work was performed to investgate the vibration characteristic of the specimen under thermal circumstance. The heating zone was restricted to the central section. The free–free boundary condition was selected to minimize the complexity of the test setup. This type of boundary condition alleviated the uncertainties associated with effects of heat conduction and thermally induced stress at the mounting frame interface. Figure 3 shows the heating zone and the instrumentation locations of the cylindrical shell.

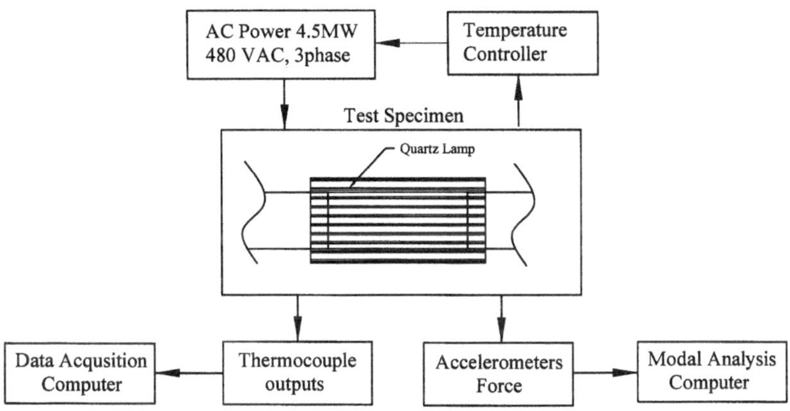

Fig. 2 Test configuration of experiment

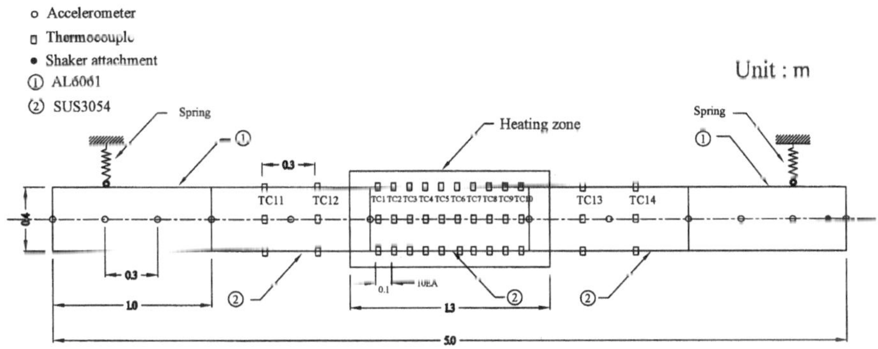

Fig. 3 Test specimen heating zone and instrumentation locations

2.2.1 Heating System

The heating system in this experiment was a radiation-heating type using quartz lamps. It consisted of quartz lamps, power controller, temperature controller and a cooling device, as shown in Table 1. This system provided closed-loop temperature control of the cylindrical shell.

Table 1 Specification of heating system

Unit	Specification
Quartz lamp	– Power: 600 W – Light length: 0.248 m – Voltage: 480 V – Total length: 0.303 m
Power controller	– Input: 480 V (45–65 Hz) – Control: input signal control – Cooling: Air cooling
Temperature controller	– Input: 18 ch. analog – Output: 12 ch. analog, 4 ch. digital
Cooling unit	– Water cooling system

2.2.2 Data Acquisition System

For temperature measurements, there were thermocouples, RTD (Resistance Temperature Detector) and non-contact type thermometer etc. We chose a K-type thermocouple that is used generally. To measure temperature gradients, a thermocouple was attached every 90 degrees in circumference direction, every 0.1 m longitudinal in direction as shown in Fig. 2, and another thermocouple attached in the non-heating sections to measure the temperature. The data sample rate was 1 Hz in the whole test and measured by the System 5,000 made by Measurement Group.

2.2.3 The Vibration Control System

Using a LDS 455 excitor, random vibration was conducted to the cylindrical shell and PCB 333A31 accelerometers were used for data acquisition. In the heating zone, PCB 357B31 high temperature accelerometers were used. The maximum operating temperatures of the accelerometers were 485°C. The vibration control system used for the experiment operation, measurement and modal analysis was the SDRC I-DEAS TDAS (Test Data Analysis System) software [14].

3 Modal Test

The cylindrical shell was uniformly heated at 100, 200 and 400°C in the middle of the cylindrical shell. The thermocouple readings were monitored to ensure that the cylindrical shell was uniformly heated. Figure 4 shows a photo of the experimental setup.

Fig. 4 Photo of experimental vibration test of thermally loaded cylindrical shell

Once vibration data acquisition was completed for a given heating temperature, frequency values for the first three modes were estimated. These modes were the first bending mode, second bending mode and third bending mode. The frequency response function at the reference point (exciting point) was selected because it contained the best response of the first third bending modes. Only the first three modes were selected because the number of accelerometers was insufficient to determine mode shapes above this number. The modal parameter estimates were obtained by fitting a second-order polynomial to each frequency peak in the selected frequency response function.

Mode shapes were generated using a single-degree-of-freedom technique. This technique extracted amplitude and phase information from each frequency response function at specified modal frequency. The information was then used for viewing animated mode shapes and static deformation plots.

4 Vibration Analysis

Vibration analysis was obtained by using ABAQUS [15]. The boundary condition was the same as that of the actual test, which was uniformly heated at 100, 200 and 400°C in the middle of cylindrical shell. Analysis had two sequences, first is the heat

Table 2 Thermal properties of the cylindrical shell [16]

Material	Young's modulus (GPa) Temp. (°C)	E	thermal conductivity (W/m °C)	Specific heat (J/kg °C)	Density (kg/m^3)	Poisson's ratio	Remark
AL6061–T6	25.0	68.9	2,163.4	1,256.0	2,712.5	0.33	Skin
	93.3	211.6					
	148.8	205.1					
	204.44	192.3					
	260.0	173.1					
	315.5	149.6					
STS304	25.0	199.9	195.2	502.4	8,027.1	0.27	Skin
	93.3	209.4					
	204.4	200.9					
	315.5	192.3					
	426.6	183.8					
	537.7	175.2					
	648.8	166.7					

transfer analysis and the second is the modal analysis. 8-node, 3-D solid elements were used for the analysis and approximately 100,000 element were used.

In addition to find out the difference between partially thermal loading condition and totally thermal loading condition, totally thermal loading conditions were applied to analysis at 100, 200, 300 and 400°C. Table 2 shows the mechanical properties of the cylindrical shell for the analysis with ABAQUS.

5 Results and Discussion

The verification and the validity of finite element analysis is conducted through the comparison with the experimental results. Figure 5 shows the mode shape of the first, second and third natural frequency comparison between experiment and finite element analysis at room temperature condition. Figure 6 shows thermocouple data of the cylindrical shell in the various thermal loading conditions. TC1 and TC10 shows half temperature of desired temperature. Because TC1 and TC10 were at the end of the thermal loading section, they were rapidly cooled rather than other thermocouples. Except TC1 and TC10, they well met the desired temperature.

Figure 7 shows a comparison of analytical and experimental frequencies for the cylindrical shell heated at uniform temperature to 100, 200, 300 and 400°C. The trend shown by the data was that the frequency decreased as the temperature increased. The tendency values between two method's results show good similarity within 4% error. Figure 8 shows the decrease from room temperature of the natural frequencies of experiment in the various thermal conditions. At 400°C, the first natural frequency had 5% decreased, the second natural frequency had 2% decreased and the third natural frequency had 3% decreases. The first natural frequency has the largest percent decrease.

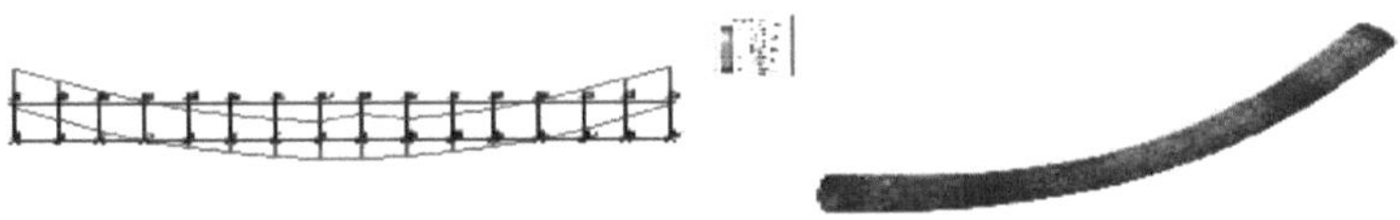

(a) First

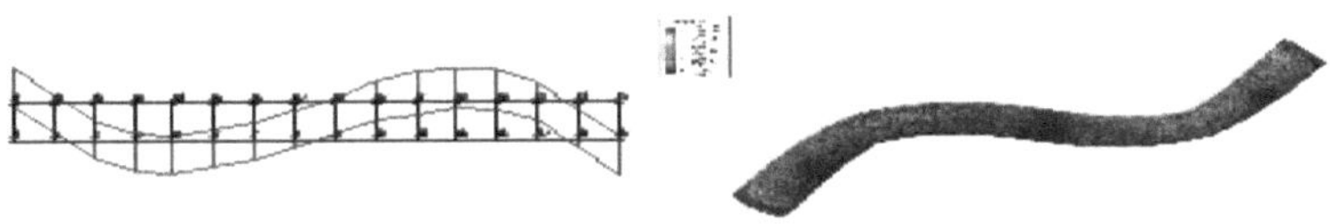

(b) Second

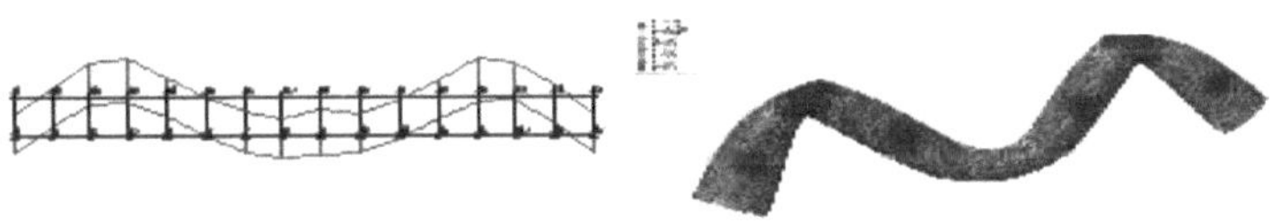

(c) Third

Fig. 5 Comparison of experimental and analytical mode shape for a room temperature condition

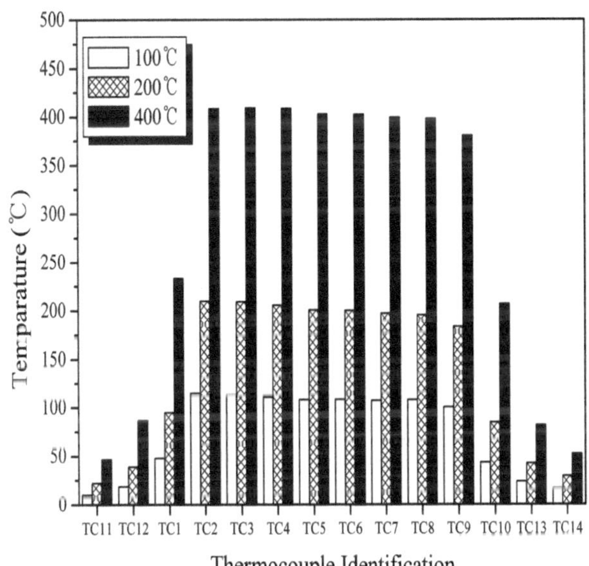

Fig. 6 Temperature distribution of the cylindrical shell

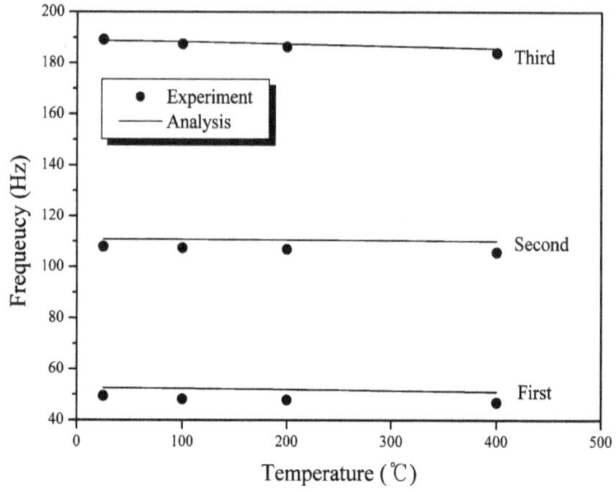

Fig. 7 Comparison of experimental and analytical frequencies for the cylindrical shell

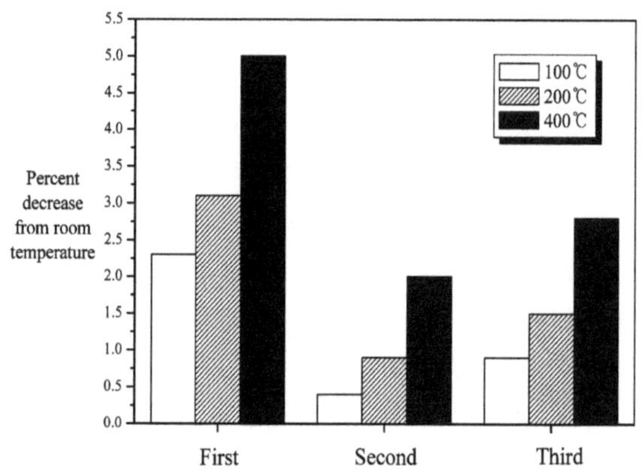

Fig. 8 Comparison of experimental frequencies changes

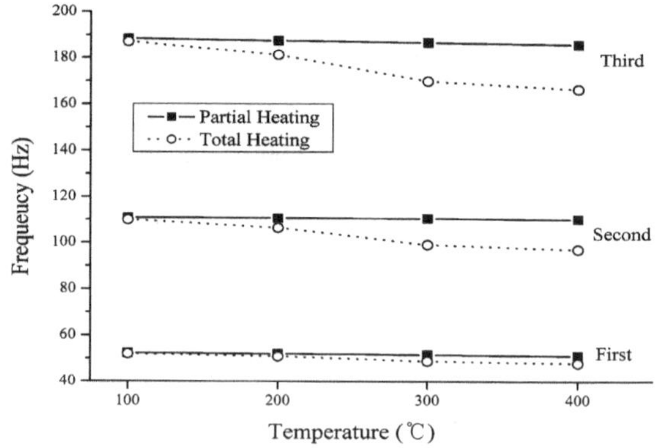

Fig. 9 Comparison of analytical frequencies on partial and total heating conditions

Figure 9 shows natural frequencies comparison of partially thermal loading condition and totally thermal loading condition. The first natural frequency has little decrease during temperature increase, but second and third natural frequencies have more than 10% decrease from room temperature condition over 300°C.

6 Conclusions

This paper presents the modal experiment of a 5 m long, 0.4 m diameter, 0.002 m thick cylindrical shell under thermal loading conditions at the center section at 100, 200 and 400°C, and finite element analysis was conducted using ABAQUS.

Comparisons between experiment and analysis showed excellent agreement on the variety of conditions.

In all conditions, the natural frequencies were decreased during temperature increase. In the experiment of the partially thermal loading condition, the first natural frequency has 5% decreased from room temperature condition. In the analysis of the totally thermal loading condition, the second and third natural frequencies have 10% decreased from room temperature condition.

References

1. B.A. Boley, BA (1956) Thermally induced vibrations of beam. Journal. of Aeronautical. Sciences. **23**: 179–181 (1956)
2. B.A. Boley, BA, A.D. Barber, AD (1957) Dynamics response of beams and plates to rapid heating. Trans. of ASME. Journal. of Appllied. Mechanics. **24**: 413–416 (1957)
3. J. Avsec J, M. Oblak M, Thermal vibrational analysis for simply supported beam and clamped beam. Journal. of Sound and Vibration. **308**: 514–525 (2007)
4. D.J. Mead, DJ (2003) Vibration and buckling of flat free–free plates under non-uniform in-plane thermal stresses. Journal. of Sound and Vibration. **260**: 141–165 (2003)
5. V. Pradeep V, N. Ganesan N (2008), Thermal buckling and vibration behavior of multi-layer rectangular viscoelastic sandwich plates. Journal. of Sound and Vibration. **310**: 169–183 (2008)
6. N. Ganesan N, V. Pradeep, V (2005) Buckling and vibration of circular cylindrical shells containing hot liquid. Journal. of Sound and Vibration. **287**: 845–863 (2005)
7. R.M. Beam, On the phenomenon of thermoelastic instability (thermal flutter) of booms with open cress section. NASA TN D-5222 (1969)
8. W.K. Michael WK, C.D. Vician, CD (1993) Correlation of analytical and experimental hot structure vibration results. NASA TM 104269 (1993)
9. J.R. Blandino, JR and E.A. Thornton EA (2001), Thermally induced vibration of an internally heated beam,. Journal. of Vibration and Acoustics. **123**: 67–75 (2001)
10. N.F. Huber NF, W.J. Bowman, WJ (1996) Longitudinal vibration effects on a copper/water heat pipe's capillary limit. Journal. of Thermophys.ics and Heat Transfer. **10**: 90–96 (1996)
11. Y.S. Lee YS, J.H. Kim JH, H.S. Kim HS, D.H. Kim DH, S.H. Gu SH, S.I. Moon SI (2006), A study on the thermal vibration analysis of the graphite disk under thermal shocks. International. Journal. of Modern . Physics. B **20**: 4105–4110 (2006)
12. K.D. Murphy KD, L.N. Virgin, S.A. LN Rizzi, SA (1997) The effect of thermal prestresses on the free vibration characteristics of clamped rectangular plates: Theory and experiment. Journal. of Vibration. and Acoustics. **119**: 243–249 (1997)

13. M. Amabili M, S. Carra, S (2009) Thermal defects on geometrically nonlinear vibrations of rectangular plates with fixed edges. Journal. of Sound and Vibration. **321**: 936–954. (2009)
14. SDRC, (2008) User guide for I-DEAS NX5 (2008)
15. Dassautt Systems Simulia Corp., *(2008) ABAQUS/Standard User's Manual*, (Version 6.8-2) (SIMULIA, Providence, RI, 2008)
16. Department of Defense, (2003) Handbook: Metallic materials and elements for aerospace vehicle structures. MIL-HDBK-5 J Superseding MIL-HDBK-5H (2003)

Model of Large Displacements in Static Analysis of Shell

Domagoj Matešan, Jure Radnić, and Alen Harapin

1 Introduction

Geometric non-linearity of a structure is a consequence of:

(i) the change in its geometry (displacements of a structure), and
(ii) non-linearity of the deformation-displacement relationship.

A model of geometric linearity does not take into account the impact of a change in the structure's geometry. Namely, in that model the equilibrium equations are satisfied on the initial non-deformed system. That approach is acceptable for structures with small displacements. When displacements are large, equilibrium equations shall be satisfied on a deformed system because application of a model of geometric linearity gives wrong results. An example of the impact of selected geometric model on cantilever internal forces is shown in Fig. 1. Application of a small displacements model in calculations of slender constructions can result in wrong conclusions in terms of their bearing capacity and safety. Sometimes those models are on the lesser safety side (if longitudinal compressive forces prevail), and sometimes they are on the greater safety side (if longitudinal tensile forces prevail).

A linear relationship between deformations and displacements is acceptable for small deformations problems. For large deformations problems, a non-linear deformation-displacement relationship shall be used. In the analysis of concrete and steel structures, where exploitation deformations and deformations at failure are small, a linear deformation-displacement relationship is satisfactory.

There are numerous papers dealing with the problem of numerical analyses of structures, which include the effects of large displacements. Some of them can be found in [1–10].

A model of geometric non-linearity in static analysis of concrete shells, which includes the effects of large displacements and small deformations, is presented.

D. Matešan (✉)
Faculty of Civil Engineering and Architecture, 21000 Split, Croatia
e-mail: domagoj.matesan@gradst.hr

A. Öchsner et al. (eds.), *Materials with Complex Behaviour*, Advanced Structured Materials 6, DOI 10.1007/978-3-642-12667-3_10, © Springer-Verlag Berlin Heidelberg 2010

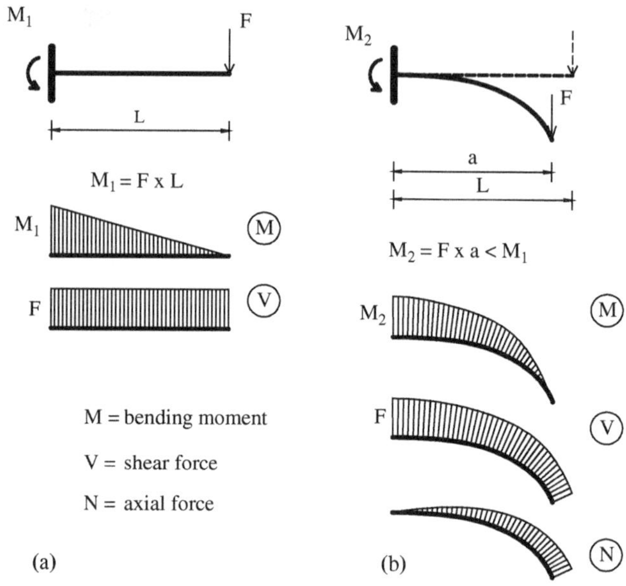

Fig. 1 An example of the geometric model impact on cantilever internal forces. (**a**) Small displacements model. (**b**) Large displacements model

A simultaneous use of the material and geometric non-linearity models is possible. Namely, a model of geometric non-linearity in the framework of previously developed model [11–13] for material non-linearity analysis of concrete shells is presented. The model was verified on the results of three experimentally tested very slender steel cantilevers, with elastic behaviour of material for all applied loads.

2 Large Displacements Model

The problem was solved using the updated Lagrange procedure, according to Nagtegaal [14] and implemented to the shell analysis, based on the following postulates:

(i) External load is applied in increments. An iterative solution procedure is carried out for each load increment until a vector of residual forces becomes arbitrary small. When a convergence criterion is satisfied, a vector of residual forces is added to the next external load increment and iteration procedure continues.
(ii) A conventional Lagrange procedure is used in each step: states of variables are defined in reference to their states at the beginning of the respective iteration step.
(iii) At the end of the each iteration step, states of variables are updated (redefined) in comparison to their states at the end of the previous iteration step.

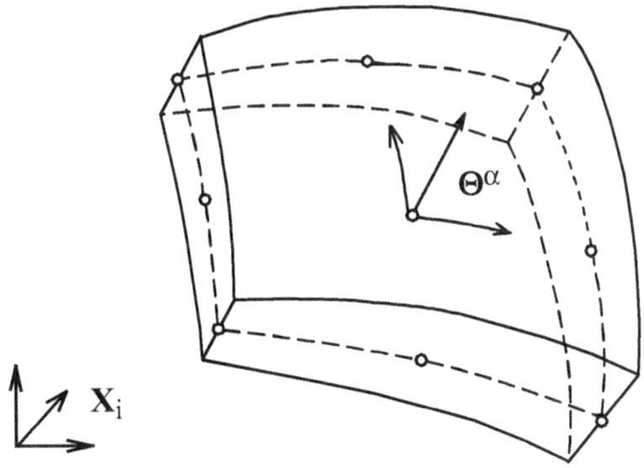

Fig. 2 Spatial ($\mathbf{X}_i$) and material ($\mathbf{\Theta}^\alpha$) coordinate systems

State at the beginning of the observed iteration step, is analyzed first. According to Lagrange procedure, spatial position of a body in the Cartesian coordinate system $\mathbf{X}_i$ is defined as:

$$\mathbf{X}_i = \mathbf{X}_i\left(\mathbf{\Theta}^\alpha\right) \tag{1}$$

where a material point of a body is uniquely defined in a curvilinear coordinate system $\mathbf{\Theta}^\alpha$ (Fig. 2).

Equilibrium of a body is expressed by the virtual work equation, with its components given in the Cartesian coordinate system:

$$\int_V \sigma_{ij}\delta\boldsymbol{\varepsilon}_{ij}dV = \int_S \mathbf{q}_i\delta\mathbf{u}_i dS \tag{2a}$$

or alternatively with its components in a respective curvilinear coordinate system:

$$\int_V \sigma^{\alpha\beta}\delta\boldsymbol{\varepsilon}_{\alpha\beta}dV = \int_S \mathbf{q}_i\delta\mathbf{u}_i dS \tag{2b}$$

In the previously given equations, σ_{ij} and $\sigma^{\alpha\beta}$ are the Cauchy stresses, $\delta\boldsymbol{\varepsilon}_{ij}$ and $\delta\boldsymbol{\varepsilon}_{\alpha\beta}$ are the small increases of deformations, V is a volume of a body and S is a body surface area. For purposes of short logs, only a surface load $\mathbf{q}_i$ is taken into account.

Since in expression (2b) integration is carried out in terms of the current state of a body, contra-variant components of the Cauchy stress $\sigma^{\alpha\beta}$ are equal to the contra-variant components of the second Piola-Kirchhoff stress $\mathbf{S}^{\alpha\beta}$. Co-variant components of deformation increase $\delta\boldsymbol{\varepsilon}_{\alpha\beta}$ are related to the Cartesian components of the displacement increase $\delta\mathbf{u}$ as:

$$\delta\boldsymbol{\varepsilon}_{\alpha\beta} = \frac{1}{2}\left(\mathbf{X}_{k,\alpha}\delta\mathbf{u}_{k,\beta} - \mathbf{X}_{k,\beta}\delta\mathbf{u}_{k,\alpha}\right) \tag{3}$$

Using the symmetry $\boldsymbol{\sigma}^{\alpha\beta}$, the virtual works equation can be written in the following form:

$$\int_V \mathbf{S}^{\alpha\beta}\mathbf{X}_{k,\alpha}\delta\mathbf{u}_{k,\beta}dV = \int_S \mathbf{q}_i\delta\mathbf{u}_i dS \tag{4}$$

At the end of the observed iteration step, spatial position of a body $\mathbf{x}_i$ is described as:

$$\mathbf{x}_i = \mathbf{x}_i\left(\boldsymbol{\Theta}^\alpha\right) = \mathbf{x}_i\left(\boldsymbol{\Theta}^\alpha\right) + \Delta\mathbf{u}_i(\boldsymbol{\Theta}^\alpha) \tag{5}$$

where, as already mentioned, $\mathbf{x}_i$ is a spatial position of a body point at the beginning of the observed iteration step, while $\Delta\mathbf{u}_i$ is a displacement increment.

At the end of the observed iteration step, the virtual work equation has the following form:

$$\int_V (\mathbf{S}^{\alpha\beta} + \Delta\mathbf{S}^{\alpha\beta})\delta\mathbf{E}_{\alpha\beta}dV = \int_S \left(\mathbf{q}_i + \Delta\mathbf{q}_i\right)\delta\mathbf{u}_i dS \tag{6}$$

where $\delta\mathbf{E}_{\alpha\beta}$ is an increase in Green (Lagrange) deformation, which can be calculated from a displacement increase using the following formula:

$$\delta\mathbf{E}_{\alpha\beta} = \frac{1}{2}\left[\left(\mathbf{X}_{k,\alpha} + \Delta\mathbf{u}_{k,\alpha}\right)\delta\mathbf{u}_{k,\beta} + \left(\mathbf{X}_{k,\beta} + \Delta\mathbf{u}_{k,\beta}\right)\delta\mathbf{u}_{k,\alpha}\right] \tag{7}$$

Note that integration in expression (6) is carried out in reference to the state of a body at the beginning of an iteration step and $\Delta\mathbf{S}^{\alpha\beta}$ is the increment of the second Piola-Kirchhoff stress.

Using the symmetry $\mathbf{S}^{\alpha\beta}$ and $\Delta\mathbf{S}^{\alpha\beta}$, equilibrium equation in the function of a displacement has the following form:

$$\int_V \left(\mathbf{S}^{\alpha\beta} + \Delta\mathbf{S}^{\alpha\beta}\right)\left(\mathbf{X}_{k,\alpha} + \Delta\mathbf{u}_{k,\alpha}\right)\delta\mathbf{u}_{k,\beta}dV = \int_S \left(\mathbf{q}_i + \Delta\mathbf{q}_i\right)\delta\mathbf{u}_i dS \tag{8}$$

Integration in (4) and (8) is carried out in reference to the same state of the body.

By reduction of the Eq. (8), the equation of incremental virtual work is obtained, which is also marked as an equation of a continuous equilibrium:

$$\int_V \left(\Delta\mathbf{S}^{\alpha\beta}\mathbf{X}_{k,\alpha} + \mathbf{S}^{\alpha\beta}\Delta\mathbf{u}_{k,\alpha} + \Delta\mathbf{S}^{\alpha\beta}\Delta\mathbf{u}_{k,\alpha}\right)\delta\mathbf{u}_{k,\beta}dV = \int_S \Delta\mathbf{q}_i\delta\mathbf{u}_i dS \tag{9}$$

The current constitutive material model can be written in the following form:

$$\Delta\mathbf{S}^{\alpha\beta} = \mathbf{C}^{\alpha\beta\gamma\delta}\Delta\mathbf{E}_{\gamma\delta} \tag{10}$$

where $\mathbf{C}^{\alpha\beta\gamma\delta}$ is the current functional relationship between the stress increase $\Delta\mathbf{S}^{\alpha\beta}$ and deformation increase $\Delta\mathbf{E}_{\gamma\delta}$. The increment of Green deformations can be calculated from the displacements decrease as:

$$\Delta\mathbf{E}_{\gamma\delta} = \frac{1}{2}\left(\mathbf{X}_{k,\gamma}\Delta\mathbf{u}_{k,\delta} + \mathbf{X}_{k,\delta}\Delta\mathbf{u}_{k,\gamma} + \Delta\mathbf{u}_{k,\gamma}\Delta\mathbf{u}_{k,\delta}\right) \tag{11}$$

Since $\mathbf{C}^{\alpha\beta\gamma\delta}$ is symmetrical in reference to the last two indexes, combination of (10) and (11) gives the expression for calculation of stress increment in the function of displacement increment:

$$\Delta\mathbf{S}^{\alpha\beta} = \mathbf{C}^{\alpha\beta\gamma\delta}\left(\mathbf{X}_{k,\gamma}\Delta\mathbf{u}_{k,\delta} + \frac{1}{2}\Delta\mathbf{u}_{k,\gamma}\Delta\mathbf{u}_{k,\delta}\right) \tag{12}$$

Expressions (9) and (12) are the basic equations of the observed iteration step. If the coordinate system is the Cartesian one, (9) and (12) are simplified:

$$\int_V \left(\Delta\mathbf{S}_{ij}\boldsymbol{\delta}_{ki} + \mathbf{S}_{ij}\Delta\mathbf{u}_{k,i} + \Delta\mathbf{S}_{ij}\Delta\mathbf{u}_{k,i}\right)\delta\mathbf{u}_{k,j}dV = \int_S \Delta\mathbf{q}_i\delta\mathbf{u}_i dS \tag{13}$$

$$\Delta\mathbf{S}_{ij} = \mathbf{C}_{ijkl}\left(\Delta\mathbf{u}_{k,l} + \frac{1}{2}\Delta\mathbf{u}_{m,k}\Delta\mathbf{u}_{m,l}\right) \tag{14}$$

where $\boldsymbol{\delta}_{ki}$ is the Kronecher symbol. States of the variables shall be updated at the end of the current iteration step.

For the curvilinear system, contra-variant components of the second stress Piola-Kirchhoff shall be corrected by a change in the volume in order to become contra-variant components of the Cauchy stress; namely:

$$\boldsymbol{\sigma}^{\alpha\beta} = \left(\mathbf{S}^{\alpha\beta} + \Delta\mathbf{S}^{\alpha\beta}\right)/\mathbf{J} \tag{15}$$

where $\mathbf{J}$ is the Jacobian of the deformation increment. If a material is almost incompressible, transformation becomes unnecessary.

In the initial Cartesian approach, the second Pioala-Kirchhoff stress shall be transformed into a real (Cauchy) stress in the Cartesian coordinate system:

$$\boldsymbol{\sigma}_{ij} = \left(\boldsymbol{\delta}_{ik} + \Delta\mathbf{u}_{i,k}\right)\left(\mathbf{S}_{kl} + \Delta\mathbf{S}_{kl}\right)\left(\boldsymbol{\delta}_{jl} + \Delta\mathbf{u}_{j,l}\right)/\mathbf{J} \tag{16}$$

Again, the Jacobian shall be equal to zero if a material is almost uncompressible.

The model for static analysis of a shell [12, 13] includes the geometric non-linearity model with the large displacements and small deformations. As already mentioned, impact of large displacements was included by transformation of components of displacements and forces (stresses) between the global and local coordinate systems. Namely, assuming the small displacement increase in each iteration step, the existence of a linear relationship between deformations and displacements can be adopted. This significantly simplifies and shortens the calculations.

Application of a local coordinate system to define the constitutive material law also simplifies the analysis in case of anisotropic material properties.

Namely, the following system shall be solved in each iteration step:

$$\mathbf{K}^{n-1}\Delta\mathbf{u}^n = \mathbf{R}^n - \mathbf{F}^{n-1} \tag{17}$$

where n is the observed iteration step; $\mathbf{K}$ is the current stiffness matrix of a construction that can include the material non-linearity (with different material models) and geometric non-linearity (large displacements, small deformations); $\Delta\mathbf{u}$ is the current displacement increment; $\mathbf{R}$ is the current vector of external nodal forces and $\mathbf{F}$ is the current vector of internal nodal forces due to material stresses. An iterative solution procedure is repeated until a vector of residual forces $\left(\mathbf{R}^n - \mathbf{F}^{n-1}\right)$ becomes arbitrary small. Then, a vector of residual forces is added to the new external load increment and iteration procedure is repeated.

3 Spatial Model and Finite Elements

A problem of geometric non-linearity of shells, simulated by 8 and 9 node curved degenerated finite elements (Fig. 3), was analyzed. The elements were free of shear and membrane locking. Each node has five unknowns: three displacements and two rotations perpendicular to the central shell plane. More detail description of adopted shell elements can be found in [11, 12].

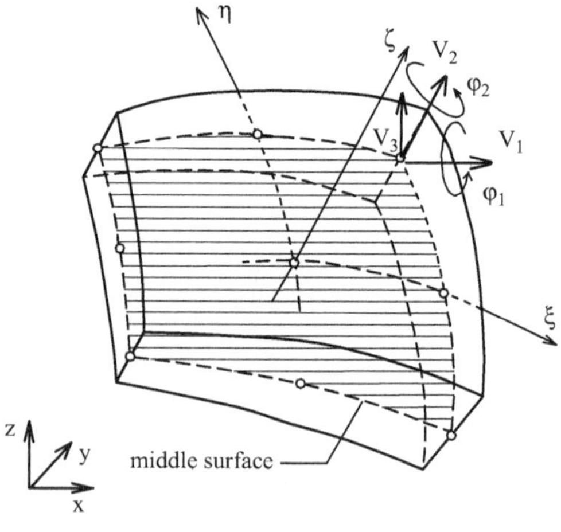

Fig. 3 Adopted degenerated shell element [11]

4 Material Model

Apart from a geometric non-linearity (large displacements), material non-linearity can also be included in the shell analysis. A conventional elastic, elasto-plastic or elasto-brittle material models can be used as well as a special model for concrete shells. A layered model along the shell thickness is used. Detailed description of some adopted material models can be found in [12, 13].

5 Some Calculation Aspects

The Newton-Raphson method is used for non-linear problem solving. A computer program provides the following possibilities for the stiffness matrix updating:

- at the beginning of the calculation only,
- at the beginning of each load increment,
- at the beginning of each iteration step of each load increment, and
- in defined load increments and defined iteration steps.

This method was proven efficient in solving of problems analyzed in Sect. 6. However, it is also possible that this method will not provide a solution for a problem with great change in geometry.

As already mentioned, the external load is given in increments. Regardless whether those are problems with pure geometric non-linearity or simultaneous geometric and material non-linearity, load increment size impacts the obtained results. The impact is more significant when the non-linearity level is greater. Figure 4 shows a scheme of a typical iteration procedure of problem solving for a particular load increment, in a 3D space. One shall note that in a real structure, solution occurs

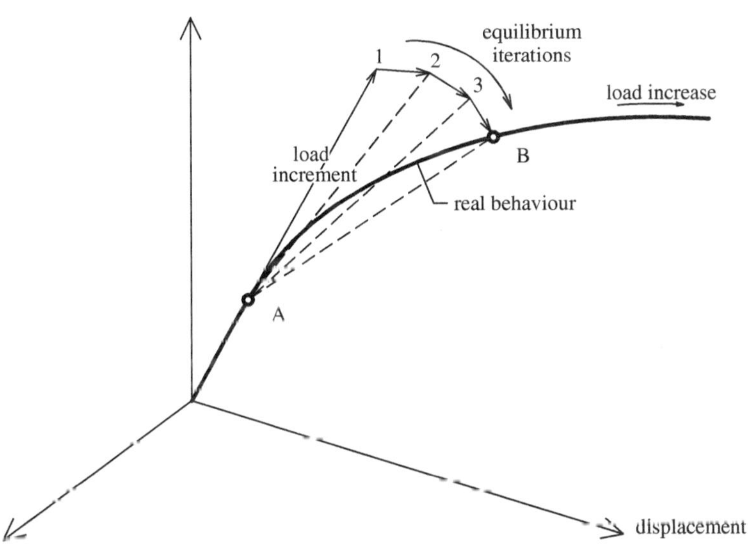

Fig. 4 Qualitative presentation of the incremental-iterative procedure for non-linear problem solving

in multi-dimensions space that corresponds to a number of degrees of freedom of a discretized system.

Note the difference between the numerical solution path A – 1 – 2 – 3 ... B and the real solution path A – B. Keep in mind that an accurate solution path is unknown; namely, solution in the last iteration step does not correspond exactly to the point B (there are always some errors depending on the selected convergence criterion). "Bypassing" of the actual path could cause inexistent material non-linearity or inexistent geometric non-linearity. Thus, sufficiently small load increments shall be used, depending on the analyzed problem. In general, smaller load increments should provide more accurate results. However, too small load increments sometimes can, not only unnecessarily extend the analyses, but also decrease accuracy of a solution due to a large scope of numerical calculations. The impact of the load increment length on the obtained results in each analyzed example shall be studied.

As already mentioned, the adopted convergence criterion directly impacts the obtained results. Thus, one shall be very cautious when selecting the convergence criterion. For purposes of procedure convergence control, a displacements increase norm in the observed iteration step was monitored in respect to the total current displacements. Too large allowable tolerance, that came handy in order to shorten duration of the analysis, can often produce a numerical state completely different than the actual behaviour of the structure. In general, smaller allowable tolerance, with the consequence of longer analysis, should give more accurate results. However, too small allowable tolerance can cause significant numerical errors and large scale deviations of numerical results from the actual results. Thus, several calculations shall be carried out using the different allowable tolerance values.

6 Verification of a Presented Numerical Model

The presented large displacements numerical model in static analysis of shells was verified on the results of three experimentally tested very slender steel cantilevers, with elastic behaviour of material for all applied loads. Cantilevers were made of high-quality steel with initial yield stress $f_s = 620$ MPa, with the elastic behaviour till just before the failure and modulus of elasticity $E_s = 205$ GPa.

In Example 1 (Sect. 6.1) the cantilever was placed horizontally and loaded by bending. In Example 2 (Sect. 6.2) the cantilever was placed vertically and loaded by longitudinal compressive force. In Example 3 (Sect. 6.3) the cantilever was placed horizontally and loaded by bending and torsion.

Cantilever displacements were measured for each increment of the applied external load F at the cantilever end. The force was a gravity one and retained its initial direction during the cantilever deformation. The force magnitude was limited in order to ensure complete elastic behaviour of the construction. Following the removal of the force F, there were no irreversible deformations left for all cantilevers.

6.1 Example 1

The basic data on experimentally tested cantilever are given in Fig. 5a. The adopted discretization for a numerical model is shown in Fig. 5b, while comparison between the experimentally and numerical displacements of a cantilever are given in Fig. 6.

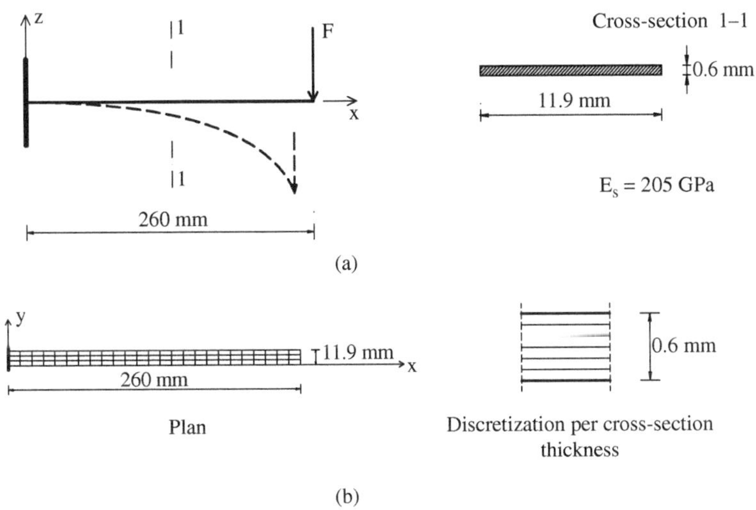

Fig. 5 Basic data on experimentally tested cantilever and adopted discretization model in Example 1. (**a**) Basic data on experimentally tested cantilever. (**b**) Discretization model

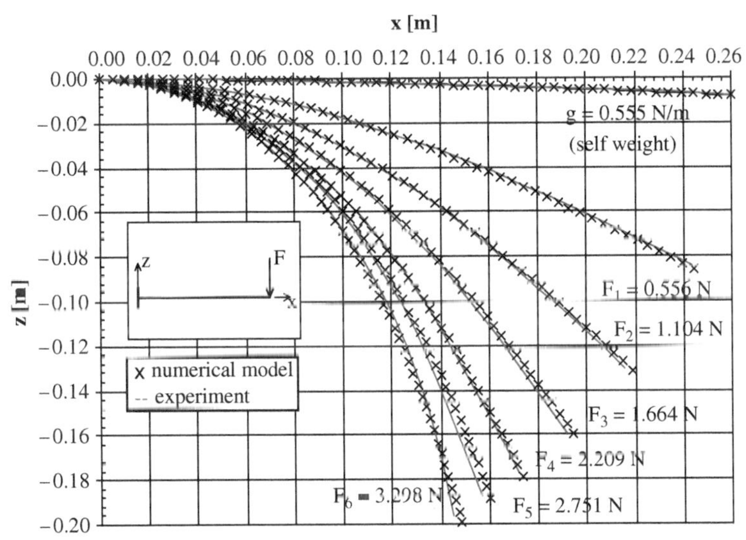

Fig. 6 Deflection of a horizontal cantilever described in Example 1

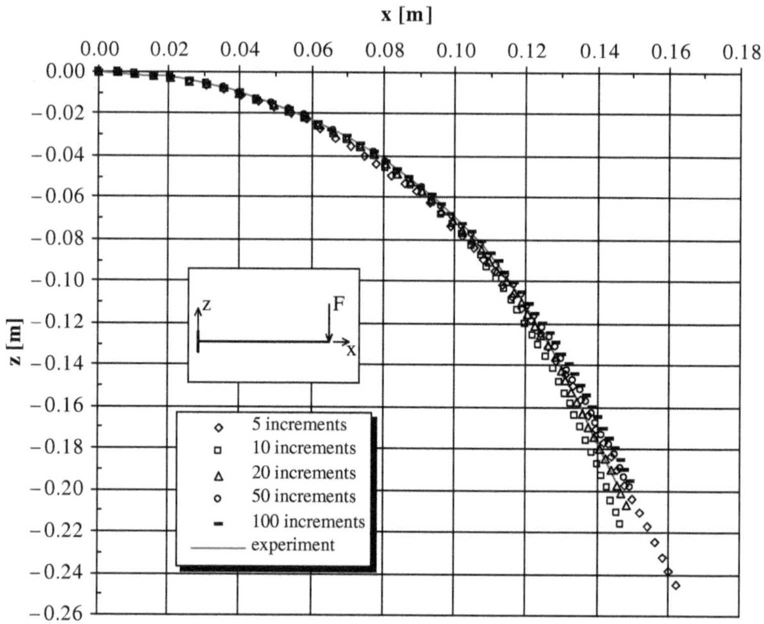

Fig. 7 Impact of the force increment size on calculated deflections of a cantilever described in Example 1 for F = 3.298 N

Note that cantilever deflections are large even due just to the dead weight. Namely, deflection of the cantilever end is several times larger than a height of the cantilever cross-section. Also, note that the cantilever stiffens gradually increase with an increase in force F. Namely, for the same force increase, cantilever displacements are decreasing. It is a consequence of larger membrane (tensile) bearing of a deformed construction. The load was applied in 0.1 F increments. The method of initial stiffness was used for the solution of a non-linear problem, with the allowed tolerance of 0.001. There is a very good correspondence between the experimentally determined and calculated displacements of a cantilever. Impact of the size of force increments on numerical results was analyzed for F = 3.298 N (Fig. 7). Its visual appearance under a load F = 1.104 N is shown in Fig. 8.

6.2 Example 2

The cantilever from Example 1 was placed vertically, with a 8 mm horizontal displacement of its top in reference to the its clamped bottom (Fig. 9). It was loaded by different values of force F at the top, as in Example 1 (Fig. 9). The cantilever displacements (Fig. 10) were relatively small for small values of force F. Exceeding of a "critical" load, leads to a sudden increase in displacements and change in the cantilever geometry. By the occurrence of membrane (tensile) bearing, the structure stiffens gradually. Note that displacements of a vertical cantilever beam for greater

Fig. 8 Cantilever subjected to a load F = 1.104 N

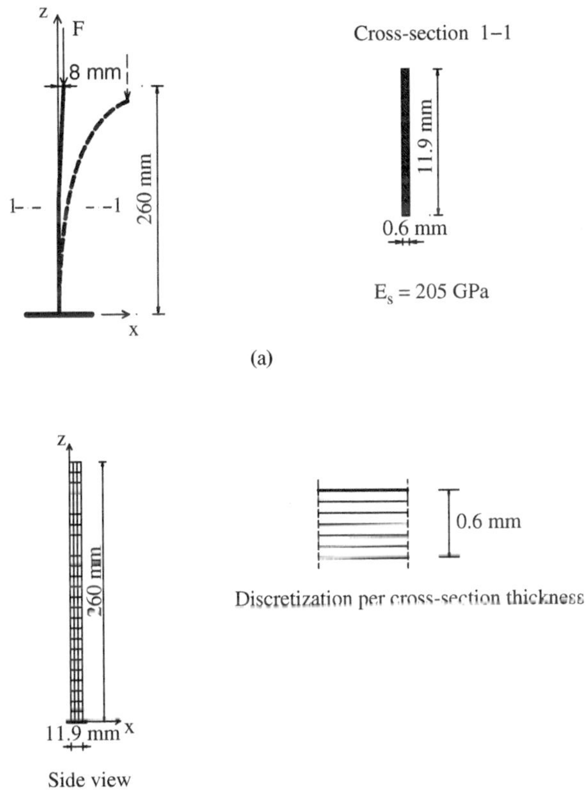

Fig. 9 Basic data on experimentally tested cantilever and adopted discretization model in Example 2. (**a**) Basic data on experimentally tested cantilever. (**b**) Model discretization

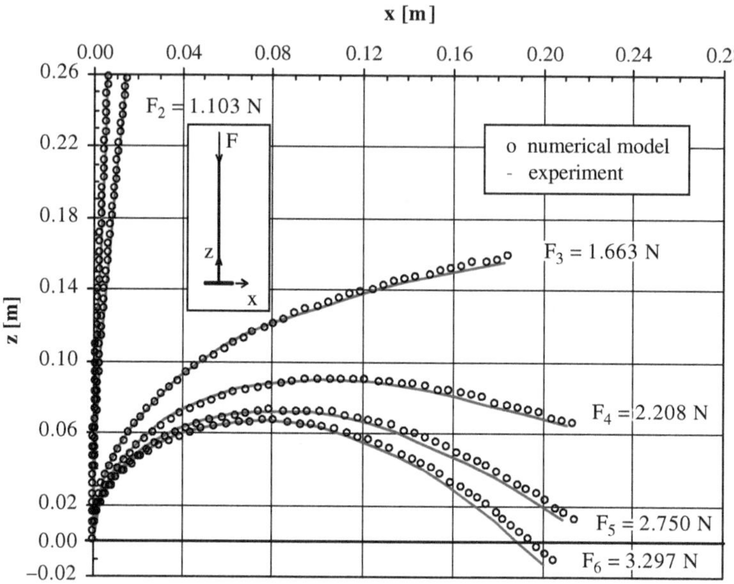

Fig. 10 Deflection of a vertical cantilever described in Example 2

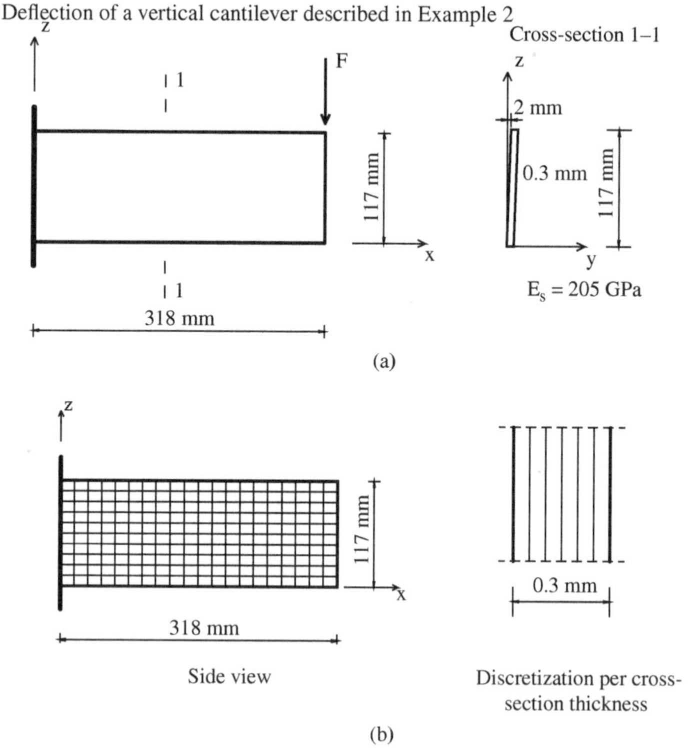

Fig. 11 Basic data on experimentally tested cantilever and adopted discretization model in Example 3. (**a**) Basic data on experimentally tested cantilever. (**b**) Model discretization

intensity of force F are greater than displacements of a horizontal cantilever in Example 2 for the same values of force F. The same discretization model was used (Fig. 9b) as in Example 1. There was a good correspondence between the experimental and calculated displacements.

6.3 Example 3

A horizontal cantilever, similar to that in Example 1, was analyzed. The difference from Example 1 was that this cantilever was very rigid to bending in its (vertical) plane and very slander to bending in perpendicular (horizontal) plane (Fig. 11a). The

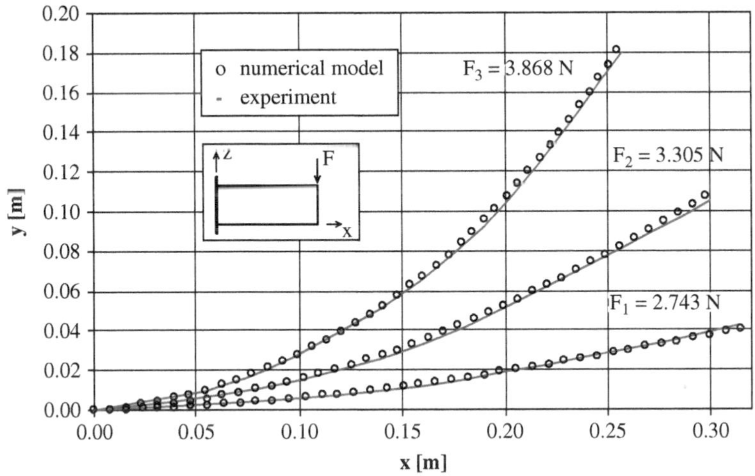

(a)

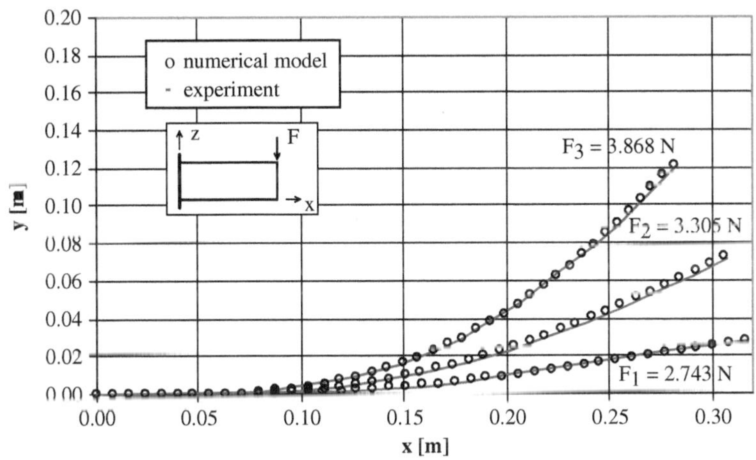

(b)

Fig. 12 Displacements of a high cantilever chords in Example 3. (**a**) Displacements of a cantilever top edge in a horizontal plane. (**b**) Displacements of a cantilever bottom edge in a horizontal plane

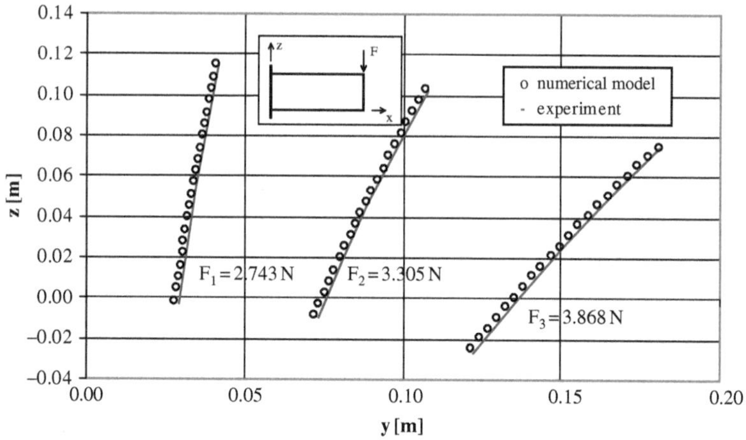

Fig. 13 Displacements of a high cantilever front in Example 3 in vertical plane

front cantilever plane deviates by about 1° from the vertical i.e. cantilever top chord is displaced from its bottom chord by 2 mm. In this example there is an impact of torsion as well as buckling of the bottom compression chord and buckling of a web.

The discretization model of cantilever is shown in Fig. 11b. An analogous numerical model as in the previous examples was used. Numerical results are shown in Figs. 12 and 13, where also good correspondence between experimental and calculated cantilever displacements can be observed.

7 Conclusion

The impact of geometric changes must be included in calculations of very slender structures because, in those cases, geometric linearity model provides either wrong or useless results. It is believed that the presented numerical model can be useful in the analyses of all slender structures that can be described well enough by shell elements and adopted material models. It is also believed that the presented numerical model is one of the most actual numerical models for static, dynamic and time-dependent analysis of reinforced and prestressed concrete shells with both material and geometric non linearity.

References

1. J.H. Argyris, P. Dunne, G. Malejannakis, D. Scharpf, On large displacement – Small strain analysis of structures with rotational degrees of freedom part II. Comput. Methods Appl. Mech. Eng. **15**, 99–135 (1978)
2. K.J. Bathe, S. Bolourchi, Large displacement analysis of three-dimensional beam structures. Int. J. Numer. Methods Eng. **14**, 961–86 (1979)

3. M. Bischoff, E. Ramm, Shear deformable shell elements for large strains and rotations. Int. J. Numer. Methods Eng. **40**, 4427–4449 (1997)
4. M.A. Crisfield, D. Tan, Large-strain elasto-plastic shell analysis using low-order elements. Int. J. Comput. Aided Eng. **18**(32), 255–286 (2001)
5. T. Hughes, E. Carnoy, Nonlinear finite element shell formulation accounting for large membrane strains. Comput. Methods Appl. Mech. Eng. **39**, 69–82 (1983)
6. K.D. Kim, Large displacement of elasto-plastic analysis of stiffened plates and shells using corotational 8-node assumed strain element. Struct. Eng. Mech. Int. J. **15**, 199–223 (2003)
7. K.D. Kim, G.R. Lomboy, G.Z. Voyiadjis, A 4-node assumed strain quasi-conforming shell element with 6 DOF. Int. J. Numer. Methods Eng. **58**, 2177–2200 (2003)
8. J.C. Simo, F. Armero, Geometrically non-linear enhanced strain mixed methods and the method of incompatible modes. Int. J. Numer. Methods Eng. **33**, 1413–1449 (1992)
9. J.C. Simo, J.G. Kennedy, On a stress resultant geometrically exact shell model. Part V, Nonlinear plasticity: Formulation and integration algorithms. Comput. Methods Appl. Mech. Eng. **96**, 133–71 (1992)
10. J.C. Simo, M.S. Rifai, D.D. Fox, On a stress resultant geometrically exact shell model. Part IV: variable thickness shells with through-the-thickness stretching. Comput. Methods Appl. Mech. Eng. **81**, 91–126 (1990)
11. H.C. Huang, *Static and Dynamic Analysis of Plates and Shells* (Springer-Verlag 1989)
12. D. Matešan, Time-dependent analysis of prestressed concrete shells. Ph.D. Thesis, University of Split, 2007
13. J. Radnić, D. Matešan, A. Harapin, Geometrical nonlinearity model in statical analysis of shells. Gra đ evinar **55**, 583–589 (2003)
14. J.C. Nagtegaal, J.E. De Jong, Some computational aspects of elasto-plastic large strain analysis. Int. J. Numer. Methods Eng. **17**, 15–41 (1981)

Nonlinear Time-Dependent Analysis of Prestressed Concrete Shells

Domagoj Matešan and Jure Radnić

1 Introduction

The paper briefly presents a model for material and geometric nonlinear analysis of prestressed concrete plates and shells under short term and long-term static load. The dominant nonlinear effects of concrete (cracking, yielding, crushing, creep, shrinkage, aging), as well as the dominant nonlinear effects related to the reinforcing steel (yielding, failure) and prestressing steel (yielding, failure, losses of the prestressing), are simulated. Detailed description of the model can be found in [1]. The model was verified on the results of performed experimental test of prestressed concrete shell under short-term and long-term static load [2]. The model is not intended for long-term load with unload.

2 Description of the Tendon Geometry

The tendon geometry is given by the following equation, namely, coordinates of tendon axis points (Fig. 1)

$$\mathbf{x}_p(s) = \sum_{k=1}^{m} \mathbf{L}_k(s)\mathbf{x}_{p,k} \tag{1}$$

where $\mathbf{x}_p(s)$ is a vector describing the tendon geometry, $\mathbf{L}_k$ are the Lagrangian inter polation functions, and $\mathbf{x}_{p,k}$ is a vector containing global coordinates from 1 to m points that define the tendon. The prestressed tendon embedded in a shell element is shown in Fig. 1.

D. Matešan (✉)
Faculty of Civil Engineering and Architecture, 21000 Split, Croatia
e-mail: domagoj.matesan@gradst.hr

A. Öchsner et al. (eds.), *Materials with Complex Behaviour*, Advanced Structured Materials 6, DOI 10.1007/978-3-642-12667-3_11, © Springer-Verlag Berlin Heidelberg 2010

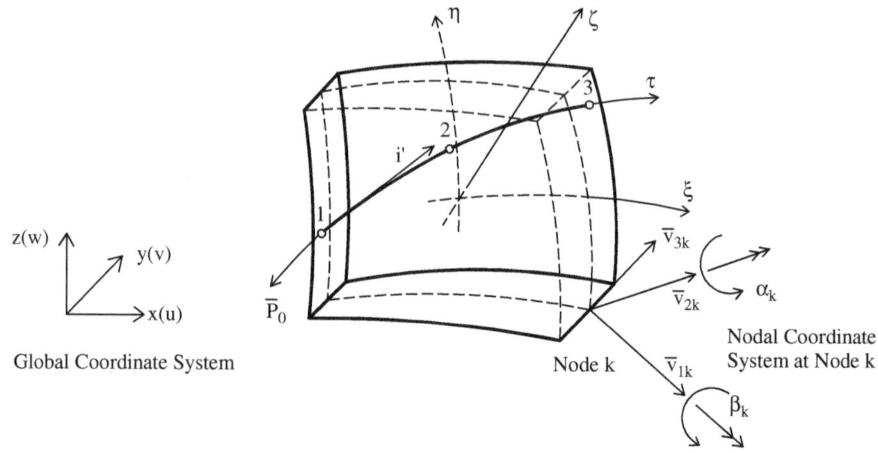

Fig. 1 Prestressed tendon embedded in a shell element

The coordinates of cable and shell element intersecting points in a natural coordinate system $(\xi, \eta, \zeta)_{p,i}$ can be calculated using the following system of nonlinear equations

$$\mathbf{x}(\xi, \eta, \zeta) = \mathbf{x}_p(s) \tag{2}$$

where $\mathbf{x}(\xi, \eta, \zeta)$ defines the geometry of intersected finite elements. For curved isoparametric shell elements, the shell geometry is explicitly described by the following equation

$$\mathbf{x}(\xi, \eta, \zeta) = \sum_{i=1}^{n} N_i(\xi, \eta)\, \mathbf{x}_i^{srednje} + \sum_{i=1}^{n} N_i(\xi, \eta)\, \frac{h_i}{2} \zeta \overline{\mathbf{v}}_{3,i} \tag{3}$$

where n is a number of nodes in an element, $N_i(\xi, \eta)$ are the shape functions for ξ = const., $\mathbf{x}_i$ are the coordinates of the central plane, h_i is a shell thickness for node i, and $\overline{\mathbf{v}}_{3,i}$ is the unit vector defining a direction of normal in node i.

The natural coordinates of the middle node of 1-D tendon element ξ, η, ζ (node 2 in Fig. 1) can be calculated using the following system of equations

$$\mathbf{x}(\xi, \eta, \zeta) = \mathbf{x}_p(s_2) \tag{4}$$

where $s_2 = (s_1 - s_3)/2$ is a length of the tendon between corresponding points.

After the natural coordinates $(\xi, \eta, \zeta)_{p,i}$ are calculated for all nodes of tendon element, then the global coordinates $\mathbf{x}_{p,j}$ can be calculated by expression (3).

The geometry of a tendon segment embedded in a shell element (Fig. 1) can be described by the equation

$$\mathbf{x}_p(\tau) = \sum_{j=1}^{3} N_j'(\tau)\,\mathbf{x}_{p,j} \tag{5}$$

where $N_j'(\tau)$ are the shape functions adopted for a one-dimensional parabolic element, and τ is a respective curvilinear coordinate.

3 Transfer of the Prestressing from the Tendon to the Concrete

A layered model along the shell thickness, with separate layers for concrete and reinforcing steel, is applied. The adopted model of prestressing is similar to the one by Figueras and Povoas [3].

In pre-tensioned structures, the bond between the prestressing steel and surrounding concrete exists due adequate adhesion. In post-tensioned structures, a bond between the tendon and concrete exists only after the grouting of tendon (the so-called bonded tendons). If tendons are not grouted, there is no bond between the tendon and surrounding concrete (the so-called unbonded tendons).

Modelling of mechanical effects due the prestressing and tendon-concrete bond requires the following steps: (i) inclusion of the contribution of prestressing to the global structure stiffness, (ii) calculation of strain increases as a result of prestressing effects acting as external load on the structure, and (iii) calculation of the actual force in the tendon and calculation of internal forces due to prestressing.

For pre-tensioned structures, there is a continue concrete-tendon bond; thus, steps (i)–(iii) are carried out in each load increment. Strain increase is calculated for each nodal force due to prestressing. Therefore, losses due to elastic shortening of the concrete are being included automatically.

Modelling of post-tensioned structures is carried out in two phases:

(a) A bond between prestressing steel and concrete is not taken into account in all prestressing phases. Strain increase in prestressed steel does not occur before tensioning of the tendon.
(b) In later phases, an "incomplete" bond between concrete and prestressing steel is considered; thus, steps (i)–(iii) are carried out successively.

Bonded tendons are contributing to the stiffness of shell elements that they are embedded into. The displacement field of bonded tendons coincides with the respective displacements of the surrounding concrete. Note that tendon elements have longitudinal strain only.

An explicit definition of the strain and stiffness matrix of un-bonded tendons is far more complex than the one for bonded tendons. Anchoring and friction effects of un-bonded tendons contribute little to the stiffness of a structure. That impact is far smaller than for bonded tendons. On the other hand, an exact calculation of a tangent stiffness matrix in incremental-iterative procedures is not necessary. Using an approximate stiffness matrix can be far more efficient in those problems. Compatibility between the displacement field of a concrete and a displacement field of un-bonded tendons only exist in the anchoring zone.

4 Numerical Modelling the Losses of the Prestressing Force

The instantaneous losses of the prestressing force are caused by friction, sliding of the anchorage and instantaneous concrete strain. The subsequent losses of the prestressing force are caused by prestressing steel relaxation, and the shrinkage and creep of concrete. A detailed calculation of the losses of the prestressing force for both pre-tensioned and post-tensioned structures (with bonded and un-bonded tendons) can be found in [1], and will be described in short hereinafter.

4.1 Losses of the Prestressing Force Caused by Friction

By tendon prestressing on anchor A (Fig. 2), tendon force $P(s)$ at distance s from the anchoring point is adopted as

$$P(s) = \alpha(s) P'_A \tag{6}$$

where

$$\alpha(s) = e^{-\mu \int_{s_A}^{s} \chi ds} \tag{7}$$

The tendon curvature χ can also include arbitrary angle deviation k.

If the tendon is prestressed on both anchors A and B, the relative displacements between the concrete and tendon will change the algebraic sign in a middle point C (Fig. 2) as follows

$$P_A(s) = P'_A e^{-\mu \int_{s_A}^{s} \chi ds} \tag{8}$$

and

$$P_B(s) = P'_B e^{-\mu \int_{s}^{s_B} \chi ds} \tag{9}$$

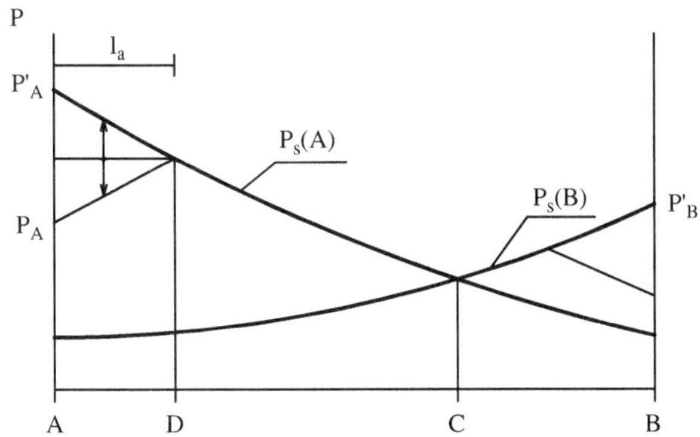

Fig. 2 Changes in prestressing force along the tendon caused by friction

Function $\alpha(s)$ is adopted either according to Eq. (8) or (9), depending on the fact which one gives the larger prestressing force.

$$\alpha(s) = e^{-\mu \int_{s_A}^{s} \chi ds} \quad \text{for } s_A \leq s \leq s_C \tag{10a}$$

$$\alpha(s) = e^{-\mu \int_{s}^{s_B} \chi ds} \quad \text{for } s_C \leq s \leq s_B \tag{10b}$$

4.2 Losses of the Prestressing Force Caused by Sliding of the Anchorage

Due to the impact of friction, the effects of the sliding of the anchorage are limited only to the distance l_a from the active anchor (Fig. 2). For the given tendon shortening Δu during sliding of the anchorage

$$\Delta u = \int_0^{l_a} \Delta\varepsilon_p(s)\, ds \tag{11}$$

or

$$\Delta u = \frac{1}{E_p A_p} \int_0^{l_a} \Delta P(s)\, ds \tag{12}$$

where $\Delta\varepsilon_p$ is a change in tendon deformation due to sliding of the anchorage, $\Delta P(s)$ is a change in prestressing force due to sliding of the anchorage, E_p is the modulus of elasticity of prestressing steel and A_p is the tendon cross-section area.

According to Fig. 2 and if $s_A = 0$, then

$$P_D(s) = P'_A e^{-\mu \int_0^{l_a} \chi ds} \tag{13}$$

and

$$P_A(s) = P_D e^{-\mu \int_0^{l_a} \chi ds} \tag{14}$$

Then

$$P_A(s) = P'_A e^{-2\mu \int_0^{l_a} \chi ds} \tag{15}$$

Tendon shortening Δu at the distance l_a due to sliding of the anchorage is defined by equation (12), where $\Delta P(s)$ is obtained as the difference between the prestressing force calculated from P'_A and P_A (Fig. 2), namely, as

$$\Delta u = \frac{P'_A}{E_p A_p} \int_0^{l_a} \left[e^{-\mu \int_0^{l_a} \chi ds} - e^{-2\mu \int_0^{l_a} \chi ds} e^{\mu \int_0^{l_a} \chi ds} \right] ds \tag{16}$$

The unknown l_a is calculated by an iterative procedure while integrals are calculated by numerical integration.

Force P_A at prestressing on one side only is given by Eq. (15). Losses caused by sliding of the anchorage at the distance l_a shall also be included in Eq. (7) or (10) for function $\alpha(s)$, which defines the tendon force due to losses caused by friction. According to that

$$\alpha(s) = e^{-2\mu \int_0^{l_a} \chi ds} e^{\mu \int_0^{l_a} \chi ds} \quad ; \quad s_A \le s \le s_D \tag{17}$$

If the length l_a in Eq. (16) is equal or greater then tendon length, Eq. (17) shall be extended to the distance from anchor A to anchor B.

4.3 Losses of the Prestressing Force Caused by Instantaneous Concrete Strain

Losses of the prestressing force in post-tensioned tendons caused by the concrete strain are calculated automatically using the numerical models for prestressing force transfer [1].

4.4 Losses of the Prestressing Force Caused by Prestressing Steel Relaxation

The model uses the following formula derived by Magura et al. [4]

$$\frac{\sigma_p}{\sigma_{pi}} = 1 - \frac{log_{10} t}{10}\left[\frac{\sigma_p}{\sigma_{pi}} - 0.55\right] \quad \text{for} \quad \frac{\sigma_p}{\sigma_{pi}} \geq 0.55 \tag{18}$$

where σ_p is a forecasted final stress for the initial σ_{pi} after t days. This formula is used only for steel subjected to constant deformations. Other cases require more complex model. Figure 3 shows the Magura model that was also used here. Let the f_{pi0} be the initial stress. After the period t_1, stress will be reduced to f_{R1} due to steel relaxation. For a modified stress f_{p1} due to instantaneous external load in time t_1, a fictional initial stress f_{pi1} is obtained and associated to a curve containing the point (t_1, f_{pi1}). That curve is used as a model for subsequent relaxation. The procedure is repeated for each load increment. The total steel relaxation till the time t_n is obtained as a sum of all separate stress decreases due to relaxation $(f_{Rk}, 1, 2, \ldots, n)$.

Losses due to steel relaxation shall be included via residual forces for the respective tendon segment

$$\Delta \mathbf{R}^u_{rlx} = -\int_{\Gamma} \mathbf{B}_p \Delta\sigma^*_p ds \tag{19}$$

which simulates the decrease in prestressing impact on the steel structure. $\Delta\sigma^*_p$ refers to a stress change due to prestressing steel relaxation in the observed time increment.

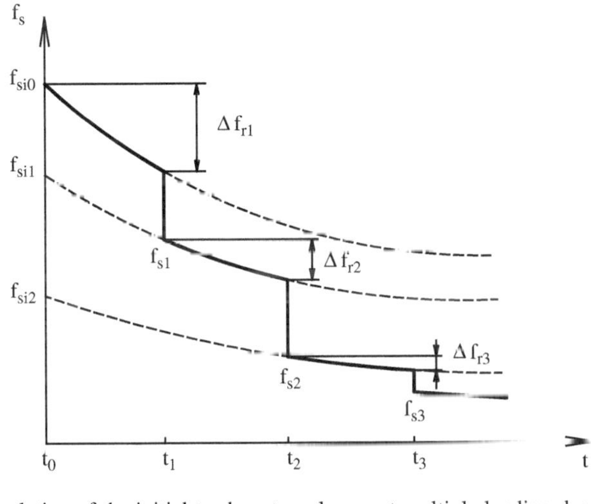

Fig. 3 Calculation of the initial tendon stress losses at multiple loading due to prestressing steel relaxation

4.5 Losses of the Prestressing Force Caused by Shrinkage and Creep of Concrete

Shrinkage and creep of concrete are included according to [5]. Shrinkage strain increment $d\varepsilon^{s}_{t_{n+1}}$ in time t_{n+1} is calculated according to

$$d\varepsilon^{s}_{t_{n+1}} = \varepsilon^{s0}\left(\beta^{s}_{t_{n+1}} - \beta^{s}_{t_n}\right) \tag{20}$$

where ε^{s0} is value of basic shrinkage and $\beta^{s}_{t_{n+1}}$, $\beta^{s}_{t_n}$ are coefficients that describe the shrinkage in time t_{n+1} and t_n.

Creep strain increment $d\varepsilon^{c}_{t_{n+1}}$ in time t_{n+1} is calculated according to

$$d\varepsilon^{c}_{t_{n+1}} = \varepsilon^{m}_{t_{n+1}}\left(\Phi_{t_{n+1},t_0} - \Phi_{t_n,t_0}\right) \tag{21}$$

where $\varepsilon^{m}_{t_{n+1}}$ is mechanical concrete strain in time t_{n+1} and Φ_{t_{n+1},t_0}, Φ_{t_n,t_0} are creep coefficients for time t_{n+1} and t_n.

Losses due to shrinkage and creep of concrete are calculated automatically via constitutive equations that are used for the description of time-dependent analysis, as described in detail in [1].

5 Numerical Procedure for Analysis of the Prestressing Structures

A detailed description of a general procedure for numerical modelling of concrete shells under long-term load, including rheologic effects of concrete, can be found in [1]. The solution for prestressing, which is similar to that described in [3], will be given hereinafter.

The basic steps of an iterative procedure are the following:

(i) Prestressing is modelled by introduction of the initial deformation ε_{p0} in the tendon, which corresponds to the tendon prestressing force. Taking into account prestressing losses caused by friction, ε_{p0} is calculated using the prestressing force P_0 before the losses

$$\varepsilon_{p0,j} = \frac{P_{0,j}}{E_p A_p} \tag{22}$$

while $P_{0,j}$ is calculated as

$$P_{0,j} = \alpha_j P'_0 \tag{23}$$

where α_j is a coefficient of losses in nodes due to friction, and P'_0 is the maximum prestressing force prior to force transfer to the concrete.

The equivalent nodal loads due to prestressing are given by

$$\sigma_{p0}(\tau) = E_p \sum_{j=1}^{3} N'_j(\tau)\varepsilon_{p0,j} \tag{24}$$

The calculated equivalent nodal loads are added to the existing nodal forces.

(ii) For the calculated displacement field $\Delta\mathbf{d}$, mechanical strain and stress are calculated in the observed tendon nodes.

(iii) The equivalent prestressing force $\bar{P}'_0$, acting on the first tendon "node" (anchor), is obtained using the integral at the distance $l_{p,l}$ of the first tendon element

$$\bar{P}'_0 = \int_{l_{p,l}} B'_{p,1}\sigma_p(\tau) A_p dl \tag{25}$$

where $B'_{p,1}$ is the 1-D deformation matrix, which relates the longitudinal tendon strain ε_p with the tangent displacement u'_1 in the first node (Fig. 1). Furthermore, $B'_{p,1}$ is calculated from the expression which relates the tendon longitudinal strain ε_p with nodal displacements $u_{p,j}$ of a 1-D element in the global coordinate system (Fig. 1)

$$\varepsilon_p = \frac{du'}{ds} = \left[a\ b\ c\right] \sum_{j=1}^{3} \frac{dN'_j}{ds} \begin{bmatrix} u_{p,j} \\ v_{p,j} \\ w_{p,j} \end{bmatrix} \tag{26}$$

The equivalent nodal forces for the first node can be expressed as

$$\begin{bmatrix} X_1 \\ Y_1 \\ Z_1 \end{bmatrix} = \int_{l_{p,1}} \left[a\ b\ c\right]^T \frac{dN'_1}{ds} \sigma_p\, A_p\, dl \tag{27}$$

where

$$\frac{dN'_1}{ds} = \frac{dN'_1}{d\tau}\frac{d\tau}{ds} = \frac{dN'_1}{d\tau}\frac{1}{v} \tag{28}$$

and

$$\bar{P}'_0 = \left[a_1\ b_1\ c_1\right] \begin{bmatrix} X_1 \\ Y_1 \\ Z_1 \end{bmatrix} \tag{29}$$

where sub-matrix $\left[a_1 \; b_1 \; c_1 \right]$ contains the components of the unit vector **a**, which is a tangent to the tendon axis at node 1 of the global coordinate system.

The equivalent prestressing force at node 1 is

$$\bar{P}_0' = \int\limits_{l_{p,1}} (a_1 a + b_1 b + c_1 c) \frac{dN_1'}{d\tau} \frac{1}{v} \sigma_p A_p dl \tag{30}$$

As derived from (30)

$$B_{p,1}' = (a_1 a + b_1 b + c_1 c) \frac{dN_1'}{d\tau} \frac{1}{v} \tag{31}$$

(iv) A convergence, i.e. the equilibrium between the initial P_0' and calculated prestressing force $\overline{P}_0'$, is calculated using the condition

$$\frac{P_0' - \overline{P}_0'}{P_0'} \leq \text{TOLER} \tag{32}$$

where TOLER is the adopted tolerance. If the condition(32) is not satisfied, the initial tendon strain is increased by

$$\Delta\varepsilon_{p0,j} = \varepsilon_{p0,j} \left(\frac{P_0'}{\overline{P}_0'} - 1 \right) \tag{33}$$

Steps (i) to (iv) shall be repeated until convergence condition(32) is satisfied, as well as the convergence condition related to nonlinear solution of the entire system [1].

(v) Following the achieved convergence, losses caused by steel relaxation are being taken into account by stress change along the tendon length l_a

$$\Delta\sigma_p(\tau) = \sum_{j=1}^{3} N_j'(\tau)\, \Delta\alpha_j \sigma_p(\tau) \tag{34}$$

where

$$\Delta\alpha_j = \Delta\alpha\left(s_j\right) = e^{-\mu \int_{s_A}^{s_j} \chi ds} - e^{-2\mu \int_0^{l_a} \chi ds} e^{\mu \int_{s_A}^{s_j} \chi ds} \tag{35}$$

The Newton–Raphson method is used for the solution of the nonlinear problem.

6 Example

The presented numerical model was verified on the results of experimental testing of prestressed concrete shell under short-term and long-term load (until unloading), described in [2]. Only one fourth of the shell was modelled due to problem symmetry. The shell geometry and its discretization by finite elements are shown in Fig. 4. Friction between the tendon and concrete is neglected.

The influence of a rigid horizontal steel tie is modelled with no lateral displacement in y-direction at shell supports. The numerical analysis has been divided in three phases (as in the experiment): prestressing, short-term load and long-term load.

Figure 5 shows the comparison between the experimental and numerical results of shell deflection at the centre of the shell. Comparison of the experiment and calculated strains of reinforcement in measuring point S-1 and S-2 are presented on Fig. 6. Figure 7 shows the comparison between the experimental and numerical results of the tendon strains. As can be observed, the obtained numerical results show good correspondence with the obtained experimental values of each phase.

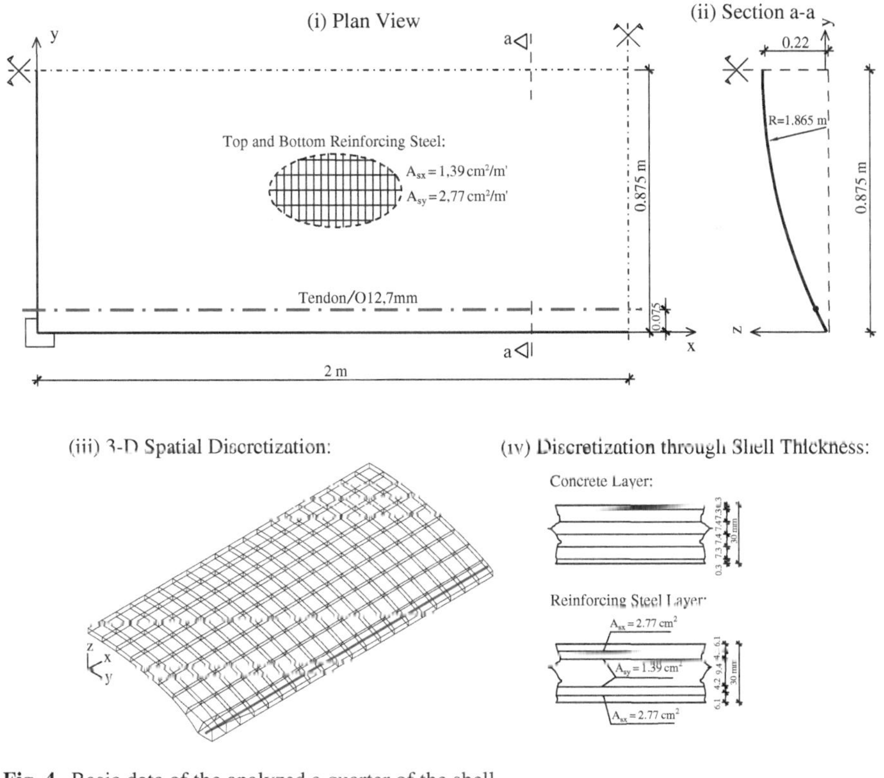

Fig. 4 Basic data of the analyzed a quarter of the shell

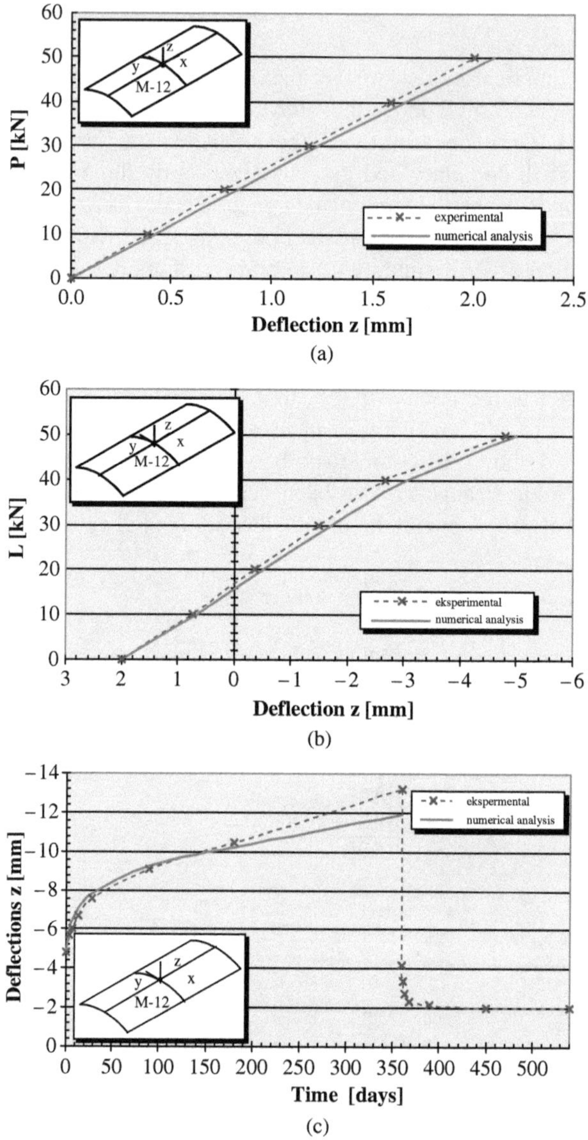

Fig. 5 Shell deflections at the measuring point M-12. (**a**) Prestressing. (**b**) Short-term load. (**c**) Long-term load (until unloading)

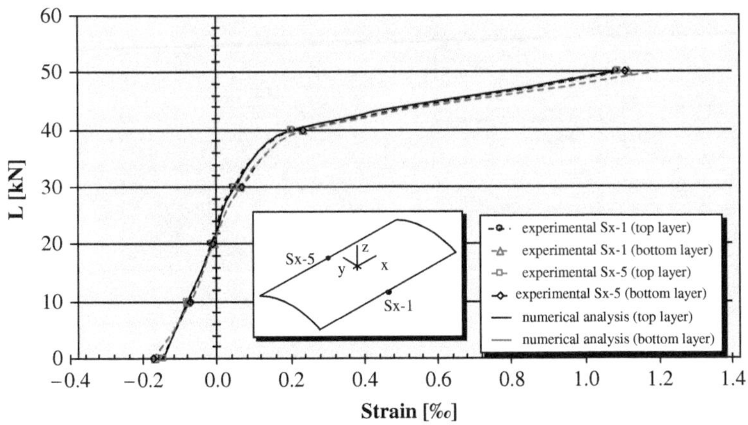

(a)

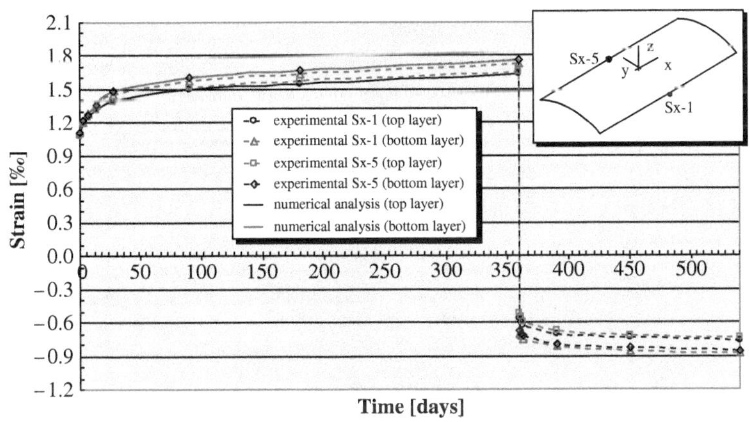

(b)

Fig. 6 Reinforcement strains at the measuring point Sx-1 and Sx-5. (**a**) Short-term load (after prestressing) – x direction. (**b**) Long-term load (until unloading) – x direction

7 Conclusion

The verification of presented numerical model on the results of performed experiments test shows good correspondence between the experimental and numerical values. Subsequent verification of the presented numerical model on several experimental tests is needed. It is believed that the presented model will be of practical use in the analysis of prestressed concrete plates and shells under short-term and long-term static load, without unload.

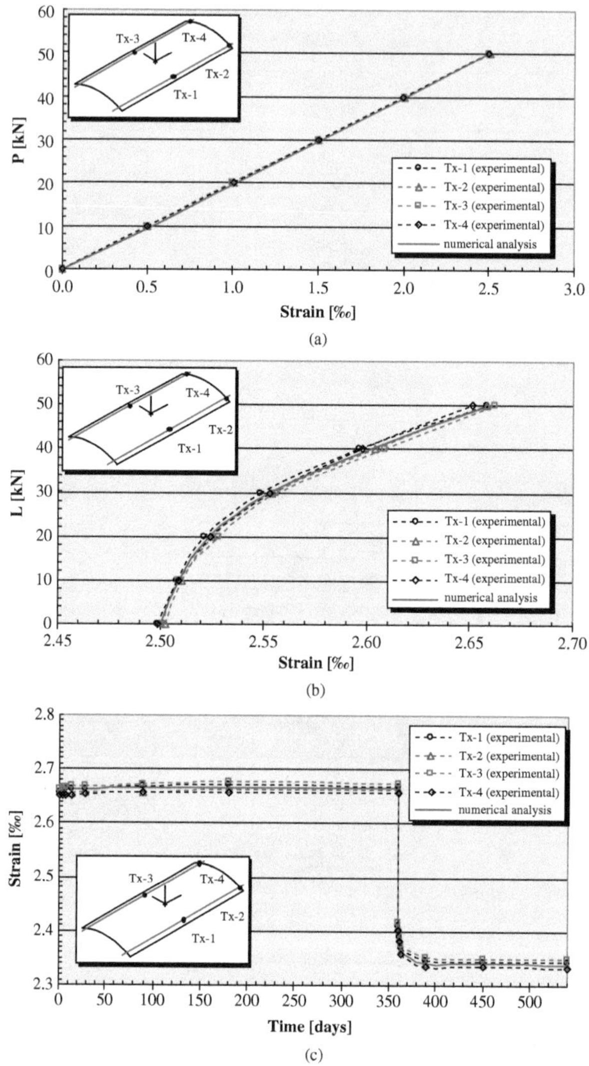

Fig. 7 Tendon strains (measuring points Tx-1, Tx-2, Tx-3 and Tx-4). (**a**) Prestressing. (**b**) Short-term load (**c**) Long-term load (until unloading)

References

1. D. Matesan, Time-dependent analysis of prestressed concrete shells. Ph.D. Thesis, University of Split, 2007
2. J. Radnić, D. Matešan, Experimental testing of prestressed concrete shell under long-term loading and unloading. 2nd International Conference on Advanced Computational Engineering and Experimenting, Abstract book ACE-X 2008, Barcelona, 2008

3. J.A. Figueiras, R.H.C.F. Povoas, Modelling of prestress in non-linear analysis of concrete structures. Comput. Struct. **53**, 173–187 (1994)
4. D. Magura, M.A. Sozen, C.P. Siess, A study of stress relaxation in prestressing reinforcement. PCI J. **9**(2), 13–57 (1964)
5. EN 1992-1, Eurocode 2: Design of concrete structures – Part 1: General rules and rules for buildings, European Standard, Brussels, 2001

DBEM and FEM Analysis of an Extrusion Press Fatigue Failure

R. Citarella, G. Cricrì, M. Lepore, and M. Perrella

1 Introduction

This paper illustrates an application of the Dual Boundary Element Method (DBEM) [1, 2], as implemented in a commercial code [3], to the simulation of a fatigue crack propagation phenomenon affecting the main cylinder of an extrusion press for aluminum sections (Figs. 1, 2, 3 and 4). Such methodology is particularly effective for simulation of three-dimensional single/multiple crack propagation under the hypothesis of isotropic and linear elastic material properties [4–8, 9].

The aforementioned crack propagates through the thickness, causing a leakage of the pressurized oil and consequent production stop after 1,310,400 cycles (the extrusion press had been working for 11 years, 240 working days per year, 5 days per week, 24 h per day, with basic cycles lasting 3 min each).

The fatigue load is induced by the pressure variation inside the cylinder from 0 to 300 bar for each cycle, as needed to push each section through the extrusion hole.

The main aim of the simulation is to assess the most probable initial crack dimensions that, after the recorded in service fatigue cycles, lead to the final crack scenario. Once assessed the initial crack scenario it is possible to understand if there was any possibility to detect such crack before the assembly phase, by using non destructive detection techniques, and if there was a rogue flaw introduced by the manufacturing process.

2 Problem Description and DBEM Analysis

The cylinder (Figs. 1, 2, 3 and 4) is made of cast iron and the related fatigue properties (Table 1), consistent with the NASGRO 3 crack propagation formula (Eq. 1), are provided in the NASGRO database where the material is indicated as [A1AC50AB1]. The cylinder is constrained along the cylinder axial direction by the

R. Citarella (✉)
Department of Mechanical Engineering, University of Salerno, 84084 Fisciano SA, Italy
e-mail: rcitarella@unisa.it

A. Öchsner et al. (eds.), *Materials with Complex Behaviour*, Advanced Structured Materials 6, DOI 10.1007/978-3-642-12667-3_12, © Springer-Verlag Berlin Heidelberg 2010

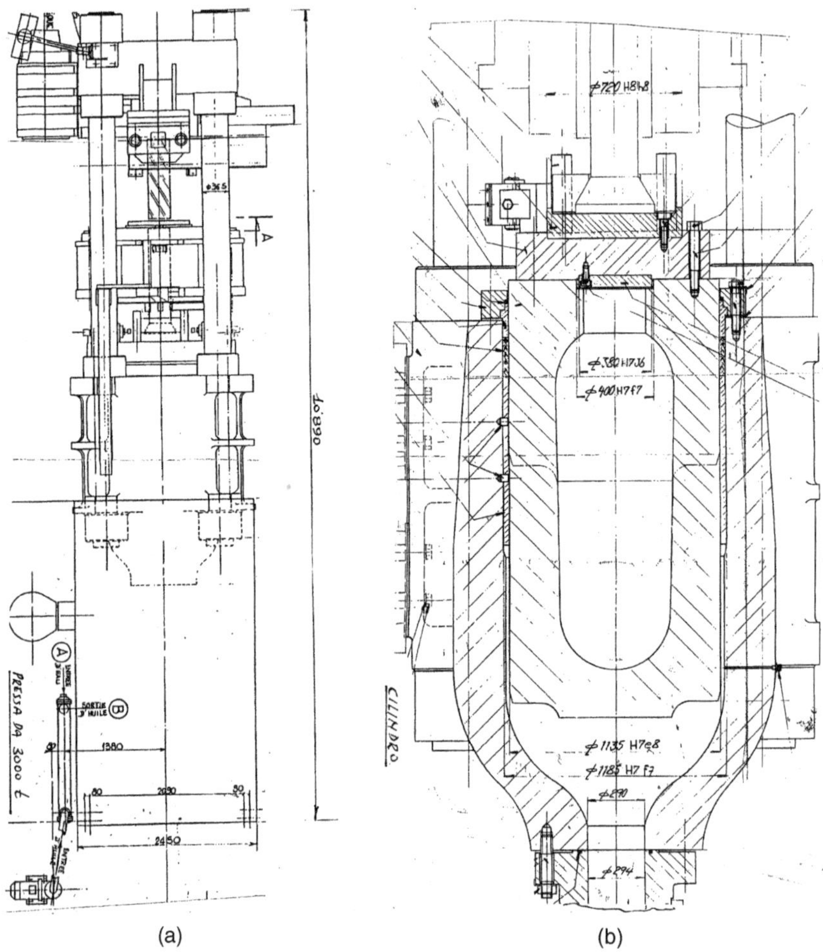

Fig. 1 (**a**) Extrusion press scheme. (**b**) Damaged cylinder scheme

axial forces applied on the rib end by the 4 columns and is clamped on the pavement by screw bolts (Figs. 1 and 2).

$$\frac{da}{dN} = \frac{C \cdot \Delta K^n \cdot \left(1 - \frac{\Delta K_{th}}{\Delta K}\right)^p}{\left(1 - \frac{\Delta K}{(1-R)\cdot K_c}\right)^q} \tag{1}$$

The in service failure is caused by a through the thickness crack that, initiated in a highly stressed area at the inner surface of the cylinder (Fig. 4), break through on the external surface.

Fig. 2 Extrusion press cylinder (front view)

Fig. 3 Extrusion press cylinder: rear view (*left*) and internal view (*right*)

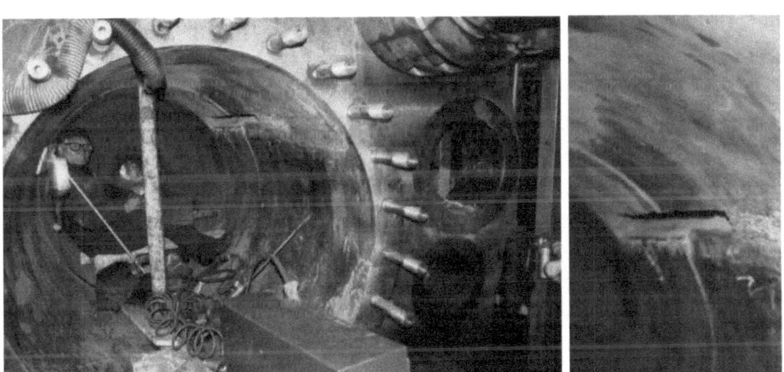

Fig. 4 Internal view of damaged cylinder (*left*) during welding repair operations and close-up of the crack under repair (*right*)

Table 1 Fatigue and mechanical properties of cast iron

Ultimate tensile strength (UTS)	5.51E+08 [N/m^2]
Yield stress (Y_s)	4.00E+08 [N/m^2]
Part through fracture toughness (K_{1e})	0.493E+08 [Nm$^{-3/2}$]
Plane strain fracture toughness (K_{1c})	0.351E+08 [Nm$^{-3/2}$]
A_k coefficient	0.75
B_k coefficient	0.5
Crack growth rate coefficient (C)	6.04E–29 [m$^{5/2}$/N]
n coefficient	2.9
p coefficient	0.5
q coefficient	0.5
Threshold SIF at R = 0 (ΔK_0)	8.76E+6 [Nm$^{-3/2}$]
C threshold positive value (C_{pth})	2.0
C threshold negative value (C_{nth})	0.1
Cut off stress ratio (RCL)	0.7
Plane stress strain constraint factor (α)	2.5
SR ratio	0.3
Young's modulus	2.1E+11 [N/m^2]
Poisson's ratio	0.3

The main driving forces for the crack propagation are provided by the tangential stresses due to the cylinder internal pressure and by the pressure on the crack faces due to the pressurized oil flowing within the crack. In order to automatically apply such pressure on the crack faces during the numerical crack propagation, an in house made routine has been developed and interfaced with the main DBEM code (BEASY).

The axial stresses produced by the pressure on the hemispheric cylinder bottom are directly balanced by the bolt reaction applied on the rib end, whereas the initial bolt pre-stress is neglected in the numerical model.

In the simulation, the crack propagation direction is based on the minimum strain energy density criterion by Sih [10].

For the numerical simulation an initial edge half-circular crack is introduced at half cylinder length, oriented in such a way to lie in an axial plane and propagated up to instability (Figs. 5 and 6), occurring when the effective SIF (K_{eff}), in at least one point of the crack front, becomes equal to the fracture toughness (K_c).

The initial crack radius ($r = 1.2$ cm) is chosen in such a way that the numerical fatigue life reproduces the experimental fatigue life, namely, with such initial crack dimension the fracture toughness is reached by the K_{eff} along the crack front, after 1,343,000 simulated cycles consistently with the experimental outcome of 1,310,400 cycles. At this stage the simulated crack becomes semielliptical with semi axes $A = 12.7$ cm (depth at crack front midpoint) and $B = 13.3$ cm (as measured at break through points). In Fig. 7 the semi axes variation during the propagation is shown.

The Stress Intensity Factors (SIFs) along the crack front are calculated with the Crack Opening Displacement (COD) and J-integral methods. In Fig. 8 they are shown at the final stage of crack propagation in which a pure mode I condition is present.

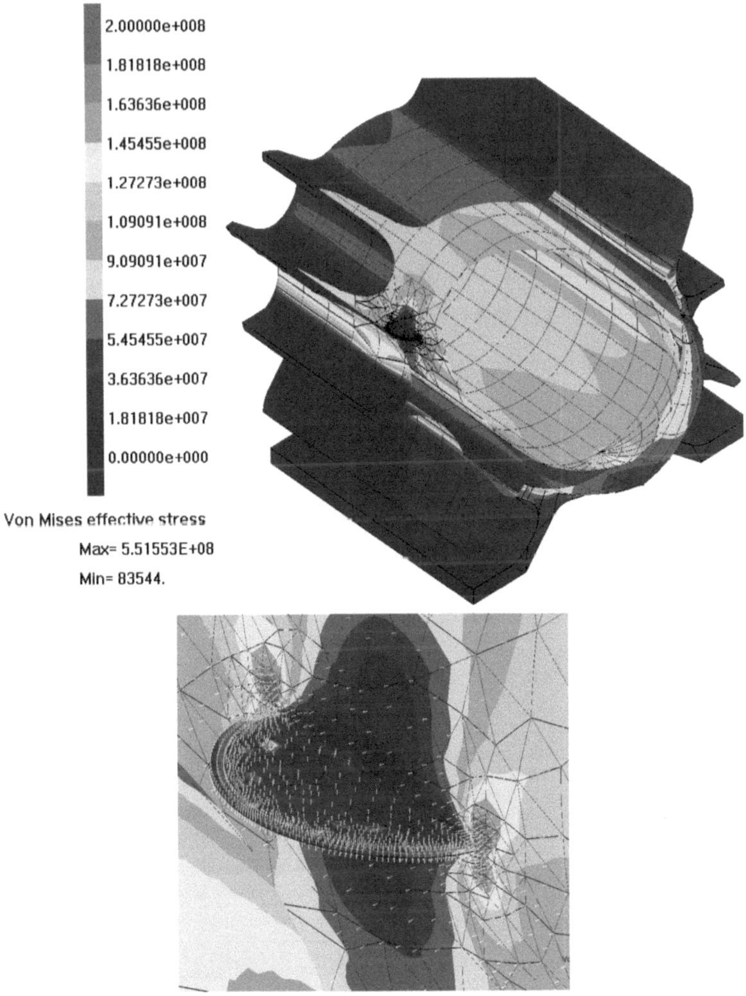

Fig. 5 DBEM cracked model in the final configuration (at crack instability), with highlight of von Mises stresses (N/m^2) and close-up of the final crack with highlight of oil pressure applied on crack faces

After propagating for 1,343,000 cycles, the crack becomes unstable as the fracture toughness is overcome and consequently the crack immediately break through the external surface, causing the pressurized oil leakage and failure of the extrusion press.

The adopted mesh is based on an initial number of 1,795 elements (before crack propagation) and a final number of 2,287 elements (end of propagation). A mixed mesh with quadratic elements (9 nodes quadrilateral) on the crack faces and reduced quadratic elements (8 nodes quadrilateral) elsewhere is adopted.

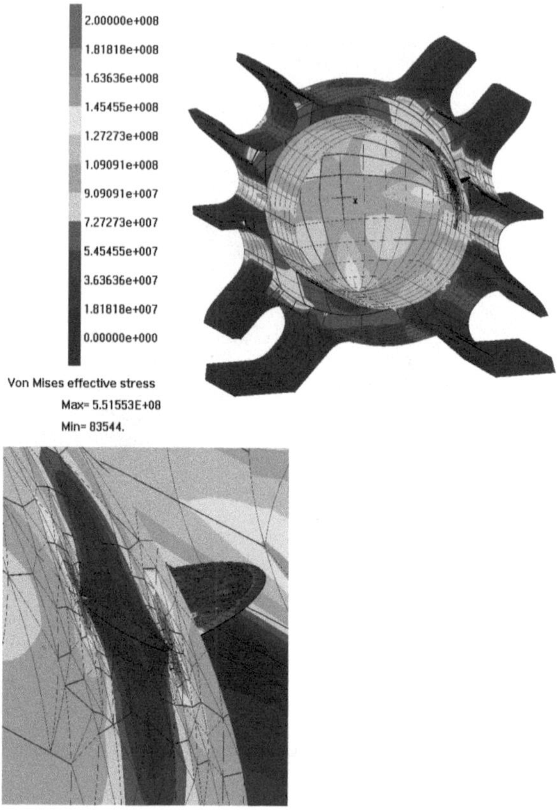

Fig. 6 DBEM cracked model in the final configuration (at crack instability), with highlight of von Mises stresses (N/m^2), and close-up of the final crack internal view

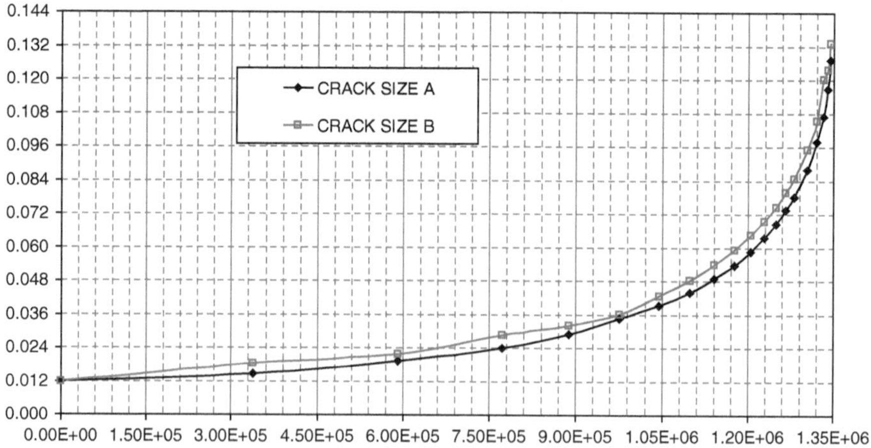

Fig. 7 Semielliptical crack axes (m) versus number of cycles

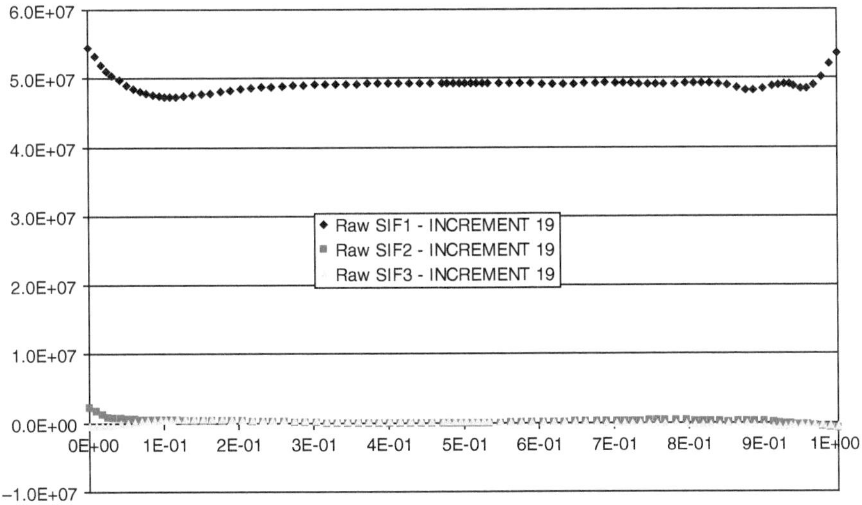

Fig. 8 SIF's values ($Nm^{-3/2}$) along the crack front, at the final stage of crack propagation (crack instability)

The availability of discontinuous elements allows to easily migrate from the very refined mesh around the cracked area to the rough mesh elsewhere where lower stress gradients are expected (Fig. 5).

3 FEM Results

The FEM simulations are performed using the ANSYS commercial code. The cylinder is modeled with about 595,000 quadratic elements (SOLID186); the details of the crack mesh are shown in Fig. 9.

Figures 10 and 11 show the von Mises stress field on the cracked cylinder, considering the initial crack (only such configuration is analyzed by FEM) and the boundary conditions previously illustrated.

SIFs are calculated by the COD method [11] as follows:

$$K_I = \Delta u(\Delta a) \cdot \frac{E}{4(1-\beta\nu)} \cdot \sqrt{\left(\frac{\pi}{2\Delta a}\right) \bigg/ \left(1 - \frac{\Delta a}{2r}\right)} \tag{2}$$

where Δa is the distance from the crack front, Δu is the corresponding crack opening, r is the crack radius (the initial crack is half circular) and β is equal to 0 or ν in case of plane strain, or plane stress respectively.

As well stated in the literature, in case of 3D cracks the plane strain hypothesis is assumed, then we set $\beta = \nu$. The distance Δa is a crucial parameter for the correct evaluation of SIFs: it should be as small as possible for the validity of Eq. (2); on the

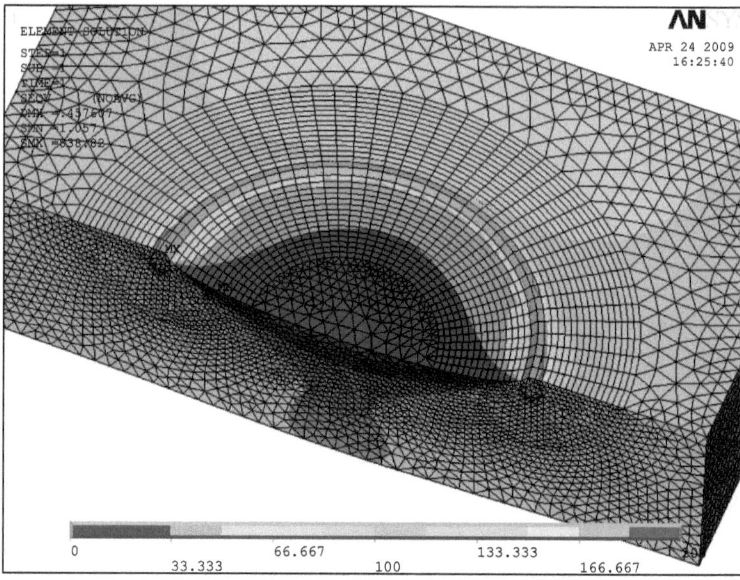

Fig. 9 Mesh details in the cracked zone for the initial cracked configuration

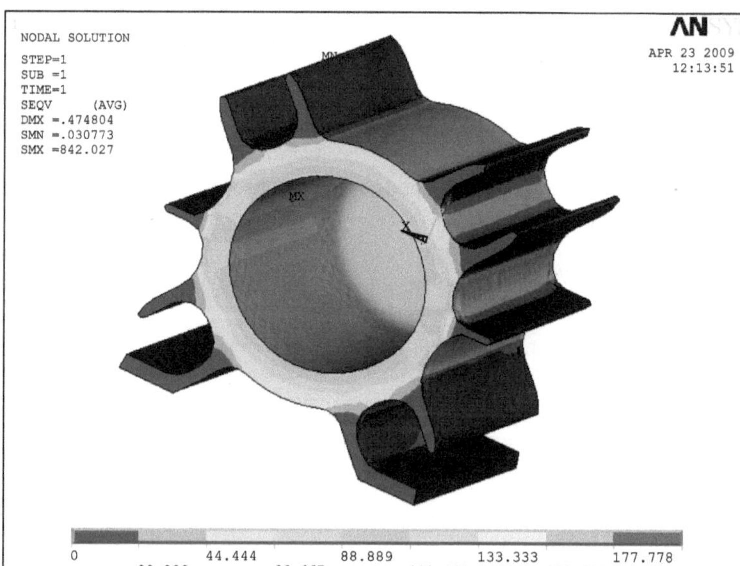

Fig. 10 FEM cracked model in the initial configuration, with highlight of von Mises stresses (N/m^2) – rear view

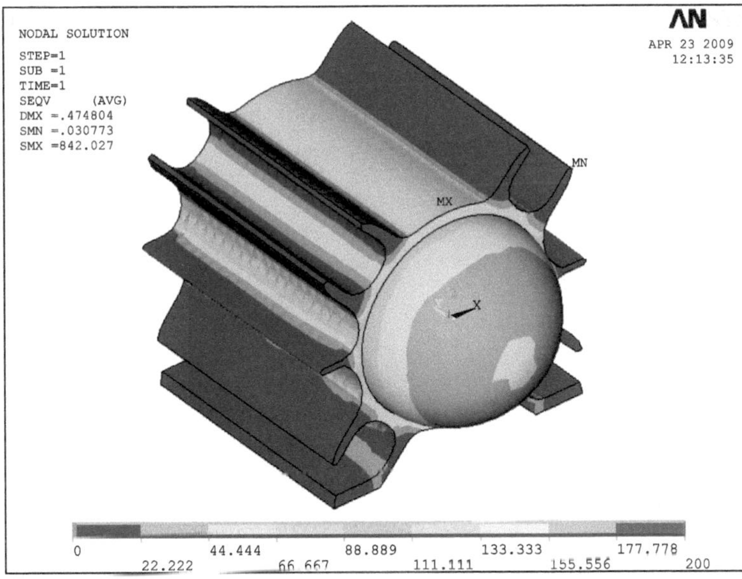

Fig. 11 FEM cracked model in the initial configuration, with highlight of von Mises stresses (N/m^2) – front view

other hand, a correct COD evaluation by FEM requires a minimum distance from the crack front, in order to avoid the mesh dependence problem. A ratio $\Delta a/r = 1/8$ coupled with an element size nearby the crack front $S = \Delta a/2$ is considered to provide sufficiently accurate evaluations.

4 Comparison of FEM and DBEM Results

In Fig. 12 comparison of FEM and DBEM von Mises stresses in the cracked zone, is provided, getting a satisfactory correlation.

The FEM and DBEM SIFs calculated along the initial crack front are shown in Fig. 13. The difference between this FEM and DBEM result ranges from –2% at the central point of the crack front and +10% at break through points.

5 Conclusions

The realized DBEM simulation shows that the initial semicircular crack, consistent with final crack scenario and with the recorded number of fatigue cycles to failure, has a radius of nearly 1.2 cm. Such relevant manufacturing defect (rogue flaw) could be detected by the manufacturer by NDE (non destructive evaluation) technique.

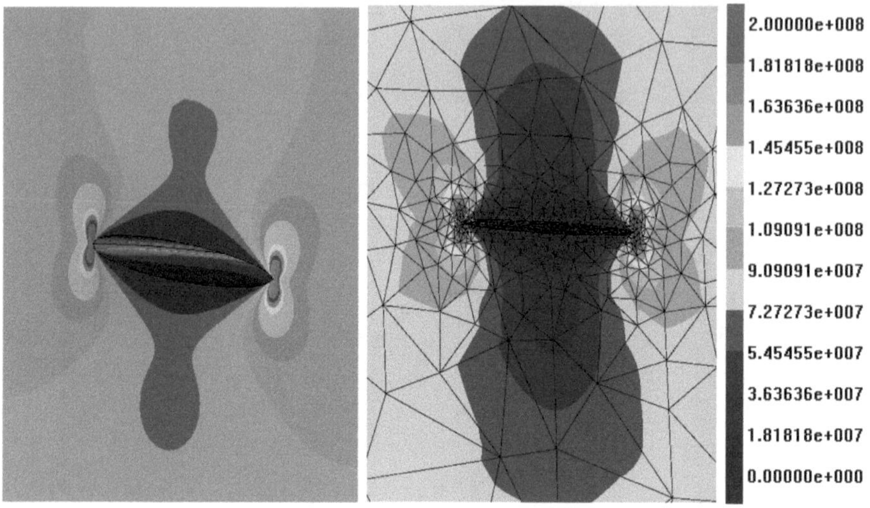

Fig. 12 Comparison of FEM and DBEM von Mises stresses (N/m^2) with reference to the initial crack configuration

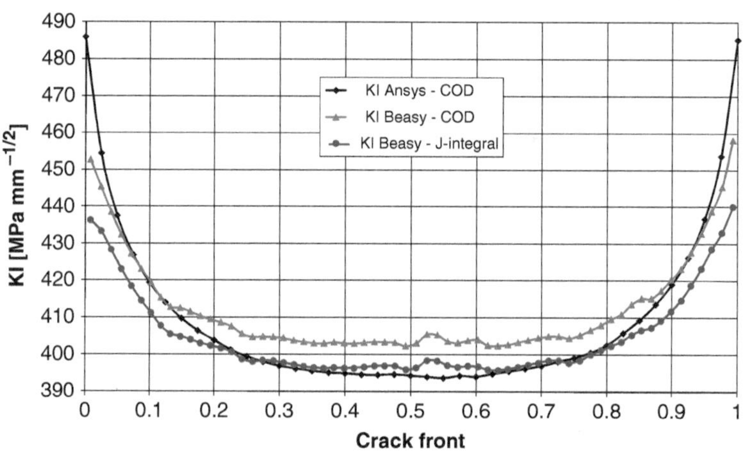

Fig. 13 K_I along the crack front at the initial stage of crack propagation by FEM (Ansys) and DBEM (Beasy)

The numerical validation, by comparison between FEM and DBEM outcomes, gave satisfactory results.

Moreover, there are some inherent advantages using DBEM for this type of analysis, including: simplified modeling of the cracked area, automatic crack introduction and propagation, reduced run times (at least for subassemblies) and accurate crack growth simulation (it is straightforward to model multiple crack propagation, load spectra effects, etc.).

References

1. Y. Mi, M.H. Aliabadi, Dual boundary element method for three dimensional fracture mechanics analysis. Eng. Anal. Bound. Elem. **10**, 161–171 (1992)
2. Y. Mi, *Three-Dimensional Analysis of Crack Growth*, Topics in Engineering, vol. 28 (Computational Mechanics Publications, Southampton, 1996)
3. C. M. BEASY Ltd., BEASY V10r10 Documentation, Southampton (2009)
4. C. Calì, R. Citarella, M. Perrella, *Three-Dimensional Crack Growth: Numerical Evaluations and Experimental Tests*, Biaxial/Multiaxial Fatigue and Fracture, ESIS Publication, vol. 31 (Elsevier, Amsterdam, 2003)
5. R. Citarella, M. Perrella, Multiple surface crack propagation: Numerical simulations and experimental tests. Fatigue Fract. Eng. Mater. Struct. **28**, 135–148 (2005)
6. R. Citarella, F.G. Buchholz, Comparison of crack growth simulation by DBEM and FEM for SEN-specimens undergoing torsion or bending loading. J. Eng. Fract. Mech. **75**, 489–509 (2008)
7. R. Citarella, F.G. Buchholz, Comparison of DBEM and FEM crack path predictions with experimental findings for a SEN-specimen under anti-plane shear loading. Key Eng. Mater. **348–349**, 129–132 (2007)
8. A. Leski, Implementation of the virtual crack closure technique in engineering FE calculations. Finite Elem. Anal. Des. **43**(3), 261–268 (2007)
9. K.S. Venkatesha, T.S. Ramamurthy, B. Dattaguru, Generalized modified crack closure integral (GMCCI) and its application to interface crack problems. Comput. Struct. **60**(4), 665–676 (1996)
10. G.C. Sih, B.C.K. Cha, A fracture criterion for three-dimensional crack problems. J. Eng. Fract. Mech. **6**, 699–732 (1974)
11. T.L. Anderson, *Fracture Mechanics: Fundamentals and Applications*, 3rd edn. (Taylor & Francis, CRC Press, Boca Raton, FL, 2005)

Damage Detections in Nonlinear Vibrating Thermally Loaded Plates

E. Manoach and I. Trendafilova

1 Introduction

The main objective of structural health monitoring (SHM) is to ascertain whether damage is present or not in a structure. Most vibration-based structural health monitoring methods (VSHM) are based on the fact that damage will alter the stiffness, mass or energy dissipation properties of a structure which in turn will alter its measured vibration response.

These methods are widely used for structural health monitoring and damage assessment purposes. Their application is somewhat limited by the need of a precise enough model of the structural vibration response. If some nonlinearities or environmental conditions (like the elevated temperature, for example) are not taken into account in the model, then a model-based VSHM method could give a false alarm due to a discrepancy between the measured and the modelled response. Temperature changes can and do affect substantially the vibration response of a structure. Thermal loads introduce stresses due to thermal expansion, which lead to changes in the modal properties. Thermal loads can also cause buckling and in some cases even lead to chaotic behaviour [1–5].

Thus, on a lot of occasions the presence of a temperature field can either mask the effect of damage or increase it, which will render a VSHM method ineffective – it might give no alarm when a fault is present or give a false alarm. This is why it is vital to be able to take into account the temperature changes when developing VSHM procedures.

Most of the previous efforts of researchers in the area of VSHM were directed towards methods based on linear modal analysis [6–10]. One of the main problems with these methods comes from the fact that in general damage starts as a local phenomenon and does not necessarily affect significantly the modal characteristics of the structure. In many cases the lower order resonance frequencies and mode shapes are not very sensitive to damage, except in cases of very large damage

E. Manoach (✉)
Institute of Mechanics, Bulgarian Academy of Sciences, 1113 Sofia, Bulgaria
e-mail: e.manoach@imbm.bas.bg

A. Öchsner et al. (eds.), *Materials with Complex Behaviour*, Advanced Structured Materials 6, DOI 10.1007/978-3-642-12667-3_13, © Springer-Verlag Berlin Heidelberg 2010

[6, 11]. Thus in reality it may be difficult to distinguish if damage is indeed the reason behind, e.g., a decrease in frequency or it is caused by environmental or operational conditions changes.

Many VSHM methods are inherently limited to linear systems – they use, for example, the superposition principle in the analysis – and cannot account for the effects of non-linearities. Another problem with a number of VSHM methods is that they rely on a linear model of the structure. As the theoretical model itself can only approximate the actual behaviour of the vibrating structure, it will introduce computational errors [6]. These errors will be greater if the non-linearities of the system are substantial. Since they are not taken into account in the model such methods might give false alarms due to a discrepancy between the measured and the modelled/expected response.

To address some of the above mentioned problems, new concepts in vibration-based monitoring have been emerging recently. These employ measured time series of the structural vibration response, or, often concomitantly, non-linear systems theory. Most of the studies in this field are devoted to the extraction of features from the structural vibration response, which can indicate the presence of damage and its location. In [12] the authors use the beating phenomenon for damage detection purposes. In [13] and [14] new attractor-based metrics are introduced as damage sensitive features. The results are promising. In [15] a panel forced by aerodynamic loads and undergoing limit-cycle oscillations and chaos is investigated. The von Kármán strain displacement relation is employed and a model of the system constituted by ordinary differential equations of motion is achieved by employing finite differences. The upstream endpoint of the panel has been considered supported by a spring of variable stiffness. Changes in the stiffness of a spring have been detected by exploring the chaotic dynamics of the panel.

In [16] a possibility for representing, interpreting and visualising the vibration response of vibrating panels using time domain measurements is investigated. The panels are thin orthotropic plates and are modelled by finite elements. It was found that the first ten resonant frequencies show low sensitivity to damage. Then the simulated vibration response of the panel is transformed and expanded in a new phase space. Preliminary results suggest that it should be possible to use the distribution of points on the attractor to extract damage sensitive features.

In our previous works [11] and [17] a numerical approach to study the geometrically non-linear vibrations of rectangular plates with and without damage is developed. A damage index and a method for damage detection and location, based on the Poincaré map of the response, have been proposed. The suggested damage assessment method shows good capability to detect and localize damage in plates.

Although the approach seems to hold a lot of potential, there is limited research addressing VSHM methods based on time series analysis and non-linear dynamics.

The main objectives of this study are twofold: (i) to study the influence of defects, elevated temperatures and their combination on the dynamic characteristics of the plate and on its geometrically nonlinear dynamic response; (ii) to test the criteria for identification of irregularities (defects) in structures proposed in [11, 17] taking into account the elevated temperature by analyzing the Poincaré map of the structural vibration response.

The application of the proposed approach is demonstrated on rectangular plates with defects at elevated temperatures. The temperature is assumed uniformly distributed over the plate surface and thickness. The plates are subjected to a harmonic loading which leads to large amplitude vibrations. The influence of damage on the time-history diagrams of the plate, as well as on the geometry of its phase-space is studied. A VSHM method is developed which applies a criterion based on features sensitive to temperature changes and damage in the same time. These features use the Poincaré maps of the structural vibration response. Taking into account the temperature influence on the extracted features allows the detection of damage and shows its location for structures subjected to temperature changes. The proposed study demonstrates the importance of taking into account the correct/exploitation temperature in a damage detection process. It is shown that in some cases of elevated temperature the Poincaré maps based criterion may be unsuitable.

2 Theoretical Model

The object of the investigation is a rectangular plate with sides a and b and thickness h, subjected to temperature changes and a dynamic loading $p(x,y,t)$ perpendicular to the plate (Fig. 1a). The geometrically nonlinear version of the Mindlin plate theory is used to model the plate behaviour, so that the shear deformation and rotatory inertia are taken into account. At each point of the middle surface of the plate, the displacements in the x, y, z directions are denoted by u, v, w, respectively, $\psi_x(x,y,t)$ and $\psi_y(x,y,t)$ are the angles of the rotation of the normal of the cross section to the plate mid-plane (see Fig. 1b).

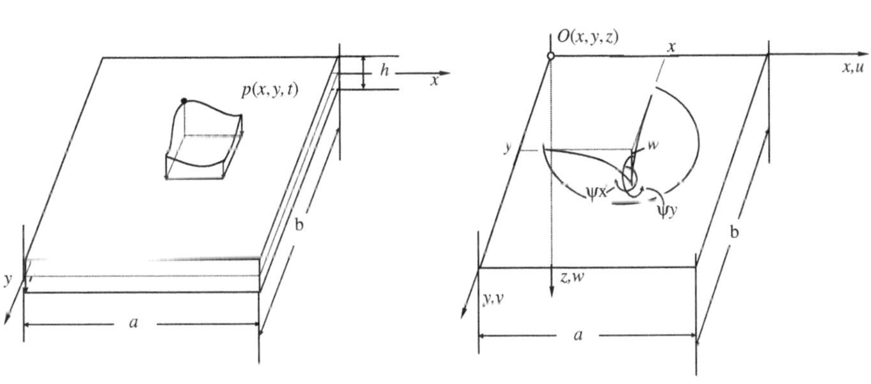

Fig. 1 Plate geometry and coordinate system. (**a**) Plate dimensions and loading. (**b**) Mid-plane of the plate and the components of the generalized displacement vector

The presence of a defect can be modelled as a reduction of the plate thickness or a stiffness reduction and therefore a variation of the flexural rigidity in the governing equations is used. The basic equations of the plate motion are described below.

2.1 Geometrical Relationships

The strain and curvature-displacements relationships associated with the mid-plane of the plate for large displacements and shear can be expressed as:

$$\varepsilon_x^0 = \frac{\partial u}{\partial x} + \frac{1}{2}\left(\frac{\partial w}{\partial x}\right)^2, \varepsilon_y^0 = \frac{\partial v}{\partial y} + \frac{1}{2}\left(\frac{\partial w}{\partial y}\right)^2, \varepsilon_{xy}^0 = \frac{\partial u}{\partial y} + \frac{\partial v}{\partial x} + \frac{\partial w}{\partial x}\frac{\partial w}{\partial y}, \quad \text{(1a–h)}$$

$$\varepsilon_{xz}^0 = \psi_x + \frac{\partial w}{\partial x}, \varepsilon_{yz}^0 = \psi_y + \frac{\partial w}{\partial y},$$
$$k_x^0 = \frac{\partial \psi_x}{\partial x}, k_y^0 = \frac{\partial \psi_y}{\partial y}, k_{xy}^0 = \frac{\partial \psi_x}{\partial y} + \frac{\partial \psi_y}{\partial x}$$

and the strain vector is given by:

$$\boldsymbol{\varepsilon} = \left\{\varepsilon_x^0 + zk_x^0, \varepsilon_y^0 + zk_y^0, \varepsilon_{xy}^0 + zk_{xy}^0, f(z)\varepsilon_{xz}^0, f(z)\varepsilon_{yz}^0\right\}^T \quad (2)$$

where $f(z)$ is a function describing the distribution of the shear strain along the plate thickness.

2.2 Constitutive Equations

Assuming that the material of the plate is linear elastic and isotropic the relations for the stress and strain components are given by:

$$\sigma_x = \frac{\boldsymbol{E(x,y)}}{1-\nu^2}\left[\varepsilon_x + \nu\varepsilon_y\right] - \frac{\boldsymbol{E(x,y)}}{1-\nu}\alpha_T \Delta \boldsymbol{T},$$
$$\sigma_y = \frac{\boldsymbol{E(x,y)}}{1-\nu^2}\left[\varepsilon_y + \nu\varepsilon_x\right] - \frac{\boldsymbol{E(x,y)}}{1-\nu}\alpha_T \Delta \boldsymbol{T}, \quad \text{(3a–d)}$$
$$\sigma_{xz} = \boldsymbol{n}^2\boldsymbol{G}\varepsilon_{xz}, \quad \sigma_{yz} = \boldsymbol{n}^2\boldsymbol{G}\varepsilon_{yz}$$

In terms of generalized stresses the above equations take the form:

$$N_x = A(\varepsilon_x^0 + \nu\varepsilon_y^0) - A\alpha_T\gamma^T, N_y = A(\varepsilon_y^0 + \nu\varepsilon_x^0) - A\alpha_T\gamma^T, N_{xy} = \frac{1-\nu}{2}A\varepsilon_{xy}^0$$

$$M_x = D(\kappa_x^o + \nu\kappa_y^o) - A\alpha_T\kappa^T, M_y = D(\kappa_y^0 + \nu\kappa_x^0) - A\alpha_T\kappa^T, M_{xy} = \frac{1}{2}(1-\nu)D\kappa_{xy}^0,$$
$$Q_x = \frac{1}{2}(1-\nu)n^2A\varepsilon_{xz}^0, \quad Q_y = \frac{1}{2}(1-\nu)n^2A\varepsilon_{yz}^0. \tag{4a–h}$$

where

$$\gamma^T(x,y) = \int_{-h/2}^{h/2} \Delta T(x,y,z)dz, \quad \kappa^T(x,y) = \int_{-h/2}^{h/2} \Delta T(x,y,z)zdz,$$
$$A(x,y) = \frac{E(x,y)h(x,y)}{1-\nu^2}, \quad D(x,y) = \frac{A(x,y)h(x,y)^2}{12} \tag{5a–d}$$

In Eqs. (3), (4) and (5) E is the Young modulus, ν is the Poison ratio, N_x, N_y and N_{xy} are the stress resultants in the mid-plane of the plate, M_x, M_y and M_{xy} are the stress couples and Q_x and Q_y are the transverse shear stress resultants, α_T is the coefficient of thermal expansion and ΔT (Kelvin) is the temperature variation (in general it can be assumed non-uniform along the plate length and thickness) with respect to a reference temperature. n^2 is a shear correction factor which is assumed equal to 5/6 throughout the paper.

2.3 Equations of Motion

The equilibrium equations may be deducted by considering the conditions for translational equilibrium in the x, y and z directions and for rotational equilibrium about x and y. They are as follows:

$$\frac{\partial N_x}{\partial x} + \frac{\partial N_{xy}}{\partial y} + \rho h\ddot{u}_x = 0$$
$$\frac{\partial N_y}{\partial y} + \frac{\partial N_{xy}}{\partial x} + \rho h\ddot{u}_y = 0$$
$$\frac{\partial M_x}{\partial x} + \frac{\partial M_{xy}}{\partial y} - Q_x + c_2\frac{\partial \psi_x}{\partial t} + \frac{\rho h^3}{12}\ddot{\psi}_x = 0$$
$$\frac{\partial M_y}{\partial y} + \frac{\partial M_{xy}}{\partial x} - Q_y + c_2\frac{\partial \psi_y}{\partial t} + \frac{\rho h^3}{12}\ddot{\psi}_y = 0 \tag{6a–e}$$
$$\frac{\partial Q_x}{\partial x} + \frac{\partial Q_y}{\partial y} + N_x\frac{\partial^2 w}{\partial x^2} + N_y\frac{\partial^2 w}{\partial y^2} + 2N_{xy}\frac{\partial^2 w}{\partial x\partial y} + c_1\frac{\partial w}{\partial t} + \rho h\ddot{w} = -p$$

Here and throughout the paper dots over variables represents derivation with respect to time, c_1 and c_2 denote the damping coefficients, and ρ is the density of the plate material.

2.4 Boundary and Initial Conditions

In the present work fully clamped plates, i.e. plates for which all their four edges are clamped and in-plane fixed, are considered. This means that all displacements u, v and w and angular rotations ψ_x and ψ_y are zero along the boundaries. The influence of the temperature variation is more essential for such plates due to the thermal expansion.

The initial conditions are accepted in the following general form:

$$\begin{aligned} &w\,(x,y,0) = w^0(x,y), \quad \dot{w}\,(x,y,0) = \dot{w}^0(x,y), \\ &\psi_x\,(x,y,0) = \psi_x^0(x,y), \, \dot{\psi}_y\,(x,y,0) = \dot{\psi}_y^0(x,y), x \in [0,a]\,, y \in [0,b] \end{aligned} \tag{7a–d}$$

3 Solution of the Problem

3.1 Reorganizing the Equations of the Plate Motion

The equation of motions (6) can be rewritten in the following form:

$$\begin{aligned}
&\frac{\partial}{\partial x}\left[A\left(\frac{\partial u}{\partial x}+\nu\frac{\partial v}{\partial y}\right)\right]+\frac{\partial}{\partial y}\left[\frac{(1-\nu)A}{2}\left(\frac{\partial u}{\partial y}+\frac{\partial v}{\partial x}\right)\right]+\rho h\ddot{u}=G_u+G_u^T \\
&\frac{\partial}{\partial y}\left[A\left(\frac{\partial v}{\partial y}+\nu\frac{\partial u}{\partial x}\right)\right]+\frac{\partial}{\partial x}\left[\frac{A(1-\nu)}{2}\left(\frac{\partial v}{\partial x}+\frac{\partial u}{\partial y}\right)\right]+\rho h\ddot{v}=G_v+G_v^T \qquad \text{(8a–e)} \\
&\frac{\partial}{\partial x}\left(D\left[\frac{\partial \psi_x}{\partial x}+\nu\frac{\partial \psi_y}{\partial y}\right]\right)+\frac{(1-\nu)}{2}\frac{\partial}{\partial y}\left(D\left[\frac{\partial \psi_x}{\partial y}+\nu\frac{\partial \psi_y}{\partial x}\right]\right) \\
&\qquad -\frac{(1-\nu^2)n^2A}{2}\left(\psi_x+\frac{\partial w}{\partial x}\right)+c_2\dot{\psi}_x+\frac{\rho h^3}{12}\ddot{\psi}_x=G_1^T \\
&\frac{\partial}{\partial y}\left(D\left[\frac{\partial \psi_y}{\partial y}+\nu\frac{\partial \psi_x}{\partial x}\right]\right)+\frac{(1-\nu)}{2}\frac{\partial}{\partial x}\left(D\left[\frac{\partial \psi_y}{\partial x}+\nu\frac{\partial \psi_x}{\partial y}\right]\right) \\
&\qquad -\frac{(1-\nu^2)n^2A}{2}\left(\psi_y+\frac{\partial w}{\partial y}\right)+c_2\dot{\psi}_y+\frac{\rho h^3}{12}\ddot{\psi}_y=G_2^T \\
&\frac{(1-\nu)n^2}{2}\left\{\frac{\partial}{\partial x}\left(A\left[\psi_x+\frac{\partial w}{\partial x}\right]\right)+\frac{\partial}{\partial y}\left(A\left[\psi_y+\frac{\partial w}{\partial y}\right]\right)\right\}+c_1\dot{w}+\rho h\ddot{w} \\
&\qquad = -p+G^L+G_3^T
\end{aligned}$$

where

$$G_u=-0.5\frac{\partial}{\partial x}\left\{A\left[\left(\frac{\partial w}{\partial x}\right)^2+\left(\frac{\partial w}{\partial y}\right)^2\right]\right\}-0.5\frac{\partial}{\partial y}\left\{A(1-\nu)\left(\frac{\partial w}{\partial x}\frac{\partial w}{\partial y}\right)\right\}$$

$$G_v = -0.5\frac{\partial}{\partial y}\left\{A\left[\left(\frac{\partial w}{\partial x}\right)^2 + \left(\frac{\partial w}{\partial y}\right)^2\right]\right\} - 0.5\frac{\partial}{\partial x}\left\{A(1-\nu)\left(\frac{\partial w}{\partial x}\frac{\partial w}{\partial y}\right)\right\} \tag{9a–d}$$

$$G_1^T = A(1+\nu)\alpha_T\frac{\partial \kappa_T}{\partial x}, \quad G_2^T = A(1+\nu)\alpha_T\frac{\partial \kappa_T}{\partial y}, G_3^T = A\alpha_T\gamma_T\left(\frac{\partial^2 w}{\partial x^2} + \frac{\partial^2 w}{\partial y^2}\right)$$

$$G^L(x,y,t) = -\left(N_x\frac{\partial^2 w}{\partial x^2} + N_y\frac{\partial^2 w}{\partial x^2} + 2N_{xy}\frac{\partial^2 w}{\partial x \partial y}\right)$$

In this work, only a uniformly distributed temperature field along the plate length and thickness will be considered. Also, it is assumed that the plate gets the elevated temperature instantly. This assumptions leads to settings $G_1^T = 0$, $G_2^T = 0$.

3.2 Numerical Approach

The pseudo-load mode superposition method (PLMS) [2, 11, 18–21] is applied to solve the problem for nonlinear vibration of plates. It will be only briefly presented here.

The widely accepted assumption for transversally loaded clamped plates that mid-plane inertia effects are negligible is assumed, i.e. $\rho h\ddot{u}_x = \rho h\ddot{u}_y = 0$. The finite element method is used to discretize the plate equations with respect to the space variables and by using the PLMS they are transformed in the frequency domain. Then an iterative procedure with respect to time is applied for the solution of the obtained system of ordinary differential equations. It is out of the scope of this paper to concentrate on the details of the solution method and the reader is referred to the above mentioned papers [2, 18–21] where the method is applied for undamaged plates and in [11] – for damaged ones. Thus the solution procedure will be presented only in brief:

Assuming G^u and G^v are known functions, Eq. (8a–b) form a linear system of PDEs which can be solved numerically. The left hand sides of Eq. (8c–e) contain only linear terms and therefore the mode superposition method can be used for their solution. Thus, the generalized displacements vector $\mathbf{U} = \{\beta\psi_x, \beta\psi_y, w\}^T$ ($\beta = h^2/12$) is expanded as a sum of the product of the vectors of the pseudo-normal modes $\mathbf{U}_n$ and the time dependent functions $q_n(t)$ as follows:

$$\mathbf{U} = \sum_{n=1}^{N_f} \mathbf{U}_n(x,y)q_n(t). \tag{10}$$

Substituting Eq. (10) into Eq. (8c–e), multiplying by $\mathbf{U}_m(x,y)$, integrating the product over the plate surface, invoking the orthogonallity condition, and assuming "proportional damping" in the sense $\iint \left(c_2\left(\psi_{xn}^2+\psi_{yn}^2\right)+c_1 w_n^2\right)dxdy = 2\xi_n\omega_n$, the equations for $q_n(t)$ will be "uncoupled" in the form:

$$\ddot{q}_n(t)+2\xi_n\omega_n\dot{q}_n+\omega_n^2 q_n(t)=F_n(t), \tag{11}$$

where ω_n are the natural frequencies of the linear elastic (undamped) Mindlin plate, ξ_n are the modal damping parameters and

$$F_n(t)=\iint \mathbf{U}_n^T[\mathbf{P}(x,y,t)+\mathbf{G}_L(x,y,t)+\mathbf{G}_T(x,y,t)]dxdy, \tag{12a–b}$$

$$\mathbf{P}(x,y,t)=(0,0,-p)^T,\quad \mathbf{G}_L(x,y,t)=(0,0,G_3^L)^T,\quad \mathbf{G}_T(x,y,t)=(0,0,G_3^T)^T.$$

The initial conditions defined by Eq. (7) are transformed also in terms of $q_n(0)$ and $\dot{q}_n(0)$:

$$q_n(0)=q_n^0,\quad \dot{q}_n(0)=\dot{q}_n^0, \tag{13a–d}$$

$$q_n^0=\iint\left(w^0 w_n+\beta\psi_x^0\psi_{xn}+\beta\psi_y^0\psi_{yn}\right)dxdy,$$

$$\dot{q}_n^0=\iint\left(\dot{w}^0\dot{w}_n+\beta\dot{\psi}_x^0\dot{\psi}_{xn}+\beta\dot{\psi}_y^0\dot{\psi}_{yn}\right)dxdy$$

Using the methodology developed by Kukreti and Issa [18] the pseudo-load vector {**P**+**G**} is interpolated by a quadratic time dependent polynomial, i.e.

$$\mathbf{P}(x,y,\tau)+\mathbf{G}(x,y,\tau)=\mathbf{A}(x,y)+\mathbf{B}(x,y)\tau+\mathbf{C}(x,y)\tau^2,\ 0\le\tau\le L_t \tag{14}$$

Where $L_t = t_{i+1}-t_i$ represents the time increment, and τ which is defined as $\tau = t-t_i$, identifies a new time origin for each time increment.

Denoting

$$\begin{aligned}&\mathbf{P}_0(x,y)=\mathbf{P}(x,y,0), \mathbf{P}_1(x,y)=\mathbf{P}(x,y,mL_t),\ \mathbf{P}_2(x,y)=\mathbf{P}(x,y,L_t),\\&\mathbf{G}_0(x,y)=\mathbf{G}(x,y,0),\ \mathbf{G}_1(x,y)=\mathbf{G}(x,y,mL_t),\ \mathbf{G}_2(x,y)=\mathbf{G}(x,y,L_t),\\&0<m<1,\ 0<x<a,\ 0<y<b\end{aligned} \tag{15}$$

the expressions for the constant vectors **A**, **B** and **C** are derived in terms of $\mathbf{P}_\mathrm{i}$ and $\mathbf{G}_\mathrm{i}$ (i = 1 to 3). The general solution of Eq. (11) is given by:

$$q_n(\tau)=E_{1n}q_n^0+E_{2n}\dot{q}_n^0+F_{1n}a_n+F_{2n}b_n+F_{3n}c_n \tag{16}$$

where E_{1n}, E_{2n}, F_{1n}, F_{2n}, F_{3n} denote complicated mathematical expressions containing ω_n, ξ_n and τ (see [19]) and

$$a_n=\iint \mathbf{U}_n^T\mathbf{A}_n dxdy,\ b_n=\iint \mathbf{U}_n^T\mathbf{B}_n dxdy,\ c_n=\iint \mathbf{U}_n^T\mathbf{C}_n dxdy \tag{17}$$

The iteration procedure applied to solve the above Eq. (11) is identical to the ones for circular plates and beams given in [21].

4 Damage Identification Technique

There are a lot of techniques to treat the nonlinear structural vibration response in the time domain. The state (phase)-space representation of the structural vibration response is a suitable and powerful tool for studying the dynamic behaviour of a structure. A standard technique for dealing with phase space $(w, \dot{w}, t)$ of periodically driven oscillators is to study the projection of $(w, \dot{w})$ at moments in time t, where t is a multiple of the period $T = 2\pi/\omega$. Here ω can be the frequency of the excitation of the mechanical system, an eigen frequency of the structure, or its multiple, and T is a period of the forcing function, an eigen period of the system, or its multiple. The result of inspecting the phase projection $(w, \dot{w})$ only at specific times $t = kT$ is a sequence of dots, representing the so-called Poincaré map. The steady-state converging trajectories, which represent the attractor, are usually formed in the phase space and in many cases of nonlinear systems they are very sensitive to any changes in the system.

In papers [11, 17] the following damage index based on the analysis of the Poincaré map was introduced:

$$I_i^d = \frac{S_i^u - S_i^d}{S_i^u}, \tag{18}$$

where

$$\begin{aligned} S_i^u &= \sum_{j=1}^{N_p-1} \sqrt{\left(w_{i,j+1}^u - w_{i,j}^u\right)^2 + \left(\dot{w}_{i,j+1}^u - \dot{w}_{i,j}^u\right)^2} \\ S_i^d &= \sum_{j=1}^{N_p-1} \sqrt{\left(w_{i,j+1}^d - w_{i,j}^d\right)^2 + \left(\dot{w}_{i,j+1}^d - \dot{w}_{i,j}^d\right)^2} \end{aligned} \tag{19a,b}$$

In these equations $I = 1,2\ldots N_{\text{nodes}}$, N_{node} is the number of nodes, N_p is the number of points in the Poincaré map and $\left(w_{ij}^u, \dot{w}_{ij}^u\right)$ and $\left(w_{ij}^d, \dot{w}_{ij}^d\right)$ denote the jth point on the Poincaré maps of the undamaged and the damaged states, respectively.

A small (close to 0) damage index will indicate no damage, while a big damage index will indicate the presence of a fault at the corresponding location. The above damage index depends on the location of the point on the plate, and consequently it is a function of the plate coordinates x and y. One can expect that the maxima of the surface $I^d(x_d, y_d)$ (18a) will represent the locations of the damage, i.e. $I_{\max}^d(x_d, y_d) = \max_i \left\{I_i^d\right\}$.

The damage criterion based on this index presumes setting a threshold value T^d for the damage index and if

$$I^d(x, y) > T^d \tag{20}$$

then one can conclude that the plate is damaged and the areas of points (x,y) for which Eq. (20) is fulfilled, form the damaged area (areas).

In the present work we shall use the same damage index and damage criterion but taking into account the temperature changes as well, $I^d = I^d(x, y, \Delta T)$. This suggestion presumes that the damage index defined by Eqs. (18) and (19) is calculated for equal values of ΔT for the healthy and damaged plate.

5 Results and Discussions

Numerical calculation of the vibrational displacements of the healthy and the damaged rectangular plates subjected to mechanical and thermal loading were performed.

The damage was modelled as a reduction (up to 50%) of the plate thickness in small parts of the plate.

First example considers the same plate as the one in [1]. The plate has the following dimensions and material properties: $a = 0.25$ m, $b = 0.24$ m, $h = 0.00027$ m, $E = 198.10^9$ Pa, $\rho = 7{,}850$ kg/m^3, $\nu = 0.3$ and $\alpha_T = 17.3\times10^{-6}$K^{-1}. This very thin plate is subjected to harmonic loading with frequency of excitation $\omega_h = 172$ rad/s $(0.7\omega_{1,1})$ and amplitude $p = 0.3$ N The time domain response of the plate center is shown in Fig. 2. The amplitudes of oscillations are very close to the ones shown in Fig. 9 in [1], so the verification of the present results is satisfactory.

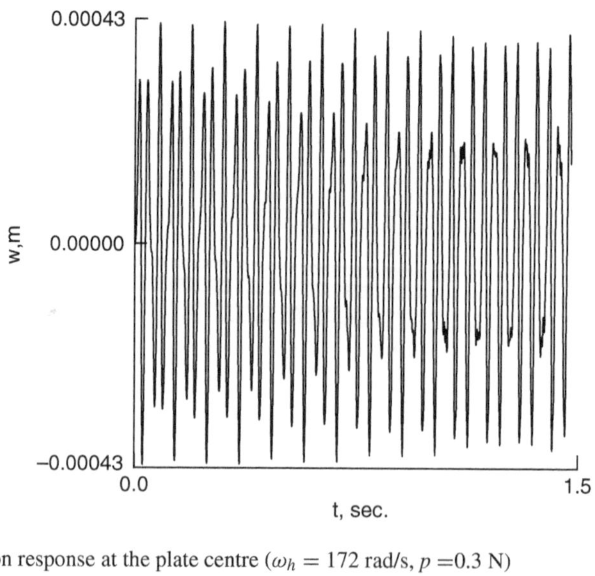

Fig. 2 Vibration response at the plate centre ($\omega_h = 172$ rad/s, p =0.3 N)

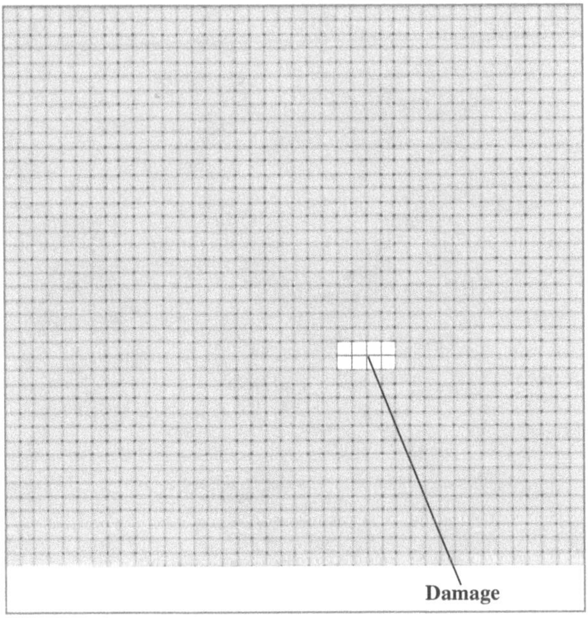

Fig. 3 Finite element discretization and damaged area (*white colour*) of the plate

Then the same plate but with increased thickness $h = 0.0005$ m (case B from [1]) was subjected to thermal and dynamic loading. For this plate two cases were considered: (1) undamaged plate and (2) plate with reduced thickness in a small part of the plate – the white area from the plate shown in Fig. 3.

It was shown in [1] that the buckling temperature for this plate is $\Delta T = 0.9$ K . It is clear that the attempt to inspect such a plate for damage without considering the temperature changes is condemned to fail.

In Fig. 4 the time-history diagrams of the healthy and the damaged plate subjected to a harmonic loading $p = 0.9$ N applied in the plate centre with frequency of excitation $\omega_h = 319$ rad/s. ($\omega_{1,1} = 455.6$ rad/s) are shown. Inspecting the time history it is visible that at the beginning the introduced small defect doesn't influence essentially the response of the plate but small changes in the eigen frequencies and modes lead to phase shift and the differences between the two responses increase with time. The phase shift can be clearly seen on the small figure in Fig. 4 where a short interval from the response is shown. The Poincaré maps of the responses of the healthy and the damaged plate in the plate centre (Fig. 5a) and in the centre of the defect are shown in (Fig. 5b), respectively. The Poincaré plots shown are obtained as a projection of $(w, \dot{w})$ at moments t, where t is a multiple of the period $T = 2\pi/\omega_h$. The damage doesn't change essentially the form of the Poincaré plot. As can be expected the difference between the two responses is larger at the points with reduced thickness. A contour plot of the damage index obtained by using Eq. (18) is plotted in Fig. 6 where a threshold value $T^d = 0.06$ is used. The contour plot is a graphical technique for representing a 3-dimensional surface by plotting

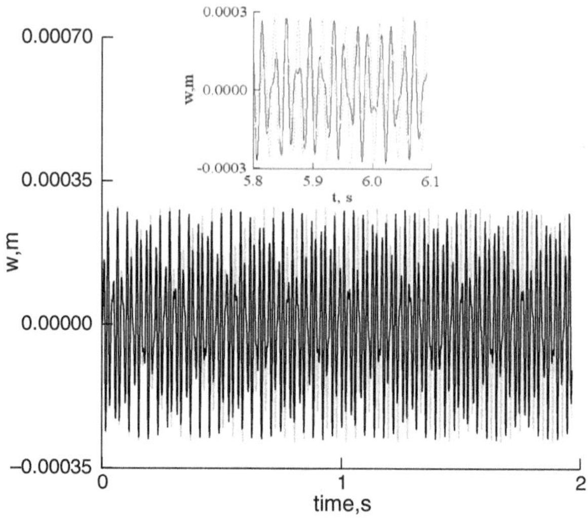

Fig. 4 Time history diagram of the plate centre response. $\omega_h = 319$ rad/s, $p = 0.9$ N. Undamaged plate (*gray line*); damaged plate (*gray line*)

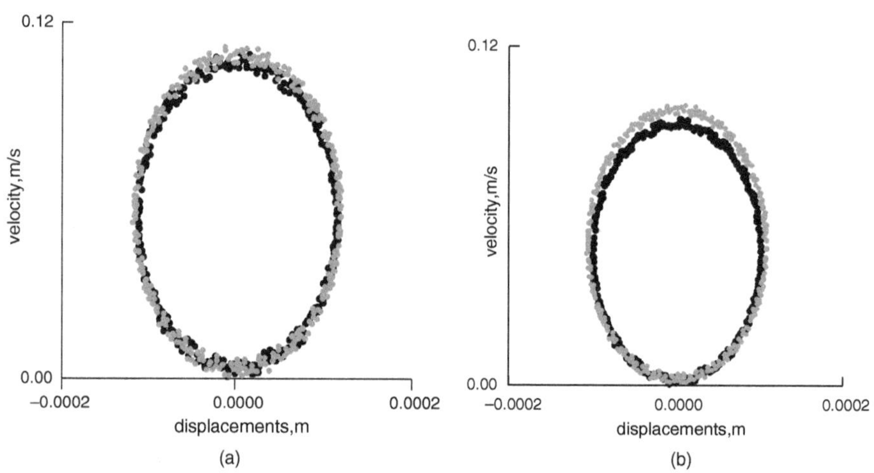

Fig. 5 (**a**) Poincaré map at the plate centre. Undamaged plate (*black dots*); damaged plate (*gray dots*). (**b**) Poincaré map at the centre of the defect. Undamaged plate (*black dots*); damaged plate (*gray dots*)

constant z slices, called contours, on a 2-dimensional plane. That is, for a given value of z, lines are drawn that connect the (x,y) coordinates which correspond to this particular value of z. The contour plot is compared to the FE model of the plate where the damaged area is coloured in white. As can be seen the damage criterion in this case works quite well and predicts rather precisely the damage location despite of the fact that the damage indexes have low values.

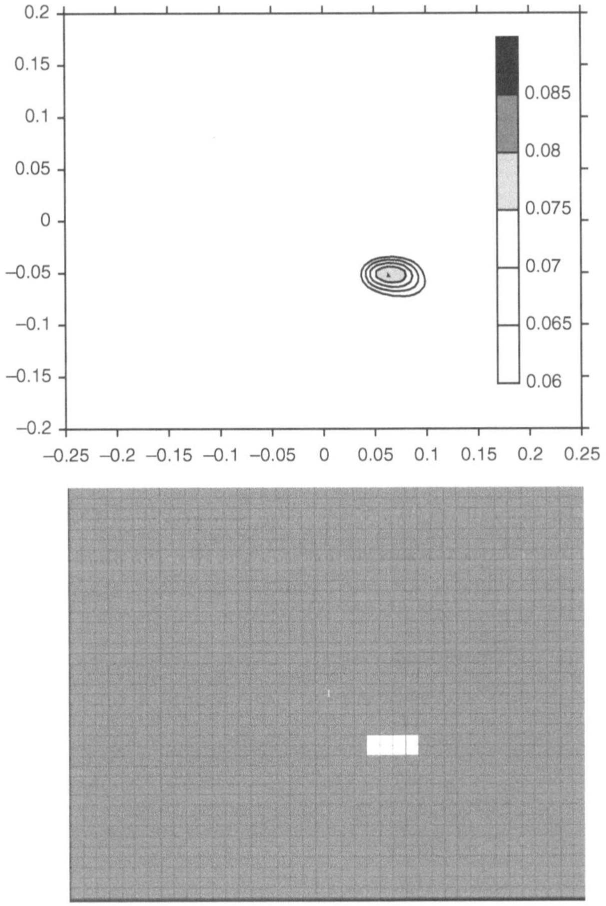

Fig. 6 Contour map of the damage index (unheated plate) and comparison with the damage location

Then the same plates were considered at elevated temperature namely $\Delta T = 0.7$ K. This temperature leads to increased amplitudes of vibrations of the plates (see Fig. 7). Again, the differences in the plate history diagrams are visible but they are not very large in the beginning of the time histories. However the Poincaré plots for the damaged and the undamaged plate have very different shapes, as can be seen from Fig. 8. This phenomenon may indicate that for these loading parameters the dynamic system changes its position in the basin of attractions moving from one region to another. This observation agrees with the fact that the plate buckles at $\Delta T = 0.9$ K [1]. The shapes of the Poincaré plots at the damaged nodes are similar. Obviously, in such case the damage criterion (20) is not appropriate and doesn't give satisfactory results for the damage location (not shown here). As can be expected neglecting the temperature influence is impossible for the damage detection purpose and leads to wrong results.

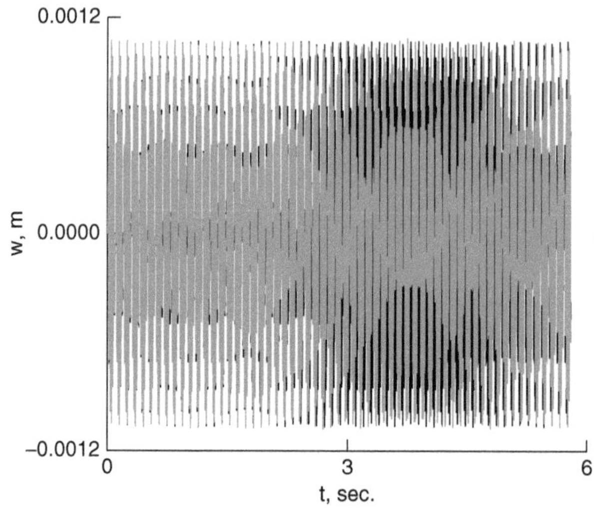

Fig. 7 Time history of the thermally loaded plate. Undamaged plate (*black line*); damaged plate (*gray line*). $\Delta T = 0.7$ K

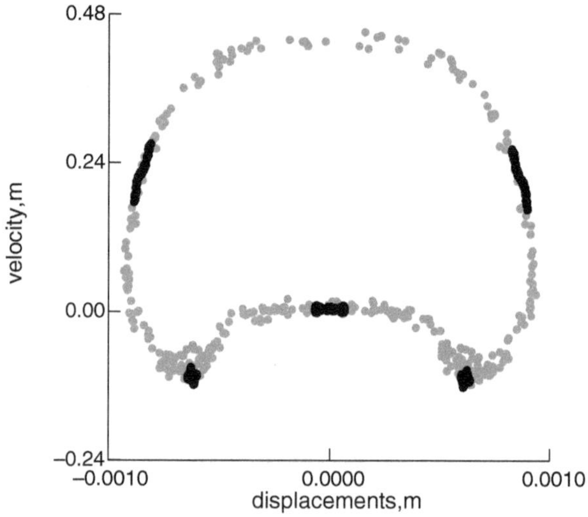

Fig. 8 Poincaré map of the response of the plate centre of heated undamaged (*black dots*) and damaged (*gray dots*) plates. $\Delta T = 0.7$ K

The second numerical example concerns a thicker rectangular plate with the following geometrical and material properties: $a = 10$ m, $b = 2.5$ m, $h = 0.05$ m, Young modulus $E = 7.10^{10}$ N/m^2, Poison ratio $\nu = 0.34$, density $\rho = 2{,}778$ kg/m^3. The damping coefficient $c_1 = c_2\frac{12}{h^2}$ in Eq. (8) was chosen to be 0.00075 $\frac{\mathrm{N\,s}}{\mathrm{m}^3}$. The

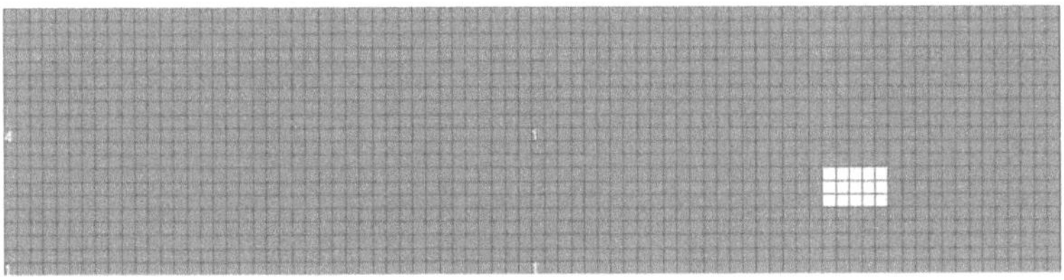

Fig. 9 Finite element mesh of the plate with defect

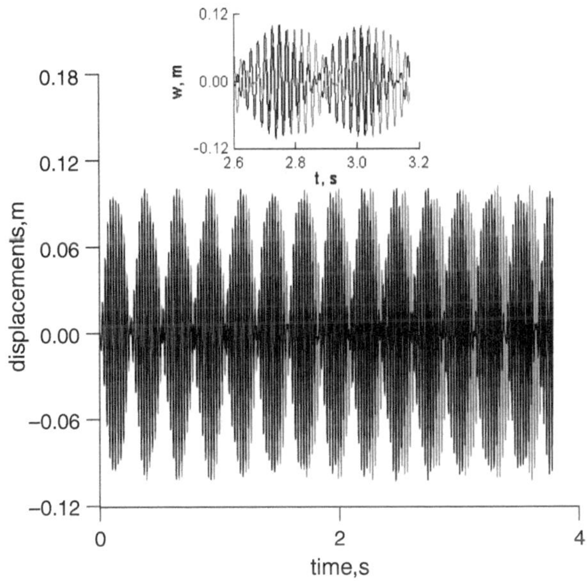

Fig. 10 Time history diagram of the plate centre, $p = 500$ N, $\omega_h = 260$ rad/s, black line – undamaged plate, gray line – damaged plate

finite element discretization and the damage area are shown in Fig. 9. Again, the damaged area has a thickness $h_{\text{damaged}} = h/2$. The plate is fully clamped and the applied harmonic load $p = 500$ N is uniformly distributed over the whole plate surface. The time history diagrams of the plate centre of the plate with a defect and without defect are shown in Fig. 10. The same time history diagrams but in the case of elevated temperatures of the plates are shown in Fig. 11. The excitation frequency is 260 rad/s, which is only 7% less than the first eigen frequency of the healthy plate. A strong beating can be observed in the responses of the healthy and damaged plates. The phase of the response of the damaged plate shifts and the difference between the responses increases with the time. The same conclusion applies in the case of the rectangular plate at elevated temperature. The

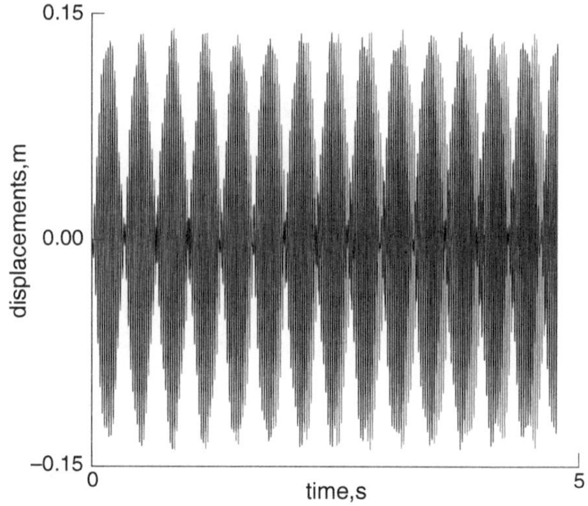

Fig. 11 Time history diagram of the plate center of heated plate, $p = 500$ N, $\omega_h = 260$ rad/s, $\Delta T = 50$ K, black line – undamaged plate, gray line – damaged plate

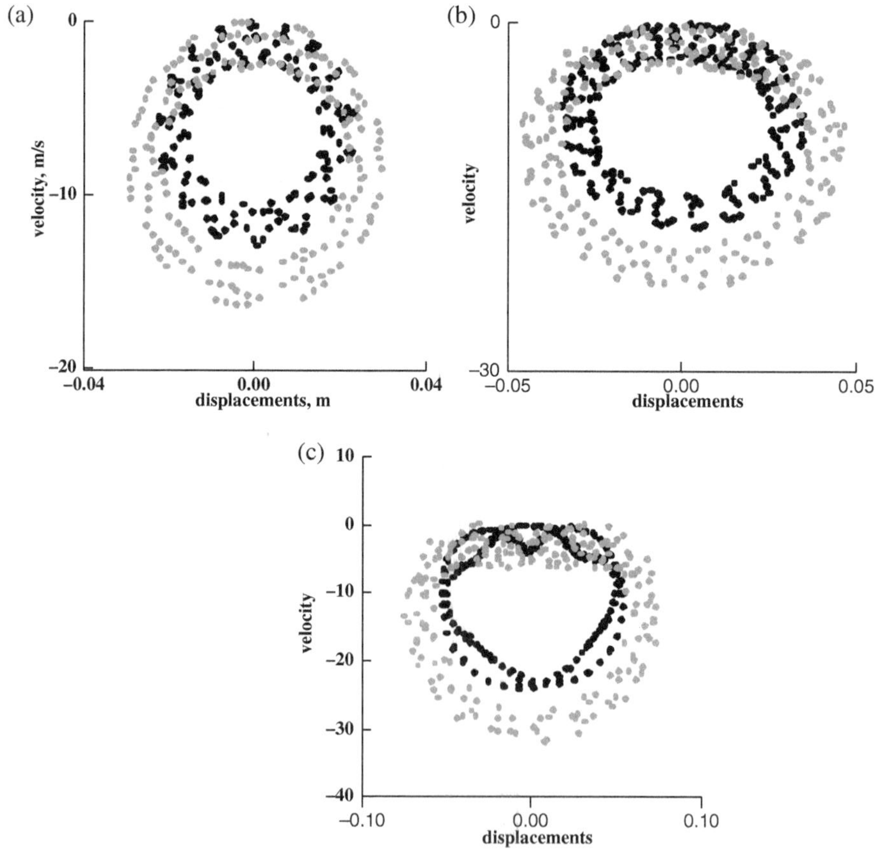

Fig. 12 Poincaré map at the centre of the defect for: (**a**) unheated plate, (**b**) heated plate $\Delta T = 50$ K, (**c**) heated plate $-\Delta T = 100$K. Undamaged plate (*black dots*); damaged plate (*gray dots*)

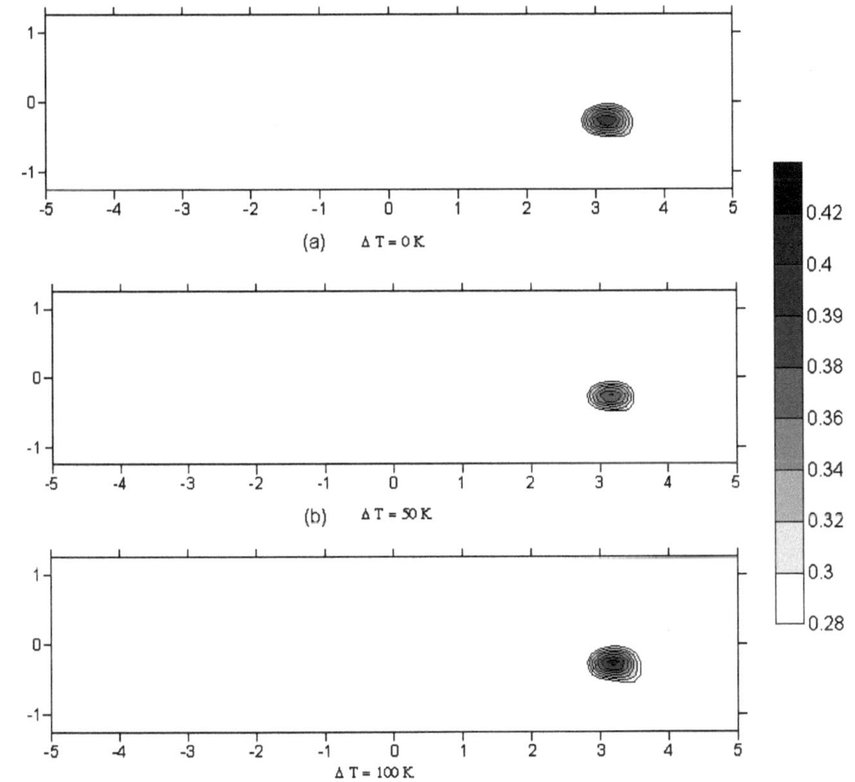

Fig. 13 Contour maps of the damage index for unheated and heated rectangular plate with damage

elevated temperature leads to larger values of the vibration amplitude. Again, the differences between the Poincaré plots of the heated and unheated plates are largest for the points from the damaged areas (see Fig. 12a–c). Accordingly, the damage indexes corresponding to the damaged area have the biggest values, which gives the possibility to locate the damage. The contour plots of I_i^d corresponding to three different temperatures are shown in Fig. 13. It can be seen that the damage location is predicted very precisely in the case of the unheated plate as well as in the cases of the heated plate with two different temperatures $\Delta T = 50$ K and $\Delta T = 100$ K. The threshold value T^d is set to 0.28 for all cases and the maximal value of I^d is almost the same ($I^d = 0.4$ for $\Delta T = 0$, $\Delta T = 50$ K and $I^d = 0.42$ for $\Delta T = 100$ K).

If, however one calculates, for example the damage index of the healthy unheated plate and the one for the damaged but heated plate then the damage location cannot be predicted precisely. This is due to the temperature change which is not taken into account for the healthy plate. The vibration responses of the healthy and the damaged plates should be compared for the same temperatures.

6 Conclusions

In this paper the computed time domain vibration responses are used to analyse the dynamic behaviour of plates in the intact condition and in the case when defects are present taking into account the temperature changes. A damage assessment method is suggested which is based on the phase space representation of the time domain nonlinear vibration response of the plate and uses the analysis of its Poincaré map. It has been demonstrated that damage as well as elevated temperatures can influence substantially the time domain response of the plate and its Poincaré maps. It can be concluded that: 1) The influence of the temperature changes is essential and can change substantially the nonlinear dynamic response of the plate and this is why temperature changes should be taken into account when developing a damage assessment procedure; 2) Temperature loadings which lead to either buckling or chaotic behaviour of the plate, might render the damage criterion suggested by Eqs. (18), (19) and (20) inappropriate. This is because even small damage, resulting in stiffness reduction of the plate, could lead to dramatic changes in the Poincaré maps of the response and consequently to unreliable results.

The potential, the sensitivity and the applicability of the developed method still have to be tested for real measurements and for more structures, defects and loading conditions.

Acknowledgments The first author wishes to thank the Bulgarian Research Fund for the partial support through grant TN-1518/2005. The authors thank to RSE for the partial support of this work.

References

1. M. Amabili, S. Carra, Thermal effects on geometrically nonlinear vibrations of rectangular plates with fixed edges. J. Sound Vib. **321**, 936–954 (2009)
2. E. Manoach, Dynamic large deflection analysis of elastic-plastic Mindlin circular plates. Int. J. Nonlinear Mech. **29**, 723–735 (1994)
3. E. Parloo, P. Verboven, P. Guillaume, M. van Overmeire, Autonomous structural health monitoring – Part II: Vibration-based in-operation damage assessment. Mech. Syst. Signal Process. **16**, 659–675 (2002)
4. P.F. Rizos, N. Aspragathos, A.D. Dimarogonas, Identification of crack location and magnitude in a cantilevered beam from the vibration modes. J. Sound Vib. **138**, 381–388 (1990)
5. I. Trendafilova, E. Manoach, M.P. Cartmell, W. Ostachowicz, A. Zak, An investigation on damage detection in aircrafts panels using nonlinear time series analysis. Key Eng. Mater. **347**, 213–218 (2007)
6. I. Trendafilova, E. Manoach, Vibration based damage detection in plates by using time series analysis. Mech. Syst. Signal Process. **22**, 1092–1106 (2008)
7. P. Verboven, E. Parloo, P. Guillaume, M. van Overmeire, Autonomous structural health monitoring – Part I: Modal parameter estimation and tracking. Mech. Syst. Signal Process. **16**, 637–657 (2002)
8. Y. Zou, L. Tong, G.P. Steven, Vibration based model-dependent damage (delamination) identification and health monitoring for composite structures – A review. J. Sound Vib. **230**, 357–378 (2000)
9. H.T. Banks, D.J. Inman, D.J. Leo, Y. Wang, An experimentally validated damage detection theory in smart structures. J. Sound Vib. **191**, 859–880 (1996)

10. L. Moniz, J.M. Nichols, C.J. Nichols, M. Seaver, S.T. Trickey, M.D. Todd, L.M. Pecora, L.N. Virgin, A multivariate, attractor-based approach to structural health monitoring. J. Sound Vib. **283**, 295–310 (2005)
11. P. Ribeiro, E. Manoach, The effect of temperature on the large amplitude vibrations of curved beams. J. Sound Vib. **285**, 1093–1107 (2005)
12. E. Manoach, in *Dynamic Large Deflection Analysis of Elastic-Plastic Beams and Plates*, ed. by N.S. Ferguson, H.F. Wolfe, M.A. Ferman, S.A. Rizzi. Proceedings of 7th International Conference Recent Advances on Structural Dynamics, vol. 1 (The Institute of Sound and Vibration Research University, Southampton, 2000), pp. 389–400
13. J. Cattarius, D.J. Inman, Time domain analysis for damage detection in smart structures. Mech. Syst. Signal Process. **11**, 409–423 (1997)
14. B.I. Epureanu, L.S. Tang, M.P. Päidoussis, Exploiting chaotic dynamics for detecting parametric variations in aeroelastic systems. AIAA J. **42**, 728–735 (2004)
15. E. Manoach, in *Coupled, Large Amplitude Vibrations of Circular Plates Subjected to Mechanical and Thermal Loading*. ed. by D.H. van Campen, M.D. Lazurko, W.P.J.M. van den Oever. Proceedings of ENOC-2005, Fifth EUROMECH Nonlinear Dynamic Conference, Eindhoven, The Netherlands (2005), pp. 2548–2557
16. A.R. Kukreti, H.I. Issa, Dynamic analysis of nonlinear structures by pseudo-normal mode superposition method. Comput. Struct. **19**, 653–663 (1984)
17. P. Ribeiro, Thermally induced transitions to chaos in plate vibrations. J. Sound Vib. **299**, 314–330 (2007)
18. Thorton, E.A, *Thermal Structures for Aerospace Applications*, AIAA Education Series (AIAA: Washington, DC, 1996)
19. M. Todd, J.M. Nichols, L.M. Pecora, L. Virgin, Vibration-based damage assessment utilizing state space geometry changes: Local attractor variance ratio. Smart Mater. Struct. **10**, 1000–1008 (2001)
20. E. Manoach, P. Ribeiro, Coupled, thermoelastic, large amplitude vibrations of Timoshenko beams. Int. J. Mech. Sci. **46**, 1589–1606 (2004)
21. E. Manoach, I. Trendafilova, Large amplitude vibrations and damage detection of rectangular plates. J. Sound Vib. **315**, 591–606 (2008)

Macroscopic Stability Analysis in Periodic Composite Solids

Fabrizio Greco, Paolo Lonetti, Paolo Nevone Blasi, and Girolamo Sgambitterra

1 Introduction

Solids with heterogeneous microstructures, such as cellular solids, particle and fiber-reinforced materials, layered materials and honeycombs, are increasingly adopted in many engineering applications since their microstructure can be designed to optimize their macroscopic properties according to the specific application. Consequently, it is extremely important to accurately predict the macroscopic material properties of such materials in terms of the microscopic behavior of its constituents. However the analysis of the macroscopic response of heterogeneous materials taking into account a precise description of its microstructure, may involve an enormous increase of computational cost, since constitutive properties of the material change within a micro-length scale which is several orders of magnitude smaller than the characteristic dimensions of the structure.

A common approach is to replace the heterogeneous material by an equivalent "homogeneous" material. The methodologies used to calculate these macroscopic properties are usually termed "homogenization" techniques.

For linear elastic solids, after some pioneering contributions [1, 2] rigorous mathematical approaches have been developed on the basis of the mathematical procedure of multi-scale perturbation assuming a periodic model for the microstructure [3, 4].

In the case of non-linear heterogeneous solids additional complications arise since nonlinear effects related to the microgeometry and the local constitutive law must be incorporated in the mathematical model. As a matter of fact the macroscopic behavior of micro heterogeneous material at finite strains, obtained by means of a classical homogenization procedure, may be often not representative of the microscopic behavior of its constituents due possible instability phenomena occurring at the micro scale. From the mathematical point of view the main difficulty arising in the homogenization of nonlinear elastic composites is related to the non-convexity

F. Greco (✉)
Department of Structural Engineering, UNICAL, Cosenza, Italy
e-mail: f.greco@unical.it

A. Öchsner et al. (eds.), *Materials with Complex Behaviour*, Advanced Structured Materials 6, DOI 10.1007/978-3-642-12667-3_14, © Springer-Verlag Berlin Heidelberg 2010

of the microscopic strain energy density function. In the case of convex microscopic strain energy functions the homogenization problem can be successfully solved by minimizing the averaged strain energy density with respect to fluctuation field periodic over a unit cell [5], leading to define the representative volume of the microstructure as one periodic cell. Unfortunately actual materials do not have convex energy functions and convexity is a too strong requirement from the physical point of view [6, 7]). For non-convex microscopic strain energy functions, by virtue of the notion of Γ-convergence [8, 9] has given an abstract formula of the homogenized strain energy density function for heterogeneous periodic microstructures which corresponds to the minimization of the averaged strain energy density with respect to admissible fluctuation fields that are periodic over an a-priori unknown ensemble of periodic cells (possibly infinite).

On the other hand, the stability analysis plays a fundamental role in the study of solids with heterogeneous microstructure, since microscopic failure mechanisms in these materials are often induced by instability phenomena and in view of the fact that the stability analysis of the microstructured solid establishes the region of validity of the standard homogenization procedure based on unit cell calculations. In fact, the limit of validity for the homogenized models of the heterogeneous solid, can be determined only by means of comparisons between the onset of the primary instability in the real microstructured solid and the corresponding instability estimated by using the homogenized model of the solid.

For solids with generic microstructures, such as composite materials, the analysis must consider both classical buckling type instability modes dominated by the geometric microstructural configuration when the stress state is prevalently negative, and constitutive-dominated instabilities occurring when tangent moduli of the material reduce greatly taking eventually negative values and in presence of a positive stress state (for additional details about the above kind of instabilities see [10]). Instabilities of the former kind arise frequently in laminated microstructures loaded primarily in compression due to fiber micro-buckling [11, 12], whereas cellular and particle-reinforced microstructures may exhibit the latter kind of instability when loaded prevalently in tension [13].

In the sake of computational efficiency, the stability analysis of elastic composite solids with periodic microstructure is carried out in terms of their macroscopic properties. However, a stability analysis based on the homogenized constitutive properties may be not able to provide an accurate prediction of microscopic instability mechanisms and often a direct analysis of the heterogeneous solid, including all microstructural details and thus involving a greater computational cost, must be carried out in order to determine the exact microstructural instability mechanism. Consequently, an appropriate analysis of the interrelations between instabilities on the macro and micro scales plays an essential role to validate a stability investigation based on the homogenized composite properties.

By using the results of [9, 14] proved the connection between microscopic bifurcation and loss of macroscopic strict rank-one convexity in the framework of functional analysis, for arbitrary solids with periodic microstructures. In this work paper it also was proven that if the wavelength of the bifurcation primary eigenmode is much larger compared to the unit cell size, the onset of the corresponding instability of the periodic principal solution can be detected as a loss of ellipticity of the corresponding one-cell homogenized tangent moduli of the solid.

From the computational point of view, accurate numerical determinations of the region of microscopic stability and of the region of macroscopic stability, intended as the region where the strong ellipticity condition for the homogenized moduli tensor still holds, for a microstructural model have been carried out in [13, 15].

From the above literature survey it emerges that a fundamental measure of macroscopic stability, based on homogenized constitutive properties, for an heterogeneous solid with periodic microstructure is that based on the strong ellipticity condition of the homogenized moduli tensor. The strong ellipticity condition is able to exactly predict the onset of microscopic instability of the periodic principal solution along a monotonic loading process when the microscopic instability mode is global in nature, i.e. its wavelength is much larger in comparison with the unit cell size (this circumstance occurs frequently, for instance, in fiber reinforced composite materials loaded prevalently in compression). On the other hand, an unconservative estimation of the primary microscopic instability load is obtained by the above mentioned macroscopic stability measure in the more general case when the instability mode is local in nature, namely its wavelength is comparable to the unit cell size, since the homogenized moduli tensor remains strongly elliptic at the onset of the primary microscopic instability. The latter kind of instabilities may occur, for instance, in cellular solids or in particle reinforced matrix materials.

In order to investigate alternative macroscopic conditions able to obtain accurate prediction of the microscopic instability mechanisms in composite solids with periodic microstructure, a stability analysis on the micro and macro scales is here carried out, both from a theoretical and numerical point of views. Firstly, theoretical details about the homogenization problem and the stability conditions for an heterogeneous solid with periodic microstructure undergoing deformations at finite strains are provided. Then alternative macroscopic stability measures are defined. After the description of the numerical implementation of the proposed method, numerical applications, devoted to cellular and particle-reinforced composite microstructures with specific hyperelastic constituents, are provided.

2 Theoretical Analysis

In order to study the stability problem of heterogeneous solids with periodic microstructure subjected to finite-strains loading conditions, the basic concepts of the homogenization procedure will be firstly introduced in this section and the macroscopic properties of a periodic microstructure will be discussed. Then the concept of macroscopic stability measures will be introduced. The microscopic stability problem is introduced which requires the examination of all perturbations of the equilibrium fluctuation field periodic over a unit cell.

2.1 Basic Equations of Homogenization: Microscopic and Macroscopic Variables

Let us consider an heterogeneous solid with a periodic microstructure defined by a unit cell occupying the domain V_i in the stress-free undeformed configuration (see Fig. 1). According to a classical assumption of the homogenization approach, the scale at which the microstructure is defined (microscopic scale, l_{micro}) is small enough for the heterogeneities to be identified. On the other hand, the homogenized continuum is defined at a macroscopic scale, l_{macro}, large enough compared to its microscopic counterpart for the heterogeneities to be 'smeared-out'. The initial and current position vectors of a material point of the microstructure are denoted by $\boldsymbol{X}$ and $\boldsymbol{x}$. The nonlinear deformation of the micro-structure is defined by $\boldsymbol{x}$ $(\boldsymbol{X})$: $V_{(i)} \to V$, mapping points $\boldsymbol{X}$ of the initial configuration $V_{(i)}$ onto points $\boldsymbol{x}$ of the actual configuration V of the micro-structure. The displacement field at $\boldsymbol{X}$ is denoted by $\boldsymbol{u}(\boldsymbol{X})$, where $\boldsymbol{u} = \boldsymbol{x}(\boldsymbol{X}) - \boldsymbol{X}$, and the deformation gradient at $\boldsymbol{X}$ is denoted as $\boldsymbol{F}(\boldsymbol{X})$, where $\boldsymbol{F} = \partial \boldsymbol{x}(\boldsymbol{X})/\partial \boldsymbol{X}$.

The material model that governs the response at a microscopic point $\boldsymbol{x}$, is assumed to be rate independent and is specified by the following incrementally linear constitutive law:

$$\dot{\boldsymbol{T}}_R = \boldsymbol{C}^R\,(F, X)\,[\dot{\boldsymbol{F}}] \tag{1}$$

where $\dot{\boldsymbol{T}}_R$ is the rate of the first Piola-Kirchhoff stress tensor, $\dot{\boldsymbol{F}}$ is the deformation gradient rate and $\boldsymbol{C}^R$ is the corresponding fourth-order tensor of nominal moduli, a $V_{(i)}$-periodic function of $\boldsymbol{X}$. Since only quasistatic loading conditions will be considered here, the rate of a field quantity correspond to its derivative with respect to a time-like parameter which increases monotonically with the evolution of the loading process. It is further assumed that the nominal moduli tensor possesses major symmetry condition, i.e. $C^R{}_{0ijkl} = C^R{}_{0klij}$. When the microscopic constitutive behavior is hyperelastic, the nominal moduli tensor and the nominal stress tensor can be defined

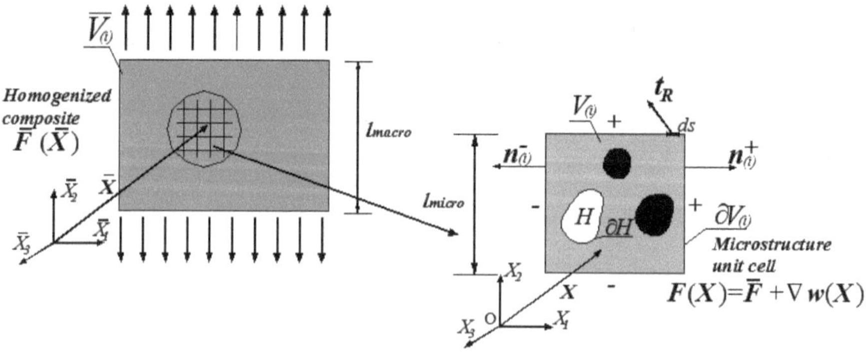

Fig. 1 Periodic unit cell attached to a generic material point $\bar{X}$ of the corresponding homogenized continuum, occupying the region $\bar{V}_{(i)}$ in its undeformed configuration

as $\boldsymbol{C}^R(\boldsymbol{X},\boldsymbol{F}) = \frac{\partial^2 W(\boldsymbol{X},\boldsymbol{F})}{\partial \boldsymbol{F} \partial \boldsymbol{F}}$, $\boldsymbol{T}_R = \frac{\partial W(\boldsymbol{X},\boldsymbol{F})}{\partial \boldsymbol{F}}$, where W is the strain energy density function that is a non-convex function of the deformation gradient F, $\partial^2 W(\boldsymbol{X},\boldsymbol{F})/\partial \boldsymbol{F} \partial \boldsymbol{F}$ denotes a second order tensor whose components are $C^R_{ijkl} = \partial^2 W(\boldsymbol{X},\boldsymbol{F})/\partial F_{ij} \partial F_{hk}$ and $\partial W(\boldsymbol{X},\boldsymbol{F})/\partial \boldsymbol{F}$ a first order tensor whose components are $T^R_{ij} = \partial W(\boldsymbol{X},\boldsymbol{F})/\partial F_{ij}$.

The macroscopic first Piola-Kirchoff stress and the macroscopic deformation gradient tensors are defined in terms of boundary data of the traction field $\boldsymbol{t}_{\mathrm{R}}$ and the deformation field $\boldsymbol{x}(\boldsymbol{X})$ according to [16], respectively as:

$$\bar{\boldsymbol{T}}_R = \frac{1}{|V_{(i)}|} \int_{\partial V_{(i)}} \boldsymbol{t}_R(\boldsymbol{X}) \otimes \boldsymbol{X} ds_{(i)}, \quad \bar{\boldsymbol{F}} = \frac{1}{|V_{(i)}|} \int_{\partial V_{(i)}} \boldsymbol{x}(\boldsymbol{X}) \otimes \boldsymbol{n}_{(i)} ds_{(i)}, \tag{2}$$

where $\otimes$ is the tensor product and $\boldsymbol{n}_{(i)}$ denotes the outward normal at $\boldsymbol{X} \in \partial V_{(i)}$.

The microscopic deformation can be assumed to be a function of the macro-deformation gradient $\bar{\boldsymbol{F}}$ and can be expressed as the sum of a linear part, representing a homogeneous deformation, and of a correction part $\boldsymbol{w}(\boldsymbol{x})$ associated to a non-homogeneous deformation. As a consequence the microscopic deformation field and its gradient assume the following expressions

$$\boldsymbol{x}(\boldsymbol{X}) = \bar{\boldsymbol{F}}\boldsymbol{X} + \boldsymbol{w}(\boldsymbol{X}), \; \boldsymbol{F}(\boldsymbol{X}) = \bar{\boldsymbol{F}} + \nabla \boldsymbol{w}(\boldsymbol{X}) \tag{3}$$

where $\boldsymbol{w}(\boldsymbol{X})$ is usually referred to as the fluctuation field.

Application of Eq. $(2)_1$ implies that the fluctuation field must comply with the following constraint:

$$\int_{\partial V_{(i)}} \boldsymbol{w} \otimes \boldsymbol{n}_{(i)} ds_{(i)} = \boldsymbol{0}, \tag{4}$$

which can satisfied for periodic fluctuation fields on the unit cell boundary (see Fig. 2):

$$\boldsymbol{w}(\boldsymbol{X}^+) = \boldsymbol{w}(\boldsymbol{X}^-) \; on \; \partial V_{(i)}. \tag{5}$$

By periodicity of the field $\boldsymbol{w}(\boldsymbol{X})$ it is understood that all components of $\boldsymbol{w}(\boldsymbol{X})$ take identical values at points on opposite sides of the boundary $\partial V_{(i)}$, $\partial V_{(i)}{}^+$ and

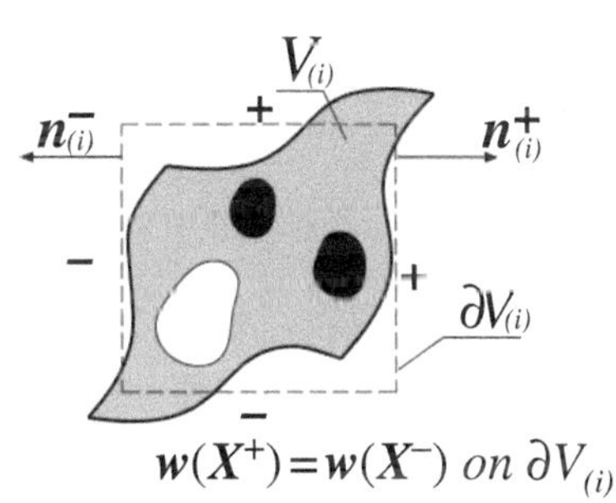

Fig. 2 Periodic boundary constraints for the fluctuation field

$\partial V_{(i)}^{-}$, with outwards normals $\boldsymbol{n}_{(i)}^{+} = -\boldsymbol{n}_{(i)}^{-}$ at two associated points $\boldsymbol{X}^{+} \in \partial V_{(i)}^{+}$ and $\boldsymbol{X}^{-} \in \partial V_{(i)}^{-}$, which are deduced by translation parallel to the directions of the periodicity vectors spanning $V_{(i)}$. Due to the assumed $V_{(i)}$–periodic distribution of material and geometrical properties of the heterogeneous solid with respect to a unit cell, the local boundary conditions Eq. (2) allows to completely determine its mechanical response and ensure a periodic distribution of the stress and strain field quantities. As will be detailed in the following section, the mechanical response of the heterogeneous solid at a given macro-deformation gradient, can be determined by means of computations performed over one unit cell if the equilibrium solution of the microstructure is stable, otherwise an assembly of unit cells must be considered.

The abbreviated notation # appended to a region will be used to denote the periodic properties of a field on the boundary of the region. Similarly, the antiperiodicity of a field quantity, implying that the field quantity takes opposite values at points on opposite sides of the boundary $\partial V_{(i)}$ of the unit cell, will be denoted by –#.

2.2 Incremental Macroscopic Response

In order to obtain the incremental homogenized response of the solid, suppose that the microstructure at a generic stage of a quasi-static loading path $\bar{\boldsymbol{F}}(\beta)$ (with the load parameter $\beta \geq 0$ increasing monotonically with increasing macroscopic load) beginning from the initial configuration associated to the region $V_{(i)}$ (i.e. with $\bar{\boldsymbol{F}}(\beta) = 1$ when $\beta = 0$) occupies the region V defined by the deformation $\boldsymbol{x}(\boldsymbol{X}) = \bar{\boldsymbol{F}}(\beta)\boldsymbol{X} + \boldsymbol{w}_{\bar{\boldsymbol{F}}(\beta)}(\boldsymbol{X})$ driven by the macroscopic load $\bar{\boldsymbol{F}}(\beta)$. The deformed configuration is assumed to correspond to a known equilibrium solution, defined in term of the fluctuation solution $\boldsymbol{w}_{\bar{\boldsymbol{F}}}(\boldsymbol{X})$ at the given macro-deformation gradient, of the unit-cell deformation problem:

$$\int_{V_{(i)}} \boldsymbol{T}_R\left(\bar{\boldsymbol{F}} + \nabla \boldsymbol{w}_{\bar{\boldsymbol{F}}}, \boldsymbol{X}\right) \cdot \nabla \delta \boldsymbol{w} dV_{(i)} = 0 \quad \forall \delta \boldsymbol{w} \in H^{1,p}(V_{(i)\#}), \tag{6}$$

where $H^{1,p}(V_{(i)\#})$ denotes the usual Sobolev space of vector valued functions periodic over the unit cell $V_{(i)}=[0,1]^{N}$. The corresponding Euler-Lagrange equations of the above variational equation subjected to periodic constraints are:

$$\begin{cases} Div\boldsymbol{T}_R = \boldsymbol{0} \ in \ V_{(i)} \\ \left(\boldsymbol{T}_R\boldsymbol{n}_{(i)}\right)^{+} = \left(\boldsymbol{T}_R\boldsymbol{n}_{(i)}\right)^{-} \ on \ \partial V_{(i)}, \\ \boldsymbol{T}_R\boldsymbol{n}_{(i)} = \boldsymbol{0} \ on \ \partial H_{(i)} \end{cases} \tag{7}$$

where $()^{+}$ and $()^{-}$, respectively, denote variables evaluated at two associated points $\boldsymbol{X}^{+} \in \partial V^{+}$ and $\boldsymbol{X}^{-} \in \partial V^{-}$, and $\partial H_{(i)}$ denotes the boundary of the eventual hole part of the unit cell. When the equilibrium solution for the unit-cell deformation problem is unique along the macroscopic loading path, it is referred to as the "principal solution path". It is worth noting that the equilibrium solution is determined by Eq. (7) up to a possible rigid body motions which must be excluded by imposing artificial constraints.

Let us consider the incremental equilibrium problem induced at the microscopic scale by an incremental change in the macroscopic deformation gradient $\dot{\bar{F}}(\beta)$. Once the microscopic distribution of the nominal moduli tensor $C^R\left(\bar{F}+\nabla w_{\bar{F}}, X\right)$ is known, the local problem defined by the following variational expression:

$$\int_{V_{(i)}} C^R\left(\bar{F}, X\right)[\dot{\bar{F}}+\nabla \dot{w}_{\dot{\bar{F}}}]\cdot\nabla\delta\dot{w}dV_{(i)}=0 \quad \forall\delta\dot{w}\in H^{1,p}(V_{(i)\#}), \tag{8}$$

can be solved, where $\dot{w}_{\dot{\bar{F}}}$ is the incremental fluctuation field induced by $\dot{\bar{F}}(\beta)$ determining the microstructure incremental equilibrium solution with antiperiodic incremental tractions $\dot{t}_R \in V_{(i)-\#}$, i.e. $\dot{t}_R\left(X^+\right)=-\dot{t}_R\left(X^-\right)$ on $\partial V_{(i)}$, vanishing tractions $\dot{t}_R\left(X\right)=0$ on the boundary $\partial H_{(i)}$of eventual hole parts, and periodic boundary constraints, $\dot{w}\left(X\right)\in V_{(i)\#}$. Incremental rigid body displacements of the unit cell must be avoided by introducing artificial constraints.

When the local incremental problem solution is known, the macroscopic constitutive response can be determined as

$$\dot{\bar{T}}_R=\bar{C}^R\left(\bar{F}\right)[\dot{\bar{F}}], \tag{9}$$

where $\bar{C}^R\left(\bar{F}\right)$ is by definition the homogenized tangent moduli tensor. Considering that the increment of the macroscopic stress tensor is equal to the macroscopic incremental stress tensor, i.e. $\dot{\bar{T}}_R=\dot{\bar{T}}_R$ and by virtue of Eq. (9), we obtain that the homogenized tangent modulus tensor of the solid can be determined through the following relation:

$$\bar{C}^R\left(\bar{F}\right)[\dot{\bar{F}}]=\frac{1}{\left|V_{(i)}\right|}\int_{V_{(i)}} C^R\left(\bar{F}, X\right)[\dot{\bar{F}}+\nabla \dot{w}_{\dot{\bar{F}}}]dV_{(i)}. \tag{10}$$

From Eq. (10) the components of the homogenized moduli tensor can be determined as

$$\bar{C}^R_{ijhk}\left(\bar{F}\right)=\frac{1}{\left|V_{(i)}\right|}\int_{V_{(i)}} C^R_{ijmn}\left(\bar{F}, X\right)[I^{hk}_{mn}+\nabla\dot{w}_{hk}]dV_{(i)}, \tag{11}$$

in terms of the incremental fluctuation field induced by unit value for each component of the macroscopic deformation increment, namely $\dot{w}_{hk}$ is the solution of the incremental boundary value problem for $\dot{\bar{F}}=I^{hk}$, where $I^{hk}_{mn}=\delta_{mh}\delta_{nk}$.

It is worth noting that the above definition of the homogenized moduli tensor based on computations over one unit cell is strictly valid only when the equilibrium configuration of the microstructure is incrementally stable, otherwise the one cell homogenization is no longer useful since a larger assembly of unit cells (possibly infinite) must be considered.

2.3 Variational Formulation for Hyperelastic Materials

For hyperelastic microscopic constituents, an homogenized strain energy function can be introduced by means of the following expression [9]:

$$\bar{W}\left(\bar{\boldsymbol{F}}\right) = \inf_{k\in} \left\{ \min_{w\in H^{1,p}(k^N V_i\#)} \left\{ \frac{1}{k^N \left|V_{(i)}\right|} \int_{k^N V_{(i)}} W\left(\boldsymbol{X}, \bar{\boldsymbol{F}} + \nabla \boldsymbol{w}\right) dV_{(i)} \right\} \right\}, \tag{12}$$

where the homogenized strain energy function $\bar{W}\left(\bar{\boldsymbol{F}}\right)$ is determined as the minimum volume average of the microscopic strain energy function with respect to admissible fluctuation fields, belonging to the Sobolev space of vector valued functions periodic over all possible ensemble of $\mathrm{k}^\mathrm{N} = [0,\mathrm{k}]^\mathrm{N}$ unit cells ($\mathrm{N} = 2$ or $\mathrm{N} = 3$ for two- or three-dimensional problems, respectively) with k an arbitrary integer, namely $\boldsymbol{w}(\boldsymbol{X}) \in H^{1,p}(\mathrm{k}^\mathrm{N} V_{(i)}\#)$. The strain energy function W is assumed to be objective, i.e. $W(\boldsymbol{QF}) = W(\boldsymbol{F})$ for all proper orthogonal $\boldsymbol{Q}$ and arbitrary deformation gradients $\boldsymbol{F}$. This implies that the homogenized strain energy function, as its microscopic counterpart, is unaffected by a superposed macroscopic rigid body motion after deformation.

A formal calculation of the first and second derivatives of the homogenized strain energy function based on Eq. (12), which makes use of Eq. (6), shows that Eq. (12) defines a macro-stress potential and the macro-stress and homogenized moduli tensor are defined in terms of the first and second derivatives of the macro-stress potential with respect to the macro-deformation gradient:

$$\bar{T}_R = \frac{\partial \bar{W}}{\partial \bar{\boldsymbol{F}}}, \quad \bar{\boldsymbol{C}}^R = \frac{\partial^2 \bar{W}}{\partial \bar{\boldsymbol{F}} \partial \bar{\boldsymbol{F}}}. \tag{13}$$

When the microscopic strain energy function $W(\boldsymbol{F})$ is a convex function of $\boldsymbol{F}$, the computation of the macroscopic energy function can be reduced to a computation on a unit cell, and the one-cell homogenized strain energy function can be defined by the following minimization problem:

$$\bar{W}^1\left(\bar{\boldsymbol{F}}\right) = \min_{w\in H^{1,p}(V_i\#)} \left\{ \frac{1}{\left|V_{(i)}\right|} \int_{V_{(i)}} W\left(\boldsymbol{X}, \bar{\boldsymbol{F}} + \nabla w\right) dV_{(i)} \right\}, \tag{14}$$

namely $\bar{W}\left(\bar{\boldsymbol{F}}\right) = \bar{W}^1\left(\bar{\boldsymbol{F}}\right)$.

The minimization principles Eqs. (12) and (14) defines the equilibrium state of the hyperelastic microstructure according to Eq. (7) except rigid body motions which must be opportunely excluded. As a matter of fact, the minimum principles Eqs. (12) and (14) imply the variational Eq. (6) written with reference to an assembly of unit cells or to a unit cell region, respectively.

Practical materials are not characterized by convex strain energy functions since convexity is a too strong and physically unacceptable restriction [6, 7]. On the contrary non convex micro-energy functions do not preclude nonuniqueness phenomena such as buckling on the micro-scale. Due to these phenomena it is possible that lower values for the homogenized strain energy function can be obtained by minimization over domains containing several unit cells. In this circumstance, Eq. (12) determines the current fluctuation field and defines the size of the representative volume of the microstructure, which is a priori unknown, which captures the minimizing micro-buckling mode. Moreover it results that $\bar{W}\left(\bar{\boldsymbol{F}}\right) \leq \bar{W}^1\left(\bar{\boldsymbol{F}}\right)$ and the equality holds only when the minimizing fluctuation field based on one unit cell computation is also the minimizing fluctuation field for any possible unit cells assembly. Equation (12) is accompained by a notable difficulty associated to the infinity of the required domain and requires a full space investigation on the micro-scale. Therefore, it should be preferable to perform the homogenization on the basis of Eq. (14) since it involves a much simpler calculation, which gives the correct results in terms of macroscopic stress and homogenized moduli through Eq. (13) only in the region of the macroscopic strain space where $\bar{W}\left(\bar{\boldsymbol{F}}\right) = \bar{W}^1\left(\bar{\boldsymbol{F}}\right)$, namely the region of validity of the one-cell homogenization. As will be shown in the following the region of validity of the one-cell homogenization can be determined by means of a microscopic stability analysis.

2.4 Stability Analysis of the Microstructure

The response of the microstructure to infinitesimal deformations from the current equilibrium configuration $\mathrm{k}^{\mathrm{N}}V$, taken as reference and assumed known, at the macroscopic strain $\bar{\boldsymbol{F}}$ and the corresponding fluctuation solution $\mathrm{w}_{\bar{\mathrm{F}}}$, is analyzed. To this end an additional microscopic displacement field $\boldsymbol{u}(x,\tau)$ is considered as superimposed on the reference configuration and compatible with the essential boundary conditions on the boundary of $\mathrm{k}^{\mathrm{N}}V$ (thus corresponding to a periodic field on $\mathrm{k}^{\mathrm{N}}V$). This demands respectively $\boldsymbol{u}(\boldsymbol{x},\tau=0)=0$ on $\partial\,\mathrm{k}^{\mathrm{N}}V$ and $\boldsymbol{u}(\boldsymbol{x}^+,\tau)=\boldsymbol{u}(\boldsymbol{x}^-,\tau)$ on $\partial\mathrm{k}^{\mathrm{N}}V$, where τ is a time-like parameter with $\tau \geq 0$. As shown in Fig. 3, the microscopic displacement field u deforms the microstructure from the reference configuration $\mathrm{k}^{\mathrm{N}}V$ to the generic configuration $\mathrm{k}^{\mathrm{N}}V(\tau)$. The gradient of the microscopic deformation field relative to the reference configuration V, is denoted by $\boldsymbol{F}_{(0)}(\boldsymbol{x},\tau)$.

For sufficiently small values of the time-like parameter τ, the additional displacement field represents an infinitesimal deformation (also called incremental) from the current configuration B and the displacement field and the gradient of the additional deformation can be expressed as

$$\boldsymbol{u}(\boldsymbol{x},\tau)=\dot{\boldsymbol{u}}_0(\boldsymbol{x})\tau+o(\tau),\ \boldsymbol{F}_{(0)}(\boldsymbol{x},\tau)=\boldsymbol{I}+\boldsymbol{L}(\boldsymbol{x})\tau+\boldsymbol{O}(\tau). \tag{15}$$

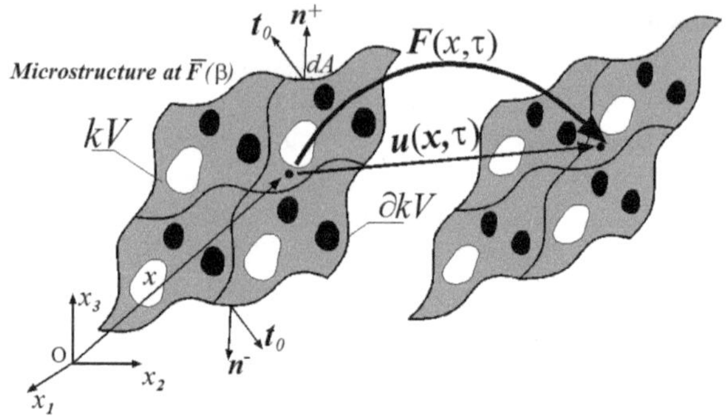

Fig. 3 Incremental deformations superimposed on the current equilibrium configuration of the microstructure

The displacement rate field $\dot{\boldsymbol{u}}_0(\boldsymbol{x})$ can be considered as the "quasi-static" velocity field, whereas its gradient $\boldsymbol{L}(\boldsymbol{x})$ the velocity gradient. The microscopic deformation defined in terms of the additional displacement field $\boldsymbol{u}(\boldsymbol{x},4)$ is coupled with the macroscopic deformation gradient at a typical point $\bar{\boldsymbol{X}}$ of the macro-continuum by virtue of Eq. $(3)_1$. Hence, the following relationship holds:

$$\boldsymbol{u}(\boldsymbol{x},\tau) = \dot{\boldsymbol{u}}_0(\boldsymbol{x})\tau + \boldsymbol{o}(\tau) = \left(\dot{\bar{\boldsymbol{F}}}_{(0)}\boldsymbol{x} + \dot{\boldsymbol{w}}(\boldsymbol{x})\right)\tau + \boldsymbol{o}(\tau) = \dot{\boldsymbol{w}}(\boldsymbol{x})\tau + \boldsymbol{o}(\tau), \quad (16)$$

where $\dot{\bar{\boldsymbol{F}}}_{(0)}$ is the rate of the macroscopic deformation gradient relative to configuration $k^N V$, evaluated in the current configuration, which vanishes due to the assumed periodicity of the additional displacement field, and $\dot{\boldsymbol{w}}(\boldsymbol{x})$ is the fluctuation field velocity. It follows that $\boldsymbol{L} = \nabla\dot{\boldsymbol{w}}(\boldsymbol{x}) = \partial\dot{\boldsymbol{w}}(\boldsymbol{x})/\partial\boldsymbol{x}$.

The incremental version of the classical criterion of stability of the current equilibrium configuration $k^N V$ [17], is here used according to the incremental constitutive law adopted for the material and the current configuration is taken as the reference one. The second order approximation of the difference between the internal deformation work D

$$D = \int_0^t \int_{k^N V} \left[\boldsymbol{T}_{R(0)}(\boldsymbol{x},\tau)\cdot\dot{\boldsymbol{F}}_{(0)}(\tau)\right] dV d\tau = \left\{\int_{k^N V} \left[\boldsymbol{T}_{R(0)}(\boldsymbol{x},\tau)\cdot\dot{\boldsymbol{F}}_{(0)}(\tau)\right]_{\tau=0} dV\right\} t +$$
$$\left\{\int_{k^N V} \left[\boldsymbol{T}_{R(0)}(\boldsymbol{x},\tau)\cdot\dot{\boldsymbol{F}}_{(0)}(\tau)\right]_{\tau=0} dV\right\} \frac{t^2}{2} + o\left(t^2\right)$$

and the work done by the antiperiodic tractions $\boldsymbol{t}_{R(0)}$ acting in the examined equilibrium configuration

$$L = \int_0^t \left[\int_{\partial k^N V} \boldsymbol{t}_{R(0)}\cdot\dot{\boldsymbol{u}} ds\right] d\tau = \int_{\partial k^N V} \boldsymbol{T}_0\boldsymbol{n}\cdot\left(\boldsymbol{v}\,t + \ddot{\boldsymbol{u}}_0\frac{t^2}{2}\right) ds + o\left(t^2\right)$$

during the additional deformation from $k^N V$ to $k^N V(\tau)$ can be expressed as:

$$D - L = \left(\int_{k^N V} \dot{\boldsymbol{T}}_{R(0)} \cdot \boldsymbol{L} dV \right) \frac{t^2}{2} + o\left(t^2\right), \qquad (17)$$

where $\boldsymbol{T}_{R(0)}$ is the first Piola-Kirchhoff stress tensor tensor based on the configuration $k^N V(\tau = 0)$ and $\boldsymbol{T}_0$ is the Cauchy stress tensor in the reference configuration $k^N V$ which corresponds to $\boldsymbol{T}_{R(0)}(\boldsymbol{x},\tau = 0)$. The first-order terms vanish due to equilibrium in the examined configuration $k^N V$ and dead loading is assumed on $\partial\, k^N V$, namely $\boldsymbol{t}_{R(0)}$ is independent on τ. In the above equations dV and ds indicate the reference volume and area elements, respectively. On the other hand, due to the assumed periodicity for displacement field and antiperiodicity of surface tractions, L is identically zero.

The structural stability condition of the microstructure at the macroscopic deformation $\bar{\boldsymbol{F}}$ is based on the positive definiteness of the functional shown in Eq. (17), quadratic in the fluctuation rate field, referred to as the stability functional, for every incremental deformations satisfying the essential periodicity boundary constraints. When the incremental constitutive law written with the reference configuration coincides with the current one:

$$\dot{\boldsymbol{T}}_{R(0)} = \boldsymbol{C}_0^R(\boldsymbol{F})[\boldsymbol{L}], \qquad (18)$$

where $C_0{}^R$ is the fourth-order tensor of nominal instantaneous moduli, the stability functional becomes

$$\int_{k^N V} \boldsymbol{C}_0^R(\boldsymbol{x}, \boldsymbol{F})[\nabla \dot{\boldsymbol{w}}(\boldsymbol{x})] \cdot \nabla \dot{\boldsymbol{w}}(\boldsymbol{x}) dV. \qquad (19)$$

Therefore, a deformed state of the microstructure characterized by the fluctuation field $w(x)$ induced by the macroscopic load $\bar{\boldsymbol{F}}$, is stable if the minimum eigenvalue of the stability functional is positive when the minimum is taken over all admissible incremental fluctuations periodic on the $k^N V$ ensemble of unit cells:

$$\Lambda\left(\bar{\boldsymbol{F}}\right) = \inf_{k \in} \left\{ \min_{\dot{\boldsymbol{w}} \in H^{1,p}(k^N V\#)} \left\{ \frac{\int_{k^N V} \boldsymbol{C}_0^R\left(\boldsymbol{x}, \bar{\boldsymbol{F}} + \nabla \boldsymbol{w}_{\bar{\boldsymbol{F}}}\right)[\nabla \dot{\boldsymbol{w}}] \cdot \nabla \dot{\boldsymbol{w}} dV}{\int_{k^N V} \nabla \dot{\boldsymbol{w}} \cdot \nabla \dot{\boldsymbol{w}} dV} \right\} \right\} > 0. \qquad (20)$$

The assumed major symmetry of the microscopic moduli tensor ensures that all eigenvalues and corresponding eigenmodes of the quadratic functional are real. The Euler Lagrange equations and surface conditions corresponding to the above eigenvalue problem are:

$$\begin{cases} div\left\{\boldsymbol{C}_0^R\left(\boldsymbol{x},\bar{\boldsymbol{F}}+\nabla\boldsymbol{w}_{\bar{\boldsymbol{F}}}\right)[\nabla\dot{\boldsymbol{w}}]-\Lambda\left(\bar{\boldsymbol{F}}\right)\nabla\dot{\boldsymbol{w}}\right\}=\boldsymbol{0}\ in\ k^N V \\ \left(\left\{\boldsymbol{C}_0^R\left(\boldsymbol{x},\bar{\boldsymbol{F}}+\nabla\boldsymbol{w}_{\bar{\boldsymbol{F}}}\right)[\nabla\dot{\boldsymbol{w}}]-\Lambda\left(\bar{\boldsymbol{F}}\right)\nabla\dot{\boldsymbol{w}}\right\}\boldsymbol{n}\right)^+= \\ -\left(\left\{\boldsymbol{C}_0^R\left(\boldsymbol{x},\bar{\boldsymbol{F}}+\nabla\boldsymbol{w}_{\bar{\boldsymbol{F}}}\right)[\nabla\dot{\boldsymbol{w}}]-\Lambda\left(\bar{\boldsymbol{F}}\right)\nabla\dot{\boldsymbol{w}}\right\}\boldsymbol{n}\right)^-\ on\ \partial k^N V \\ \left\{\boldsymbol{C}_0^R\left(\boldsymbol{x},\bar{\boldsymbol{F}}+\nabla\boldsymbol{w}_{\bar{\boldsymbol{F}}}\right)[\nabla\dot{\boldsymbol{w}}]-\Lambda\left(\bar{\boldsymbol{F}}\right)\nabla\dot{\boldsymbol{w}}\right\}\boldsymbol{n}=\boldsymbol{0}\ on\ \partial H \end{cases} . \tag{21}$$

where $\boldsymbol{n}$ is the outward normal to the boundary of the unit cell assembly in the current configuration $k^N V$. The second and third equations in (21) respectively represents the antiperiodicity condition and the free surface conditions on the hole boundary for the equivalent traction $\left\{\boldsymbol{C}_0^R\left(\boldsymbol{x},\bar{\boldsymbol{F}}+\nabla\boldsymbol{w}_{\bar{\boldsymbol{F}}}\right)[\nabla\dot{\boldsymbol{w}}]-\Lambda\left(\bar{\boldsymbol{F}}\right)\nabla\dot{\boldsymbol{w}}\right\}\boldsymbol{n}$.

It is worth noting that the stability functional Eq. (19) can be written in an equivalent form with reference to the initial configuration as

$$\int\limits_{k^N V_{(i)}} \boldsymbol{C}^R\left(\boldsymbol{X},\boldsymbol{F}\right)\left[\nabla_{(i)}\dot{\boldsymbol{w}}(\boldsymbol{X})\right]\cdot\nabla_{(i)}\dot{\boldsymbol{w}}(\boldsymbol{X})dV_{(i)},$$

where $\nabla_{(i)}\dot{\boldsymbol{w}}(\boldsymbol{X}) = \partial\dot{\boldsymbol{w}}(\boldsymbol{X})/\partial X$, due to the relations between the instantaneous and fixed-reference moduli and between the deformation gradient rate and velocity gradient rate:

$$C^R_{0\,ijkl} = \frac{1}{\det \boldsymbol{F}} F_{jm} F_{ln} C^R{}_{imkn}, \quad \dot{\boldsymbol{F}} = \boldsymbol{L}\boldsymbol{F}. \tag{22}$$

It follows that the stability condition (20) can be also formulated as

$$\Lambda\left(\bar{\boldsymbol{F}}\right)=\inf_{k\in}\left\{\min_{\dot{\boldsymbol{w}}\in H^{1,p}(k^N V_{(i)}\#)}\left\{\frac{\int\limits_{k^N V_{(i)}} \boldsymbol{C}^R\left(\boldsymbol{x},\bar{\boldsymbol{F}}+\nabla\boldsymbol{w}_{\bar{\boldsymbol{F}}}\right)\left[\nabla_{(i)}\dot{\boldsymbol{w}}(\boldsymbol{X})\right]\cdot\nabla_{(i)}\dot{\boldsymbol{w}}(\boldsymbol{X})dV_{(i)}}{\int\limits_{k^N V_{(i)}} \nabla_{(i)}\dot{\boldsymbol{w}}(\boldsymbol{X})\cdot\nabla_{(i)}\dot{\boldsymbol{w}}(\boldsymbol{X})dV_{(i)}}\right\}\right\}>0. \tag{23}$$

Along a deformation path $\bar{\boldsymbol{F}}(\beta)\,\beta \geq 0\ \beta = 0\ for\ \bar{\boldsymbol{F}} = \boldsymbol{I}$ starting where the stability functional is positive definite, namely $\Lambda\left(\bar{\boldsymbol{F}}(0)\right) > 0$, usually λ decreases and at some load level β_c (termed microscopic critical load parameter) the stability functional becomes positive semi-definite. At this load necessarily, the initially unique and stable principal solution ceases to be unique since an eigenmode (incremental periodic solution to the homogeneous problem) exists and the loss of microscopic structural stability occurs:

$$\Lambda\left(\beta_c\right) = 0,\ \Lambda\left(\beta\right) > 0\ for\ 0 \leq \beta < \beta_c.$$

Therefore the primary instability is detected when the minimum eigenvalue first vanishes. The microscopic stability region $\beta|\Lambda\left(\bar{\boldsymbol{F}}(\beta)\right) > 0$, inside which the fundamental periodic solution, for which all cells deform identically, is unique, establishes also the region where the one-cell standard homogenized energy is the correct one, namely $\Lambda\left(\bar{\boldsymbol{F}}(\beta)\right) > 0$ implies $\bar{W}\left(\bar{\boldsymbol{F}}\right) = \bar{W}^1\left(\bar{\boldsymbol{F}}\right)$.

2.5 Macroscopic Stability Analysis

A microscopic stability analysis along a macroscopic loading path $\bar{\boldsymbol{F}}(\beta)$, in turn leading to determine the microscopic stability region, requires a notable computational effort since it involves the examination of perturbations of the equilbrium fluctuation solution defined in a infinite domain. As a consequence, it should be preferable to carry out the stability analysis of the heterogeneous solid in terms of its macroscopic properties determined by means of calculations performed on a unit cell.

A basic macroscopic measure of the stability of the periodic solid at the load parameter β can defined as the strong ellipticity condition of the homogenized moduli tensor:

$$\bar{\Lambda}\left(\bar{\boldsymbol{F}}(\beta)\right) = \min_{\|\bar{m}\|=\|\bar{n}\|=1} \left\{\bar{\boldsymbol{C}}_0^R\left(\bar{\boldsymbol{F}},\bar{\boldsymbol{X}}\right)(\bar{\boldsymbol{m}}\otimes\bar{\boldsymbol{n}})\cdot\bar{\boldsymbol{m}}\otimes\bar{\boldsymbol{n}}\right\} > 0 \qquad (24)$$

in which the minimum is taken over all unit vectors $\bar{\boldsymbol{m}}$ and $\bar{\boldsymbol{n}}$.

As a matter of fact the strong ellipticity of the homogenized solid moduli ensures stability under Dirichlet boundary conditions when the macroscopic incremental moduli tensor is spatially constant. Moreover, when extended to arbitrary rank tensors corresponds to the the positive definiteness of the homogenized nominal moduli tensor:

$$\bar{\Lambda}^{(R)}\left(\bar{\boldsymbol{F}}(\beta)\right) = \min_{\|\bar{L}\|=1} \left\{\bar{\boldsymbol{C}}_0^R\left(\bar{\boldsymbol{F}}\right)\left[\bar{\boldsymbol{L}}\right]\cdot\bar{\boldsymbol{L}}\right\} > 0, \qquad (25)$$

too restrictive and physically unrealistic since it would imply uniqueness in corresponding boundary value problems for the homogenized solid, an unacceptable situation for nonlinear deformations. Obviously (25) implies (24).

A macroscopic primary instability load associated to the stability condition Eq. (24) can be defined as

$$\bar{\Lambda}(\beta_{cM}) = 0, \bar{\Lambda}(\beta) > 0 \, for \, 0 \le \beta < \beta_{cM},$$

where β_{cM} is the macroscopic critical paramenter and, consequently, the macroscopic stability region $\beta|\bar{\Lambda}(\beta) > 0$ can be determined. The critical load parameter $\beta_{cM}^{(R)}$ associated to Eq. (25) can be defined analogously

As proved in [14] the microscopic stability condition Eq. (20) implies the macroscopic stability condition Eq. (24), provided the microscopic material is strongly elliptic:

$$\min_{\|m\|=\|n\|=1} \left\{\boldsymbol{C}_0^R(\boldsymbol{F},\boldsymbol{X})(\boldsymbol{m}\otimes\boldsymbol{n})\cdot\boldsymbol{m}\otimes\boldsymbol{n}\right\} > 0 \Rightarrow \Lambda\left(\bar{\boldsymbol{F}}(\beta)\right) < \bar{\Lambda}\left(\bar{\boldsymbol{F}}(\beta)\right).$$

Moreover, [14] have shown that the primary microscopic instability along a monotonic macroscopic loading process can be detected as a loss of macroscopic stability Eq. (24) provided that the wavelength of the first instability is much larger than the unit cell size (global instability mode). On the contrary, when the primary microscopic instability encountered in the loading process has a wavelength comparable to the unit cell size (local instability), the macroscopic stability condition Eq. (24) still holds and the one-cell homogenized moduli remains strongly elliptic:

$$\Lambda\left(\bar{\boldsymbol{F}}\left(\beta_c\right)\right)=0 \Rightarrow \begin{cases} \textit{local instability}\ \bar{\Lambda}\left(\bar{\boldsymbol{F}}\left(\beta_c\right)\right) > 0 \\ \textit{global instability}\ \bar{\Lambda}\left(\bar{\boldsymbol{F}}\left(\beta_c\right)\right)=0 \end{cases}.$$

It follows that the macroscopic stability condition based on the strong ellipticity of the homogenized moduli tensor is able to exactly predict the onset of microscopic instability of the periodic principal solution along a monotonic loading process when the microscopic instability mode is global in nature. On the other hand, an unconservative estimation of the primary microscopic instability load is obtained in the more general case when the instability mode is local in nature, since the homogenized moduli tensor remains strongly elliptic at the onset of the primary microscopic instability.

Therefore alternative macroscopic stability measures are here introduced and their ability to obtain conservative prediction of the primary instability load of the microstructure will be investigated.

2.6 Conjugated Macroscopic Stability Measures

As an alternative to Eq. (18) the incremental material response at a point x of the microstructure can be expressed as:

$$\dot{\boldsymbol{T}}_{f(0)}=\boldsymbol{C}_0^f\left(\boldsymbol{F},\boldsymbol{X}\right)\left[\boldsymbol{D}\right], \tag{26}$$

where $\dot{\boldsymbol{T}}_{f(0)}$ and D are, respectively, the stress rate and the strain rate when the reference configuration coincides with the current one, corresponding to the work conjugate stress-strain measure pair ($\boldsymbol{T}_{\mathrm{f}}$, $\boldsymbol{F}(\boldsymbol{U})$) based on strain measures coaxial with $\boldsymbol{U}$, the right stretch tensor associated to $\boldsymbol{F}$, and having principal values $f(\lambda_i)$, with f a monotonic increasing function of the principal values λ_i of $\boldsymbol{U}$ such that $f(1)=0$ and $\mathrm{d}f/\mathrm{d}\lambda_i(1)=1$ [18]. The rate of strain $\boldsymbol{D}$ equals the symmetric part of the velocity gradient $\boldsymbol{L}=\partial\boldsymbol{v}/\partial\boldsymbol{x}$ also called the Eulerian strain rate. By using the following expression relating the rate of the first Piola-Kirchhoff stress tensor $\boldsymbol{T}_R$ to $\dot{\boldsymbol{T}}_{f(0)}$ (Ogden, 1984) evaluated when the reference configuration coincides with the current one

$$\dot{\boldsymbol{T}}_{R(0)}=\dot{\boldsymbol{T}}_{f(0)}+\frac{1}{2}\left[f''\left(1\right)-1\right]\left(\boldsymbol{T}_0\boldsymbol{D}+\boldsymbol{D}\boldsymbol{T}_0\right)+\boldsymbol{L}\boldsymbol{T}_0, \tag{27}$$

the fourth-order tensor of instantaneous moduli $\boldsymbol{C}_0^f$can be easily related to the fourth-order tensor of nominal instantaneous moduli $\boldsymbol{C}_0{}^R$.

A well-known sub-class ($\boldsymbol{T}^{(\mathrm{m})}$, $\boldsymbol{E}^{(\mathrm{m})}$) of stress-strain measure pairs can be obtained by choosing $f(\lambda_i)=(\lambda_i{}^m-1)/m$, where m is an integer (Ogden, 1984). The stress-strain pairs associated to the logarithmic ($\boldsymbol{E}^{(0)} = \ln \boldsymbol{U}$), the Green-Lagrange ($\boldsymbol{E}^{(2)} = \boldsymbol{U}^2 - \boldsymbol{I}$) the Biot strain ($\boldsymbol{E}^{(1)} = \boldsymbol{U} - \boldsymbol{I}$) measures, can be determined by taking $m \rightarrow 0$, $m = 2$ and $m = 1$, respectively. The corresponding moduli tensor are respectively denoted as $\boldsymbol{C}_0^{(0)}$, $\boldsymbol{C}_0^{(2)}$ and $\boldsymbol{C}_0^{(1)}$. The strain measure $\boldsymbol{E}^{(-2)}$ is usually attributed to Almansi.

According to the above representation of the microscopic constitutive response, the following family of macroscopic stability measures is introduced:

$$\bar{\Lambda}^f \left(\bar{\boldsymbol{F}}(\beta)\right) = \min_{\|\bar{\boldsymbol{D}}\|=1} \left\{ \bar{\boldsymbol{C}}_0^f \left(\bar{\boldsymbol{F}}\right) \left[\bar{\boldsymbol{D}}\right] \cdot \bar{\boldsymbol{D}} \right\} > 0, \tag{28}$$

where $\bar{\boldsymbol{D}}$ is a symmetric tensor and $\bar{\boldsymbol{C}}_0^{\mathrm{f}}$ is the macroscopic tensor of instantaneous homogenized moduli which relates the macro-strain rate $\bar{\boldsymbol{D}}$ to the rate of the macro-stress tensor $\dot{\bar{\boldsymbol{T}}}_{f(0)}$ and is associated to the macroscopic strain measures $\bar{\boldsymbol{F}}_{(\bar{U})}$ defined by the usual continuum relations in terms of macroscopic quantities, where $\bar{\boldsymbol{U}}$ is the macroscopic right stretch tensor in the polar decomposition of the macroscopic deformation gradient $\bar{\boldsymbol{F}} = \bar{\boldsymbol{R}}\bar{\boldsymbol{U}}$, $\bar{\boldsymbol{R}}$ denoting the macroscopic rotation tensor.

The positive definiteness inequalities Eq. (28), can be linked to the infinitesimal stability condition of a homogeneously deformed homogenized material element subjected to deformation dependent surface tractions which do not work on material rotations (namely, with the problem of so-called "material" or "constitutive" stability), thus providing a physical understanding of the restrictions imposed on the homogenized material behavior. This mechanical interpretation has been introduced in [10] for a material element of a homogeneous body.

It is worth noting that if the macroscopic stability analysis is carried out by using Eq. (28), which restrict the analysis to symmetric incremental deformation gradients, rigid rotations, irrelevant from the physical point of view can be avoided.

A specific macroscopic strain measure must be adopted in Eq. (28) in order to define a unique stability measure. If the macroscopic Biot strain tensor is used, Eq. (28) specializes to the following positiveness condition

$$\bar{\Lambda}^{(1)} \left(\bar{\boldsymbol{F}}(\beta)\right) = \min_{\|\bar{\boldsymbol{D}}\|=1} \left\{ \bar{\boldsymbol{C}}_0^{(1)} \left(\bar{\boldsymbol{F}}\right) \left[\bar{\boldsymbol{D}}\right] \cdot \bar{\boldsymbol{D}} \right\} > 0. \tag{29}$$

On the other hand, the use of the macroscopic logarithmic strain measure leads to :

$$\bar{\Lambda}^{(0)} \left(\bar{\boldsymbol{F}}(\beta)\right) = \min_{\|\bar{\boldsymbol{D}}\|=1} \left\{ \bar{\boldsymbol{C}}_0^{(0)} \left(\bar{\boldsymbol{F}}\right) \left[\bar{\boldsymbol{D}}\right] \cdot \bar{\boldsymbol{D}} \right\} > 0. \tag{30}$$

Moreover, the choice of the macroscopic Green-Lagrange strain measure leads to:

$$\bar{\Lambda}^{(2)} \left(\bar{\boldsymbol{F}}(\beta)\right) = \min_{\|\bar{\boldsymbol{D}}\|=1} \left\{ \bar{\boldsymbol{C}}_0^{(2)} \left(\bar{\boldsymbol{F}}\right) \left[\bar{\boldsymbol{D}}\right] \cdot \bar{\boldsymbol{D}} \right\} > 0. \tag{31}$$

Finally when the Almansi strain measure is adopted we obtain:

$$\bar{\Lambda}^{(-2)}\left(\bar{\boldsymbol{F}}(\beta)\right) = \min_{\|\bar{\boldsymbol{D}}\|=1}\left\{\bar{\boldsymbol{C}}_0^{(-2)}\left(\bar{\boldsymbol{F}}\right)\left[\bar{\boldsymbol{D}}\right]\cdot\bar{\boldsymbol{D}}\right\} > 0$$

A macroscopic primary instability load associated to the conjugated stability measures can be defined as

$$\bar{\Lambda}^f\left(\beta^f_{cM}\right) = 0,\ \bar{\Lambda}(\beta) > 0\, for\, 0 \le \beta < \beta^f_{cM},$$

where β^f_{cM} is the macroscopic critical parameter and, consequently, the macroscopic stability region $\beta|\bar{\Lambda}^f(\beta) > 0$ can be determined.

Since the overall stress and strain measures are based on the nominal stress tensor and deformation gradient [19], the components of the macroscopic constitutive tensor $\bar{\boldsymbol{C}}_0^f$ can be obtained in terms of the components of the macroscopic moduli tensor by using Eq. (18) and the following usual continuum relations as:

$$\bar{C}^f_{0\,ijkl} = \bar{C}^R_{0\,ijkl} - \frac{f''(1)-1}{4}\left(\bar{T}_{ik}\delta_{jl} + \bar{T}_{il}\delta_{kj} + \bar{T}_{lj}\delta_{ki} + \bar{T}_{kj}\delta_{li}\right) - \bar{T}_{lj}\delta_{ik},$$

where $\bar{\boldsymbol{L}}$ is the macro-deformation velocity gradient $\bar{\boldsymbol{L}} = \dot{\bar{\boldsymbol{F}}}\bar{\boldsymbol{F}}^{-1}$, $\bar{\boldsymbol{D}}$ its symmetric part and $\bar{\boldsymbol{T}}_{(0)} = \bar{\boldsymbol{J}}^{-1}\bar{\boldsymbol{T}}_R\bar{\boldsymbol{F}}^T$ is the macroscopic Cauchy stress tensor. It turns out that the macroscopic stability measure Eq. (28) are expressed in terms of macroscopic properties determined by means of calculations performed on a unit cell, in line with Eq. (24).

It is worth noting, incidentally, that also the macroscopic stability condition Eq. (24) can be considered as a constitutive stability condition for an homogeneous material both in the context of wave propagation and strain localization [20].

3 Numerical Results

The general theory formulated in the previous section is hereby applied to specific materials and specific microstructural models. In the first subsection the computational implementation of the stability analysis is discussed. Then a specific hyperelastic constitutive law is introduced. Then critical load parameters associated to microscopic and macroscopic onset of instability are presented for different loading paths and microgeometries.

3.1 Computational Implementation

The stability problem of a periodic microstructure stated in Sect. 2 was discretized by means of a displacement-type finite element (FE) approximation using plane strain Lagrange quadratic elements. The FE model has been developed by using the commercial software COMSOL MULTIPHISYCS™ [21].

A one-way coupled FE model is employed to compute sequentially respectively the principal solution path for the unit cell, the incremental solutions needed to determine the homogenized tangent moduli and the minimum eigenvalue of the microscopic structural stability functional. At first the finitely deformed configurations of a unit cell are determined by discretization of the variational problem Eq. (6) along the principal equilibrium path for a given macroscopic loading process $\bar{\boldsymbol{F}}(\beta)$, assuming that the loading process produces a unique response. A parametric non-linear solver based on a continuation approach, is adopted to find the solution to the sequence of nonlinear stationary PDE problems (6) arising when the load parameter β varies. A step size equal to $\Delta\beta = 10^{-3}$ is adopted to discretize the loading path.

Secondly, the incremental equilibrium problems of the unit cell for each unit incremental macroscopic deformation mode, $\dot{\bar{\boldsymbol{F}}} = \boldsymbol{I}^{hk}$ $h,k = 1,2,3$, are solved superimposed on the given finite deformation along the loading path by the discretization of the following variational equation:

$$\int_{V_{(i)}} \boldsymbol{C}^R(\beta, \boldsymbol{X})\, [\boldsymbol{I}^{hk} + \nabla \dot{\boldsymbol{w}}_{\boldsymbol{I}^{hk}}] \cdot \nabla \delta \dot{\boldsymbol{w}} dV = 0 \quad \forall \delta \dot{\boldsymbol{w}} \in H^{1,p}(V_{(i)\#}).$$

The homogenized moduli can be thus obtained by Eq. (11).

Then the load parameter associated to the lowest zero eigenvalue of the microscopic structural stability functional Eq. (20) must be computed over all possible ensemble of unit cells. From the computational point of view the linearized eigenvalue problem with a varying domain of definition is solved in the following simple way. For a fixed ensemble of unit cells the lowest value of β for which the minimum eigenvalue of the stability functional is zero is determined, namely β_c, together with the associated eigenmode by the discretization of the minimization problem Eq. (20). Then we successively enlarge the ensemble by increasing the number k. The minimum value of β_c for all currently possible instability modes then determines the optimal ensemble of unit cells. This value corresponds to the loss of microscopic stability.

Finally the macroscopic stability analysis is performed by monitoring the lowest eigenvalue of the acoustic tensor $\bar{\boldsymbol{Q}}_{0ih}(\bar{\boldsymbol{n}}) = \bar{C}^R_{0ijhk}\bar{n}_j\bar{n}_k$ for every direction of propagation $\bar{\boldsymbol{n}}$:

$$\bar{\Lambda}(\bar{\boldsymbol{F}}(\beta)) = \min_i \left[\min_{\|\bar{n}\|=1} \lambda_i^{\bar{Q}(\bar{n})}(\beta) \right] \text{ with } \lambda_i^{\bar{Q}(\bar{n})} | (\bar{\boldsymbol{Q}}_0(\bar{\boldsymbol{n}}) - \lambda_i^{\bar{Q}(\bar{n})}\boldsymbol{I})\hat{\boldsymbol{\Phi}}_i = \boldsymbol{0}.$$

The first macroscopic instability is detected when the lowest eigenvalue becomes zero.

Similarly, the onset of macroscopic instability according to the conjugated stability measures is determined by monitoring the lowest eigenvalue of Eq. (28)

$$\bar{\Lambda}^f(\bar{\boldsymbol{F}}(\beta)) = \min\left[\lambda_i^{\bar{\mathcal{C}}_0^f}(\beta)\right] \text{ with } \lambda_i^{\bar{\mathcal{C}}_0^f} |(\bar{\boldsymbol{C}}_0^f - \lambda_i^{\bar{\mathcal{C}}_0^f}\boldsymbol{I})\hat{\boldsymbol{\Phi}}_i = \boldsymbol{0},$$

where $\bar{\boldsymbol{C}}_0^f$ denotes a second order tensor defining the following eigenvalue problem

$$\left(\begin{bmatrix} \bar{\mathcal{C}}^f_{0\,1111} & \bar{\mathcal{C}}^f_{0\,1122} & 2\bar{\mathcal{C}}^f_{0\,1112} \\ \bar{\mathcal{C}}^f_{0\,1122} & \bar{\mathcal{C}}^f_{0\,2222} & 2\bar{\mathcal{C}}^f_{0\,2212} \\ 2\bar{\mathcal{C}}^f_{0\,1112} & 2\bar{\mathcal{C}}^f_{0\,2212} & 2\bar{\mathcal{C}}^f_{0\,1212} \end{bmatrix} - \lambda_i^{\bar{\mathcal{C}}_0^f}\, diag\,\{1,1,1\}\right)\begin{Bmatrix} \tilde{\Phi}_{11}^{(\iota)} \\ \tilde{\Phi}_{22}^{(\iota)} \\ \tilde{\Phi}_{12}^{(\iota)} \end{Bmatrix} = \begin{Bmatrix} 0 \\ 0 \\ 0 \end{Bmatrix}.$$

The onset of macroscopic instability is determined by computing the lowest load parameter for which the lowest eigenvalue of Eq. (28) becomes zero.

Similar considerations apply to the macroscopic condition Eq. (25) for which the corresponding eigenvalue problem is:

$$\left(\begin{bmatrix} \bar{C}^R_{1111} & \bar{C}^R_{1122} & \bar{C}^R_{1112} & \bar{C}^R_{1121} \\ \bar{C}^R_{2211} & \bar{C}^R_{2222} & \bar{C}^R_{2212} & \bar{C}^R_{2221} \\ \bar{C}^R_{1211} & \bar{C}^R_{1222} & \bar{C}^R_{1212} & \bar{C}^R_{1221} \\ \bar{C}^R_{2111} & \bar{C}^R_{2122} & \bar{C}^R_{2112} & \bar{C}^R_{2121} \end{bmatrix} - \lambda_i^{\bar{C}^R}\, diag\,\{1,1,1,1\}\right)\begin{Bmatrix} \Phi_{11}^{(\iota)} \\ \Phi_{22}^{(\iota)} \\ \Phi_{12}^{(\iota)} \\ \Phi_{21}^{(\iota)} \end{Bmatrix} = \begin{Bmatrix} 0 \\ 0 \\ 0 \\ 0 \end{Bmatrix}.$$

Periodic boundary conditions were implemented in the finite and incremental homogenization procedure by means of the extrusion coupling variable methodology, according to which the displacement field or its increment is made available on the opposite boundary faces of the unit cell. Once the displacement field or its increment is extruded from the source domain (the negative unit cell assembly faces $\partial\, k^N V_{(i)}^-$) to the destination one (the positive unit cell faces $\partial k^N V_{(i)}^+$), periodic boundary constraints are imposed as point constraints on the destination boundaries of the unit cell. In order to exclude rigid body motions, the fluctuation field can be assumed to be zero at the corner points of the unit cell, implying that the displacement field at the corner points is driven by the macroscopic deformation gradient, i.e. $\boldsymbol{u}(\boldsymbol{X}) = \left(\bar{\boldsymbol{F}} - \boldsymbol{1}\right)\boldsymbol{X}$, $\dot{\boldsymbol{u}}\,(\boldsymbol{X}) = \dot{\bar{\boldsymbol{F}}}\boldsymbol{X}$.

The evolution of the minimum eigenvalues of the microscopic and macroscopic stability conditions is managed by developing a computer code written in the COMSOLSCRIPT™ programming language, which is interfaced with COMSOL MULTIPHISICS™ [21].

3.2 Constitutive and Microgeometry Models

The theoretical formulation of this work is valid for materials characterized by an incrementally linear constitutive law. In the sake of simplicity, numerical examples are developed by adopting the experimentally based compressible [22] constitutive law for each microstructural components, with the following strain energy density:

$$W = -\frac{\mu}{2}\left[J_m \ln\left(1 - \frac{\|\boldsymbol{F}\|^2 - 3}{J_m}\right) + 2\ln J\right] + \left(\frac{k-\mu}{2} - \frac{\mu}{J_m}\right)(J-1)^2$$

where μ and κ are, respectively, the shear and bulk moduli of the solid at zero strain and J_m is a constant which calibrates the solid's strain saturation. By assuming $\mu > 0$, $\kappa > [(J_m+2)/J_m]\,\mu$, $J_m>0$ the Gent solid becomes polyconvex [6] and ensures that the strong ellipticity condition is satisfied for the microscopic material. For the numerical calculations the values of $J_m = 50$ and $\kappa/\mu = 10$ are adopted.

Two different types of macroscopic loading path are considered, namely a uniaxial and a equibiaxial loading along the reference coordinate axes:

$$\bar{\boldsymbol{F}}(\beta) = \begin{bmatrix} 1+\beta & 0 & 0 \\ 0 & 1 & 0 \\ 0 & 0 & 1 \end{bmatrix}, \quad \bar{\boldsymbol{F}}(\beta) = \begin{bmatrix} 1+\beta & 0 & 0 \\ 0 & 1+\beta & 0 \\ 0 & 0 & 1 \end{bmatrix}.$$

In the first application, a cellular microstructure with an initial square distribution of circular voids is analyzed under plane strain conditions in the X_1–X_2 plane. In the second one a particle reinforced composite microstructure is considered with a square distribution of inclusions, which can be considered as representative of a cross section of a fiber-reinforced solid with cylindrical fibers aligned in the X_3 axis direction.

In both cases the unit cell dimensions are $L_1 = L_2 = L$ and the radius of the voids and the inclusions is $R = 0.25$ L. Hence the initial porosity equal to the volume fraction of the inclusions is $\pi/16$. In the numerical calculations the shear modulus at zero strain of the matrix material has been assumed equal to $\mu = 807$ N/mm^2.

Typical meshes used to discretize a unit cell of the above mentioned microstructural models are shown in Fig. 4. These meshes have been used to determine the one-cell homogenized moduli and, consequently, the macroscopic stability analyses. For the cellular microgeometry the mesh involves 15,616 degrees of

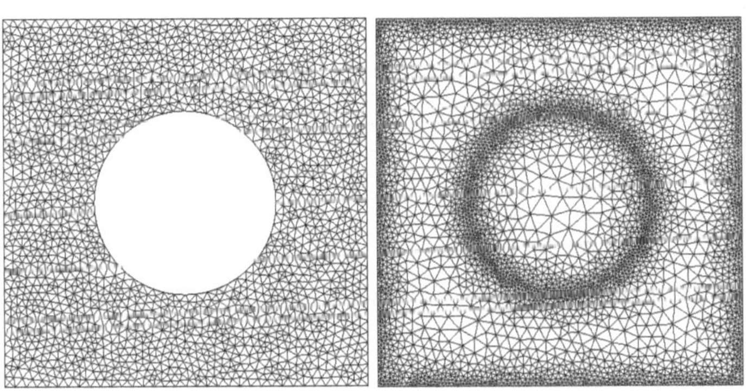

Fig. 4 Typical meshes adopted to discretize the unit cells of the examined microgeometry models

freedom and 3,810 quadratic Lagrangian triangular elements. In the case of the reinforced microgeometry 31,194 degrees of freedom and 7,598 quadratic Lagrangian triangular elements.

The microscopic stability analysis involves an increasing assembly of unit cells. In the numerical calculations the largest assembly examined corresponds to an array of 20×20 unit cells. This assembly is assumed to be a reasonable approximation of the possibly infinite domain of microscopic stability analysis and provided values of the critical load parameter in the case of a global instability mode sufficiently close to those obtained by using the macroscopic stability measure Eq. (24). In particular, the instability mode has been classified as global, when by considering increasing unit cell assemblies the lowest value of the load parameter for which the minimum eigenvalue of stability functional first vanishes decreases and approaches from above the load parameter corresponding to the macroscopic loss of ellipticity within a relative percentage error $(\beta_{cM} - \beta_c)/\ \beta_{cM} \times 100$ equal to 0.1.

3.3 Microscopic and Macroscopic Primary Instabilities

3.3.1 Homogeneous Microstructure

In order to show the main characteristics of the microscopic constitutive response and to analyze the influence of the stiffness contrast between the inclusion and the matrix at first the response of a homogeneous unit cell is determined. In this case the microscopic and macroscopic properties coincide. Firstly the biaxial and uniaxial plane strain response of the unit cell is calculated. More specifically the dimensionless First Piola Kirchhoff stress (force per unit width applied on the block) versus the load parameter under tension and compression is plotted in Fig. 5. Note that the equibiaxial has the stiffest response under tension, while the response of the uniaxial case is the stiffest in compression, except for low levels of compression.

The stability analysis carried out for the homogeneous microstructure shows that the macroscopic stability condition Eq. (24) is always satisfied. It turns out from the modified Van Hove theorem, valid for a constant strongly elliptic nominal moduli tensor and a rectangular domain [23], that the microscopic stability condition is also satisfied:

$$\text{constant } \boldsymbol{C}_0^R \Rightarrow \int_V \boldsymbol{C}_0^R \left[\nabla \dot{\boldsymbol{w}}(\boldsymbol{x})\right] \cdot \nabla \dot{\boldsymbol{w}}\,(\boldsymbol{x})\, dV > 0 \quad \forall \dot{\boldsymbol{w}}(\boldsymbol{x}) \in V\#, \nabla \dot{\boldsymbol{w}}\,(\boldsymbol{x}) \neq \boldsymbol{0}$$
$$\Rightarrow \Lambda\left(\bar{\boldsymbol{F}}\right) > 0 .$$

The stability analysis has been carried out also with reference to some representative conjugated macroscopic stability conditions. For both the uniaxial and equibiaxial cases, all the examined conjugated macroscopic stability measures are violated, except the $\bar{\Lambda}^{(2)}$and $\bar{\Lambda}^{(1)}$conditions in compression and the $\bar{\Lambda}^{(-2)}$ and $\bar{\Lambda}^{(-1)}$

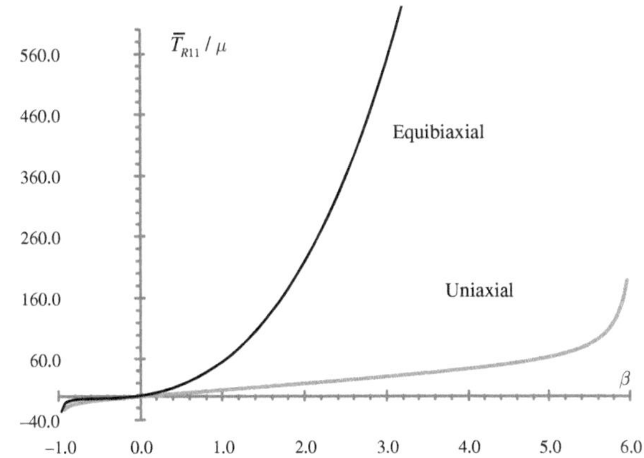

Fig. 5 Plane strain constitutive response of the Gent material under uniaxial and equibiaxial loading

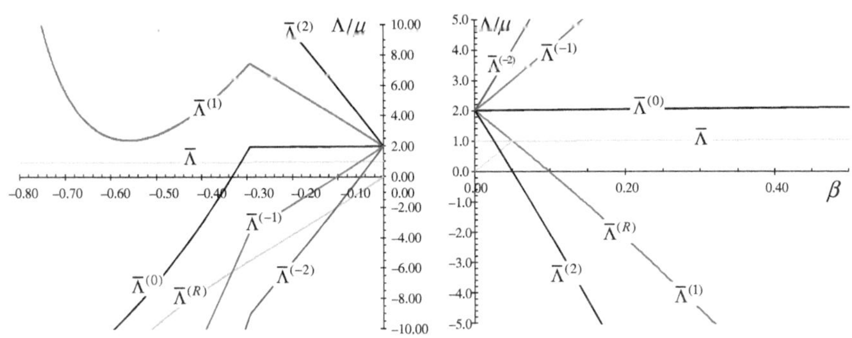

Fig. 6 Stability analysis of a homogeneous microstructure under equibiaxial loading

conditions in tension (Figs. 6 and 7). As expected the macroscopic stability measure is violated at 0^- in compression due to rotational instabilities, and before the macroscopic condition Eq. (24) in tension.

3.3.2 Cellular Microstructure

The homogenized plane strain response of the cellular microstructure, shown in Fig. 8, is determined by two factors: the non-linear effect of the constitutive law exhibiting a stiffening response with increasing strains and the effect related to the varying porosity, softening in tension. The constitutive stiffening dominates the response and as a result, the stress–strain responses in uniaxial loading is softer than in the corresponding plane strain equibiaxial case. Moreover, strain saturation in uniaxial case occurs at a higher level of strain.

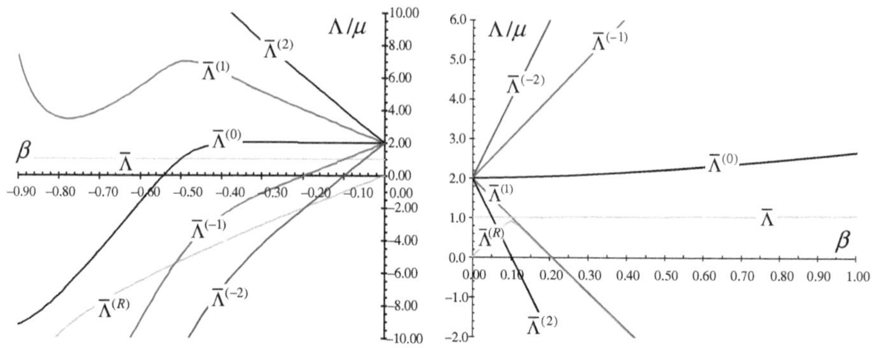

Fig. 7 Stability analysis of a homogeneous microstructure under uniaxial loading

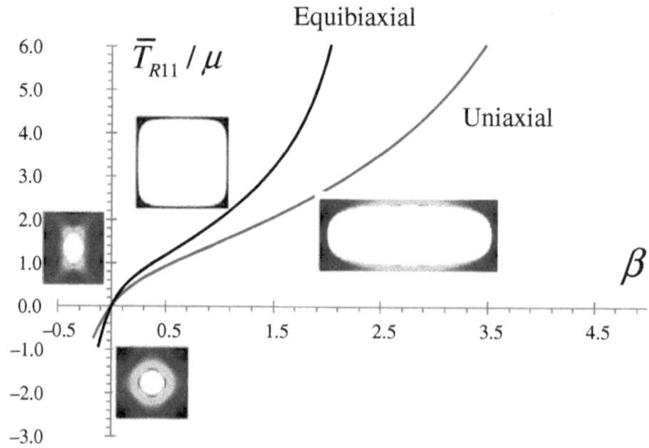

Fig. 8 Homogenized response of the cellular microstructure under uniaxial and equibiaxial loading along the principal solution path

The stability analysis, illustrated in Figs. 9 and 10, shows that in compression the onset of microscopic instability (occurring at $\beta_c^- = -0.07595$ in the biaxial case and at $\beta_c^- = -0.1435$ in the uniaxial one) always precedes the macroscopic loss of strong ellipticity (i.e. the macroscopic loss of stability according to Eq. (24) and the local microscopic instability mode is periodic on a 2×2 cell assembly. The corresponding instability modes in compression for the 2×2 cell assembly are shown in Fig. 11, where it can be noted that the bifurcated mode involves an alternation of void ovalization. In the tension case the first instability mode is global in nature as shown in Fig. 12 and, consequently, the macroscopic loss of stability according to Eq. (24) coincides with the microscopic one and occurs at $\beta_c^+ = 1.695$ in the biaxial case and at $\beta_c^+ = 3.515$ in the uniaxial case.

As far as the macroscopic conjugated stability measures are concerned, numerical calculations have shown that in tension the conditions $\bar{\Lambda}^{(2)}$ and $\bar{\Lambda}^{(1)}$ are first violated before the macroscopic loss of strong ellipticity in the equibiaxial case, whereas in the uniaxial case the $\bar{\Lambda}^{(1)}$is violated after the macroscopic loss of

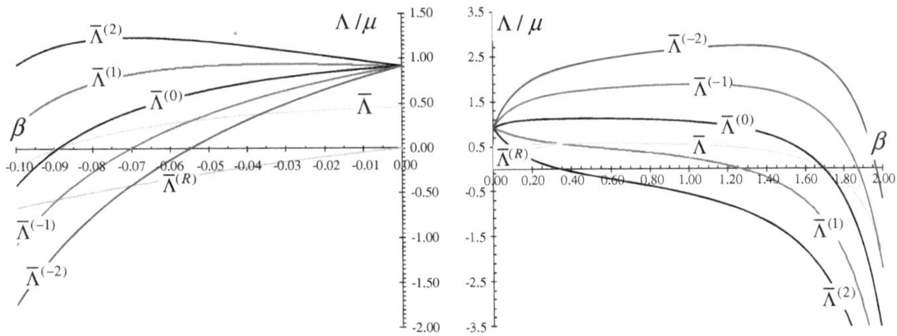

Fig. 9 Stability analysis of a cellular microstructure under biaxial loading

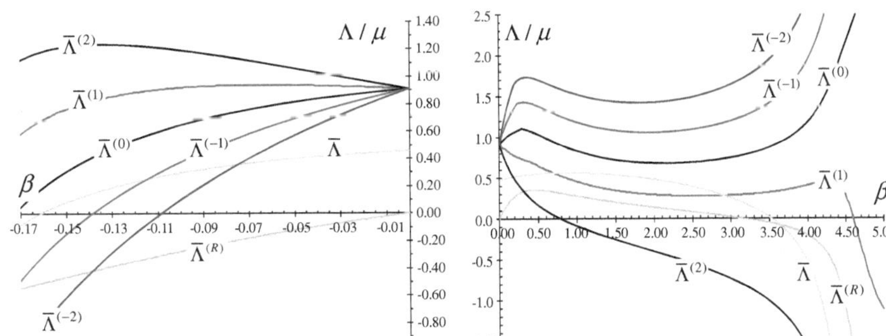

Fig. 10 Stability analysis of a cellular microstructure under uniaxial loading

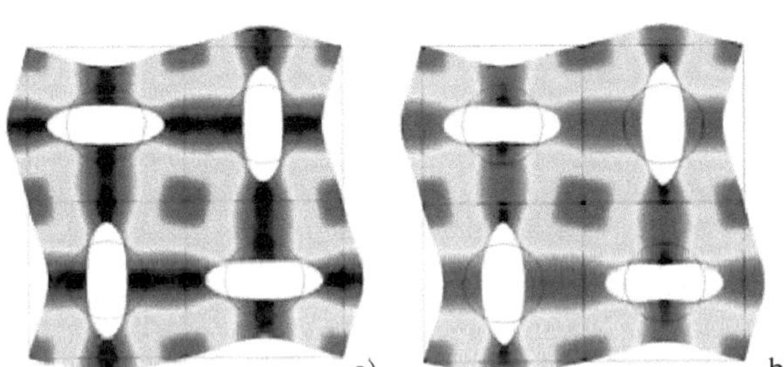

Fig. 11 Primary instability modes energy in the equibiaxial compression (**a**) and uniaxial compression (**b**) case

ellipticity. On the contrary the loss of macroscopic conditions $\bar{\Lambda}^{(-1)}$ and $\bar{\Lambda}^{(-2)}$occur after the macroscopic loss of strong ellipticity in the equibiaxial case, in the uniaxial case being always satisfied. In the compressive case, the situation is reversed since the loss of conditions $\bar{\Lambda}^{(-1)}$ and $\bar{\Lambda}^{(-2)}$ occur before the macroscopic loss of

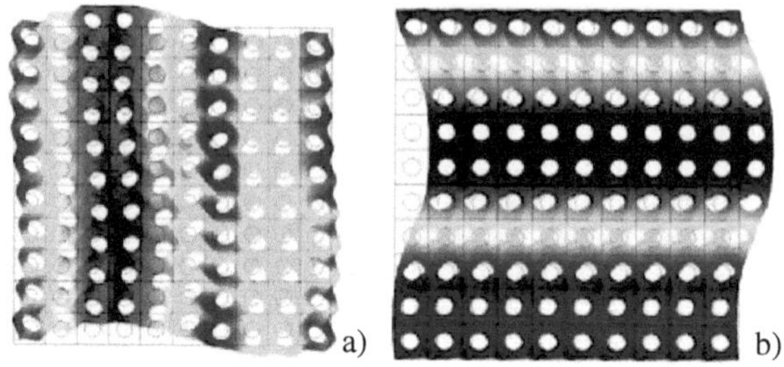

Fig. 12 Instability modes in the equibiaxial tension (**a**) and uniaxial tension (**b**) case

ellipticity, whereas the conditions $\bar{\Lambda}^{(2)}$ and $\bar{\Lambda}^{(1)}$ are first violated after the macroscopic loss of ellipticity. In the equibiaxial case the loss of stability relative to the $\bar{\Lambda}^{(0)}$ condition coincides with the macroscopic loss of strong ellipticity, both in tension and in compression. In the uniaxial case the $\bar{\Lambda}^{(0)}$ condition is not violated in the examined macrostrain range, although it seems approaching zero for larger level of compression. In the examined range of macrostrain, the conditions $\bar{\Lambda}^{(2)}$ and $\bar{\Lambda}^{(1)}$ are not violated in compression, although show a decreasing behavior and tend to have a root for larger levels of strains.

The sequences of the onset of macroscopic instabilities according to the conjugated stability measures are:

$$\text{in the equibiaxial case}: \beta_{cM}^{(2)+} < \beta_{cM}^{(1)+} < \beta_{cM}^{(0)+} = \beta_c^+ = \beta_{cM}^+ < \beta_{cM}^{(-1)+} < \beta_{cM}^{(-2)+},$$

$$\beta_{cM}^{(0)-} = \beta_{cM}^- < \beta_c^- < \beta_{cM}^{(-1)-} < \beta_{cM}^{(-2)-},$$

$$\text{in the uniaxial case}: \beta_{cM}^{(2)+} < \beta_c^+ = \beta_{cM}^+ < \beta_{cM}^{(1)+},$$

$$\beta_{cM}^- < \beta_c^- < \beta_{cM}^{(-1)-} < \beta_{cM}^{(-2)-},$$

showing that the $\bar{\Lambda}^{(2)}$ condition provides conservative microscopic instability estimates in tension whereas the conditions $\bar{\Lambda}^{(-1)}$ and $\bar{\Lambda}^{(-2)}$ provide conservative predictions in compression. Moreover, the $\bar{\Lambda}^{(1)}$ condition is able to provide conservative predictions only in the equibiaxial case. The critical value of the load parameter for the examined stability measures are shown in Table 1.

In the equibiaxial case the $\bar{\Lambda}^{(0)}$ condition gives an exact prediction of the microscopic instability, within numerical errors related to the FE discretization. On the other hand in the uniaxial case the $\bar{\Lambda}^{(0)}$ condition gives an unconservative microscopic instability load prediction.

Among the proposed conjugated stability measures, the $\bar{\Lambda}^{(1)}$ and the $\bar{\Lambda}^{(-1)}$ conditions give the less conservative prediction of the microscopic critical load parameter in equibiaxial tension and compression, respectively. In the uniaxial case the less conservative predictions are provided by the $\bar{\Lambda}^{(2)}$ and the $\bar{\Lambda}^{(-1)}$ conditions in tension and compression, respectively.

Table 1 Cellular microstructure: critical load parameter values for the microscopic and macroscopic stability measures

Tension	$\beta_{cM}^{(2)+}$	$\beta_{cM}^{(1)+}$	$\beta_{cM}^{(0)+}$	β_c^+	β_{cM}^+	$\beta_{cM}^{(-1)+}$	$\beta_{cM}^{(-2)+}$	$\beta_{cM}^{(R)+}$
Equibiaxial	0.345	1.255	1.695	1.695	1.695	1.875	1.965	1.255
Uniaxial	0.785	4.605	–	3.525	3.525	–	–	3.155
Compression	$\beta_{cM}^{(2)-}$	$\beta_{cM}^{(1)-}$	$\beta_{cM}^{(0)-}$	β_c^-	β_{cM}^-	$\beta_{cM}^{(-1)-}$	$\beta_{cM}^{(-2)-}$	$\beta_{cM}^{(R)-}$
Equibiaxial	–	–	–0.0894	–0.075	–0.0894	–0.0703	–0.0550	0
Uniaxial	–	–	–	–0.143	–0.1615	–0.1385	–0.1085	0

As expected the macroscopic stability measure (28) is violated at 0^- in compression due to rotational instabilities, and before the macroscopic condition Eq. (24) in tension. In particular, we have:

$$\beta_{cM}^{(2)+} < \beta_{cM}^{(R)+} \leq \beta_{cM}^{(1)+} < \beta_{cM}^{(0)+} < \beta_{cM}^{(-1)+} < \beta_{cM}^{(-2)+}, \; \beta_{cM}^{(R)-} = 0.$$

It follows that in tension the macroscopic stability measure (28) gives conservative estimate of the microscopic critical load parameter. This situation always occurs when the instability mode is global in nature since condition (28) always implies (27). In addition, it can be noted that in the equibiaxial case the loss of macroscopic stability according to the condition (28) occurs with a symmetric incremental deformation gradient ($L = D$) and, consequently, it coincides with the loss of macroscopic $\bar{\Lambda}^{(1)}$ condition.

Finally, the effect of porosity has a destabilizing influence both for the microscopic and the macroscopic stability analysis, with respect to the homogeneous microstructure. As a matter of fact, the macroscopic strong ellipticity condition can be violated at sufficiently high levels of strain for the cellular microstructure, whereas in the homogeneous case the material is always strongly elliptic. The situation in the case of the conjugated macroscopic stability measures is more variegated, with effects which can change from stabilizing and destabilizing according the adopted measure.

3.3.3 Particle Reinforced Microstructure

The homogenized response shows the strong effect of the constitutive law providing stiffening both in tension and compression with strain saturation at a sufficiently large level of strain (see Fig. 13). In the case of the particle reinforced matrix with a stiff inclusion ($\mu_f/\mu_m = 50$) the main difference in the stability analysis from the porous solid is that in the compression case the first microscopic instability mode is always of global type and coincides with the macroscopic instability related to the strong ellipticity condition. On the contrary, in tension the microstructure is always structurally stable for the examined range of deformations.

The stability analysis, illustrated in Figs. 14 and 15, shows that in compression the onset of microscopic instability occurs at $\beta_c^- = -0.475$ in the biaxial

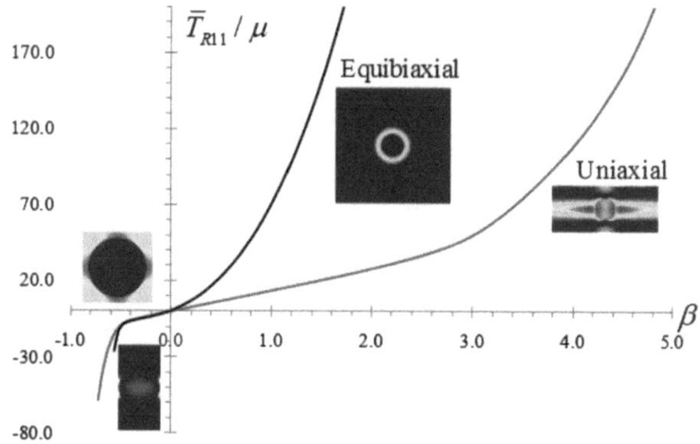

Fig. 13 Homogenized response of the reinforced microstructure under uniaxial and equibiaxial loading along the principal solution path

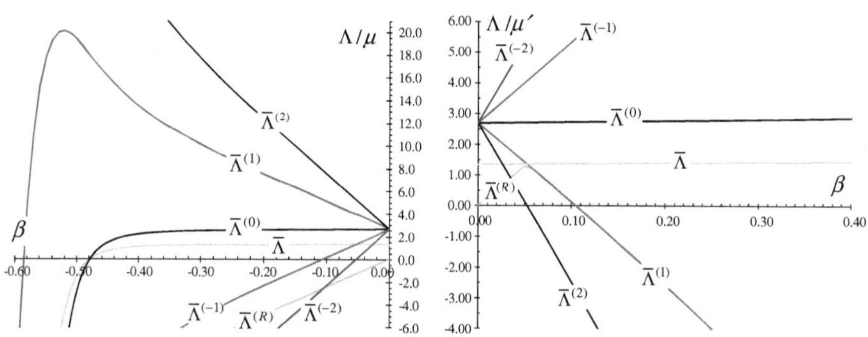

Fig. 14 Stability analysis of a particle reinforced microstructure under equibiaxial loading

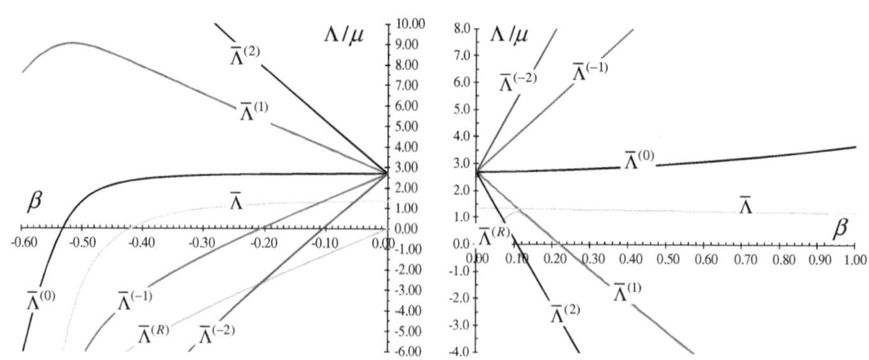

Fig. 15 Stability analysis of a cellular microstructure under uniaxial loading

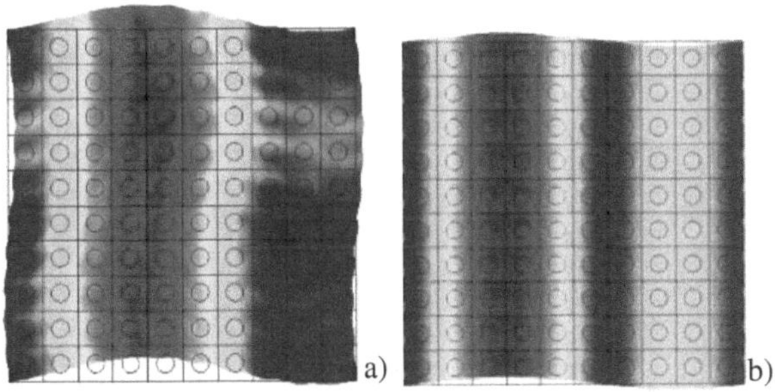

Fig. 16 Instability modes in the equibiaxial tension (**a**) and uniaxial tension (**b**) case

case and at $\beta_c^- = -0.415$ in the uniaxial one. Figure 16 shows one of the four simultaneous modes occurring in the equibiaxial compression case (a) and one of the two simultaneous modes occurring in the uniaxial compression case (b).

The sequence of eventual instabilities related to the conjugated stability measures remains the same of the cellular case in compression where the loss of conditions $\bar{\Lambda}^{(-1)}$ and $\bar{\Lambda}^{(-2)}$ occur before the macroscopic loss of ellipticity. As for the cellular microstructure in the equibiaxial compression case the loss of stability relative to the $\bar{\Lambda}^{(0)}$ condition coincides with the macroscopic loss of strong ellipticity. On the contrary, in the uniaxial case the $\bar{\Lambda}^{(0)}$ condition is eventually violated after the macroscopic loss of ellipticity both in tension and in compression.

The relationships between the onset of macroscopic instabilities according to the conjugated stability measures are:

in the equibiaxial case: $\beta_{cM}^{(2)+} < \beta_{cM}^{(1)+}$,

$$\beta_{cM}^{(1)-} < \beta_{cM}^{(0)-} = \beta_{cM}^{-} = \beta_{c}^{-} < \beta_{cM}^{(-1)-} < \beta_{cM}^{(-2)-},$$

in the uniaxial case: $\beta_{cM}^{(2)+} < \beta_{cM}^{(1)+}$,

$$\beta_{cM}^{(0)-} < \beta_{cM}^{-} = \beta_{c}^{-} < \beta_{cM}^{(-1)-} < \beta_{cM}^{(-2)-},$$

showing that the $\bar{\Lambda}^{(2)}$ and $\bar{\Lambda}^{(1)}$ conditions provide conservative microscopic instability estimates in tension whereas the conditions $\bar{\Lambda}^{(-1)}$ and $\bar{\Lambda}^{(-2)}$ provide conservative predictions in compression. It is worth noting that in the examined range of strains, for equibiaxial tension the conditions $\bar{\Lambda}^{(-1)}$, $\bar{\Lambda}^{(-2)}$and $\bar{\Lambda}^{(0)}$ always predict stability whereas for equibiaxial compression the $\bar{\Lambda}^{(2)}$ condition is always satisfied. Moreover, in the uniaxial case, contrary to the equibiaxial one, $\bar{\Lambda}$ show a decreasing behavior in tension, a symptom of possible instability at very high level of strain.

Among the proposed conjugated stability measures, the $\bar{\Lambda}^{(1)}$ and the $\bar{\Lambda}^{(-1)}$ conditions give the less conservative prediction of the microscopic critical load parameter in tension and compression, respectively. The critical value of the load parameter for the examined stability measures are shown in Table 2.

The macroscopic stability measure Eq. (25) shows the same behavior of the cellular case.

Table 2 Reinforced microstructure: critical load parameter values for the microscopic and macroscopic stability measures

Tension	$\beta_{cM}^{(2)+}$	$\beta_{cM}^{(1)+}$	$\beta_{cM}^{(0)+}$	β_{c}^{+}	β_{cM}^{+}	$\beta_{cM}^{(-1)+}$	$\beta_{cM}^{(-2)+}$	$\beta_{cM}^{(R)+}$
Equibiaxial	0.055	0.105	–	–	–	–	–	0.105
Uniaxial	0.105	0.225	–	–	–	–	–	0.225
Compression	$\beta_{cM}^{(2)-}$	$\beta_{cM}^{(1)-}$	$\beta_{cM}^{(0)-}$	β_{c}^{-}	β_{cM}^{-}	$\beta_{cM}^{(-1)-}$	$\beta_{cM}^{(-2)-}$	$\beta_{cM}^{(R)-}$
Equibiaxial	–	–	–0.475	–0.475	–0.475	–0.105	–0.055	0
Uniaxial	–	–	–0.535	–0.415	–0.415	–0.205	–0.105	0

Finally, it can be noted that the although the microscopic material is always strongly elliptic for the composite microstructure the macroscopic strong ellipticity condition can be violated in compression at sufficiently high levels of strain. The effect of a quasi-rigid inclusion, as expected, provides a stabilizing influence with respect to the cellular case

4 Conclusions and Discussion

The present work deals with the problem of prediction of failure mechanisms induced by microstructural instability phenomena in finitely strained composite materials with heterogeneous periodic microstructure, by examining a macroscopic model of the composite as obtained through an appropriate homogenization method.

This aspect is of considerable relevance since an accurate direct stability analysis of the composite solid taking into account a precise description of its microstructure, may involve severe complications due to the complexity of microstructural configuration and finite changes in constitutive and geometric microstructural properties occurring under finite strains.

Existing criteria based on the homogenized constitutive properties may be not able to provide a conservative prediction of microscopic instability mechanisms. The fundamental macroscopic condition introduced in the literature, corresponding to the strong ellipticity of the homogenized tangent moduli tensor, is able to provide an exact estimate of the microscopic instability critical load only when the instability mode has a global character. On the other hand, an unconservative estimation is obtained in the more general case when the instability mode is local in nature.

To this end a stability analysis on the micro and macro scales is here carried out for incrementally linear materials. Alternative macroscopic stability measures are defined corresponding with the positive definiteness of homogenized moduli tensor associated with a class of work conjugate stress-strain measures, and their ability to obtain conservative prediction of the primary instability load of the microstructure is then investigated.

A nonlinear finite element procedure is developed in order to solve sequentially the unit cell principal equilibrium problem, the incremental equilibrium problems giving the homogenized tangent moduli and the stability eigenvalue problem along a prescribed monotonic macrostrain path. The methodology developed is then applied to representative microstructures with hyperelastic constituents adopting a compressible Gent strain energy function, namely a composite with circular inclusions and a cellular material.

Numerical applications point out that the sequence of primary instabilities depends both on the tensile or compressive nature of the loading path and the kind of microstructure. Consequently, a conservative prediction of the microstructural instability load may be obtained by using an appropriate macroscopic stability criterion. Results show that in the tensile case a conservative predictions of the microscopic stability region can be obtained by using the positiveness condition for the homogenized moduli tensor related to the Lagrange strain measure. The same conclusion can be done with reference to the Biot strain measure, with the only exception of the cellular material under uniaxial tension. On the hand, a conservative prediction in compression in general requires the use of stability measures based on homogenized moduli tensor whose corresponding strain measures are characterized by a negative curvature of the Hill's scale function [18] smaller to that of the logarithmic strain measure (the Almansi strain measure for instance).

References

1. G. Geymonat, S. Müller, N. Triantafyllidis, Homogenization of nonlinearly elastic materials, microscopic bifurcation and macroscopic loss of rank-one convexity. Arch. Ration. Mech. Anal. **122**, 231–290 (1993)
2. COMSOL AB. COMSOL 3.4 Multiphysics User's Guide, Oct 2007
3. E. Sanchez-Palencia, in *Non-homogeneous Media and Vibration Theory*, vol 127, Lecture Notes in Physics (Springer, Heidelberg, 1980)
4. F. Greco, An investigation on static and dynamic criteria of constitutive stability. Mech. Adv. Mater. Struct. **14**(5), 347–363. Corrigendum, Mech. Adv. Mater. Struct. **15**(1), 77–78 (2007)
5. N. Triantafyllidis, B.N. Maker, On the comparison between microscopic and macroscopic instability mechanisms in a class of fiber-reinforced composites. J. Appl. Mech. **52**, 794–800 (1985)
6. R. Hill, in *Aspects of Invariance in Solid Mechanics, Advances in Applied Mechanics*, vol 18 (Academic Press, New York, NY, 1978), pp. 1–72.
7. R. Hill, On uniqueness and stability in the theory of finite elastic strains, J. Mech. Phys. Solids **5**, 229–241 (1957)
8. E.I. Ryzhak, On stable deformation of "unstable" materials in a rigid triaxial testing machine. J. Mech. Phys. Solids **30**, 234–245 (1993)
9. J.M. Ball, Convexity conditions and existence theorems in nonlinear elasticity. Arch. Ration. Mech. Anal. **63**, 337–403 (1977)
10. F. Greco, R. Luciano, Analysis of the influence of incremental material response on the structural stability. Mech. Adv. Mater. Struct. **12**(5), 363–77 (2005)
11. A. Benssousan, J.L. Lions, G. Papanicoulau, *Asymptotic Analysis for Periodic Structures* (North-Holland, Amsterdam, 1978)
12. S. Müller, Homogenization of nonconvex integral functionals and cellular elastic materials. Arch. Ration. Mech. Anal. **99**, 189–212 (1987)
13. J.C. Michel, O. Lopez-Pamies, P. Ponte Castañeda, N. Triantafyllidis, Microscopic and macroscopic instabilities in finitely strained porous elastomers. J. Mech. Phys. Solids. **55**, 900–938 (2007)

14. C. Miehe, J. Schröder, M. Becker, Computational homogenization analysis in finite elasticity. Material and structural instabilities on the micro- and macro-scales of periodic composites and their interaction. Comput. Methods Appl. Mech. Eng. **191**, 4971–5005 (2007)
15. A.N. Gent, A new constitutive relation for rubber. Rubb. Chem. Technol. **69**, 59–61 (1996)
16. R. Hill, in *On Constitutive Macro-variables for Heterogeneous Solids at Finite Strain*, Series A 326. (Proceedings of the Royal Society, London, 1972), 131–147
17. R. Hill, A self-consistent mechanics of composite materials. J. Mech. Phys. Solids **13**, 213–22 (1965)
18. Z. Hashin, S. Shtrikman, On some variational principles in anisotropic and nonhomogeneous elasticity. J. Mech. Phys. Solids **10**, 335–342 (1962)
19. S. Nemat-Nasser, Averaging theorems in finite deformation plasticity. Mech. Mater. **31**, 493–523 (1999)
20. N. Triantafyllidis, M.D. Nestorović, M.W. Schraad, Failure surfaces for finitely strained two-phase periodic solids under general in-plane loading. J. Appl. Mech. **73**, 505–515 (2006)
21. E. De Giorgi, in *Convergence Problems for Functions and Operators*, ed. by E. De Giorgi, et al. Proceedings of the International Meeting: Recent Methods in Nonlinear Analysis (Pitagora, Bologna, 1979), pp. 131–188
22. R. Hill, On constitutive inequalities for simple materials-I,II. J. Mech. Phys. Solids **16**, 229–242 (1968)
23. P. Marcellini, Periodic solutions and homogenization of nonlinear variational problems. Ann. Mat. Pura. Appl. **117**, 139–152 (1978)

Finite Element Vibration Analysis of MHSS Based on 3D Tomography Image Processing

R. Winkler, J. Schulz, M. Merkel, and A. Öchsner

1 Introduction

Cellular solids are materials based on distinct cells and from the macroscopic view they show a high porosity. The boundaries of these cells are usually made of solid ceramic, polymer or metal, while the internal regions are air cavities. The notion of porous and cellular metals first appeared in the beginning of the 1970s [1–3]. The basic idea seeks to imitate the cellular structure of high performance lightweight structures in the nature such as the human osseous structure and can therefore be related to the field of bionic research. Nowadays especially foams made of polymeric materials are widely used in all fields of technology. For example, Styrofoam® and hard polyurethane foams are widely used as packaging materials. Other typical application areas of polymeric foams are the field of heat and sound absorption. During the last few years, techniques for the manufacturing of novel cellular and porous metals have been developed [4, 5]. These materials exhibit a high potential for future oriented applications due to their specific properties. Well-known advantages of cellular metals are their high ability for energy absorption [6, 7], good damping behaviour [8–10], sound absorption [11], excellent heat insulation [12, 13] and a high specific stiffness [14, 15]. The combination of these properties opens a wide field of potential applications, e.g. in automotive, aviation or space-industry [16]. Essential limiting factors for the utilisation are unevenly distributed material parameters [7] and relatively high production costs. Less variation in the physical properties can be achieved with lattice block materials [17]. Metallic hollow sphere structures (MHSS) form a new group of advanced composite materials characterised by high geometry reproduction leading to stable mechanical and physical properties. The MHSS combine the well-known advantages of cellular metals without major scattering of their material parameters. Various joining technologies such as sintering, soldering and adhering can be used to assemble single metallic hollow spheres to interdependent structures and allow to adjust different macroscopic properties [18–20]. In this study, the dynamic response of MHSS plates is

R. Winkler (✉)
University of Applied Sciences Aalen, 73430 Aalen, Germany
e-mail: rolf.winkler@htw-aalen.de

A. Öchsner et al. (eds.), *Materials with Complex Behaviour*, Advanced Structured Materials 6, DOI 10.1007/978-3-642-12667-3_15, © Springer-Verlag Berlin Heidelberg 2010

experimentally investigated by a laser scanning vibrometer and the eigenmodes are analysed. A finite element approach is proposed to predict the dynamic behaviour of MHSS. Numerical results are compared with experimental findings.

2 Vibration Analysis

The analysis of structural vibration is necessary to determine the natural frequencies of a structure and the response to excitation. Vibration analysis is the study of the dynamic properties of structures under vibrational excitation. The experimental vibration analysis measures and investigates the dynamic response of structures under defined excitation as input. One of the interesting results are the modes. A mode is characterized by a mode shape and corresponding eigenfrequency. A mode shape describes the displacement of a surface vibrating at a particular eigenfrequency. A mode shape shows points or locations where the displacement is always zero, they are called nodes or node line. In the current work adhesively bonded MHSS plates are investigated. Figure 1 shows the MHSS sample with dimensions of 300 mm × 100 mm × 30 mm. The diameter of the hollow sphere is about 2 mm and the weigth of a complete specimen is in average 442 g.

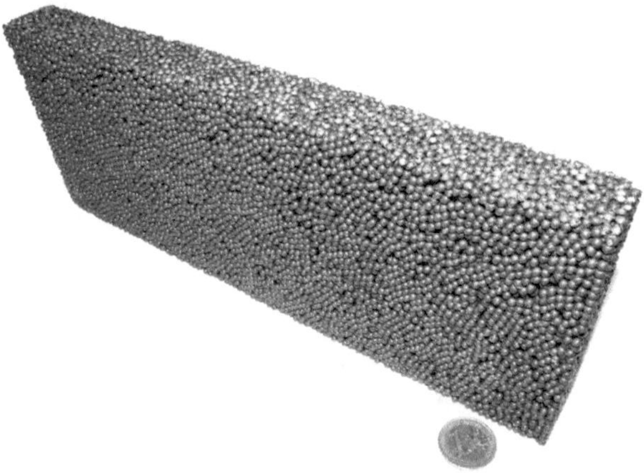

Fig. 1 MHSS sample with adhesively bonded joint between the the spheres

3 Finite Element Modeling

The governing differential equation for vibration analysis without damping can be written as:

$$\mathbf{m}\ddot{\mathbf{x}} + \mathbf{k}\mathbf{x} = \mathbf{0}. \tag{1}$$

This system of equation can be reduced to a standard eigenvalue problem as:

$$\mathbf{kx} = \omega^2 \mathbf{mx}, \tag{2}$$

where ω is the eigenfrequency of the system, $\mathbf{k}$ is the stiffness matrix, $\mathbf{m}$ is the mass matrix and $\mathbf{x}$ is the displacement matrix. There are a lot of techniques to efficiently extract the eigenvalues, forward and inverse iteration method, simultaneous iteration method and block lanczos method. In the current work, the block lanczos method was used to extract the eigenvalues. Advantages of this method are the memory efficiency and that only a few iterations are required for a good approximation. In general, the quality of a FE analysis strongly depends on the details taken into account. A good approximation is to be expected when sufficient details are reflected in the FE model.

3.1 Detailed Modeling

In Fig. 2 the basic unit of the FE model is presented. Up to now, this FE mesh has only been used for static analysis of structures with few spheres. The FE-model for the vibration analysis is built up with thousands of single spheres including joints. It is expected that this model would exceed available computer resources due to the fine mesh. A simplification of this detailed model helps to reduce computing time.

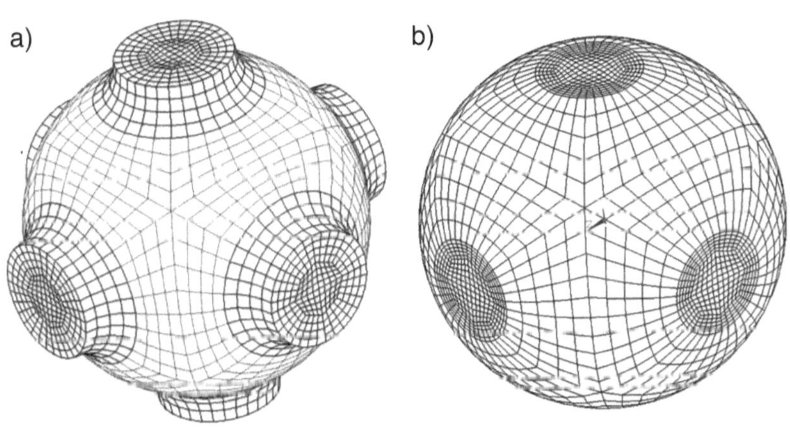

Fig. 2 Detailed FE model of an (**a**) adhesively and (**b**) sintered single hollow sphere

3.2 Homogenization and Unit Cell Method

Homogenization is a widely spread method to simplify a problem. From the point of view of micro mechanical modeling, cellular materials fall into 3 groups: honeycombs which can be studied by two-dimensional (2D) models, open cell foams, a solid scallfold which is dominated by beam-like elements and closed cell foams, in which membrane- or shell-like cell walls are present. Homogenization relations usually take the form of volume averages, so that the homogenized value $\langle f \rangle$ of a variable $f(x)$ can be written as:

$$\langle f \rangle \equiv \frac{1}{\Omega} \int_{\Omega} f(x)\, \mathrm{d}\Omega, \tag{3}$$

where x is the position vector and Ω represents the integration volume. A detailed discussion on homogenization and micro-mechanics can be found in Böhm et al. [21]. The Unit Cell method and Representative Volume Element (RVE) are respectively modeling techniques for the prediction of mechanical behavior of infinite periodic structures on both the micro-scale and macro-scale. The method assumes that the behavior of an infinite periodic structure can be represented by a model of finite size constituting a periodically repeating building block of the geometry in combination with appropriate boundary conditions, which ensure the periodicity of the structure.

3.3 Proper Model

A coarse approach is to replace the hollow sphere by a mass point and joints between spheres by beams. This idea stems from classical mass-spring-damper models used to idealise vibrating systems. The approximation has the following variables: mass, stiffness (bending and torsional) and moment of inertia (geometrical and mass). The objective is to assign proper values to those microscopic variables, so that the

Fig. 3 FE model of a single sphere in a simple cubic structure based on mass point and beams

macroscopic vibration behaviour can be described accurately. Figure 3 shows the approximated model for a single sphere as basic unit of the FE Model. In order to simplify the model, i.e. to clearly separate different effects, all mass is put to the mass point and the beams are defined without mass.

4 Modeling Based on Image Processing

4.1 Computed Tomography

Nowadays X-ray tomography is an established non-destructive material testing method in different technical areas. The three dimensional X-ray tomography (3D X-ray) gives the three dimensional design of the outer contour and inner structure. The 3D X-ray tomography is based on a X-ray source and a detector method. The sample is fixed on a rotation stage between the X-ray source and the detector that acquires the transmitted X-ray images. The X-Ray system in use is a RayScan 200 from Wälischmiller. The X-ray source is a microfocus X-ray tube operated between 10 and 225 kV and the X-Ray detector is an amorphous silicon flat panel with the resolution of 1024 $\times$ 1024 pixels. The present study has the focus on the macro architecture of MHSS. As a result of 3D X-ray scanning one gets a three-dimensional (3D) image with 16 Bit grayscale.

4.2 Gradient Based Method for Detecting Spheres

In the following, a new method is presented to localize the exact position of the spheres in the 3D X-ray images and at the same time determine each sphere radius. This sphere detection method applies the Haar-integration framework used for generic object detection and classification in 2D and 3D images. It has been introduced in [22] and extended e.g. in [23]. In this work, the fact that a sphere is defined by the set of points that have a distance r from the sphere centre, where r is the sphere radius is exploited. The probability that an arbitrary point is the centre of a sphere of radius r is estimated and those points with highest probability are selected. The probability map for all points in the 3D image gives a range of possible radii can be computed in time $O(N)$ with N being the number of voxels in the image. If the spheres all have approximately the same radius, only one run through the 3D image is needed. Besides being fast, the algorithm is extremely robust against disturbances in the sphere structure like not fully closed spheres or deformed spheres. Additionally, the method is only relying on relative gray value changes and is thus robust against variations in the gray values of the spheres and background. The mathematical background of this method is shortly introduced here. The Haar-integration framework can be applied to describe whole images or arbitrary parts of a n-D image independent of a given transformation group. A common example is to characterize objects in an image independent of their position and orientation in the 3D space. This invariance of an image X under a transformation group G is achieved by integrating a kernel function f over all group operations g applied to the image:

$$I_f(X) = \int_G f(gX)\,\mathrm{d}g. \tag{4}$$

The resulting $I_f(X)$ is identical for the image X and all its transformations under the transformation group G. Thus it serves as a transformation invariant descriptor of the image. In the special case of the detection of spherical structures in a 3D space, the rotation invariant descriptor I is computed as follows:

$$I(X, r, x_0) = \int_{O_3} R^{-1} \frac{(\nabla x)(R \cdot r - x_0)}{|(\nabla x)(R \cdot r - x_0)|} \frac{r}{|r|} \mathrm{d}R. \tag{5}$$

Hereby the value $I(X, r, x_0)$ is a direct measure for the probability that the point x_0 is the centre of a sphere with radius r. The transformation group O_3 is the group of rotations, R can be any rotation in 3D space, ∇x denotes the gradient image of X. This rotation invariant descriptor for characterizing spherical structures has been introduced in [24]. Therein the complete mathematical derivation can be found. The integration over all rotation matrices $R \in O_3$ is very time-consuming. Fortunately a fast approximation of the above rotation invariant features has been introduced in [24] and can be applied for the 3D sphere detection. It is based on a voting scheme similar to the one used in the Hough transform [25]. Every voxel x_0 in the 3D image votes for the point x_c that lies in direction of the gradient of x_0 in distance r. This originates from the fact that for any point x_0 on a sphere, the centre of the sphere is x_c. The resulting accumulator is a four-dimensional array with dimensions x, y, z and r. The subsequent application of a four-dimensional Gaussian filter

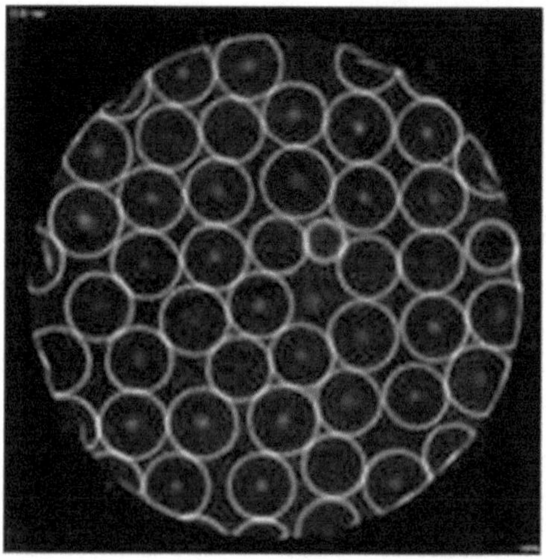

Fig. 4 X-ray image with accummulater map

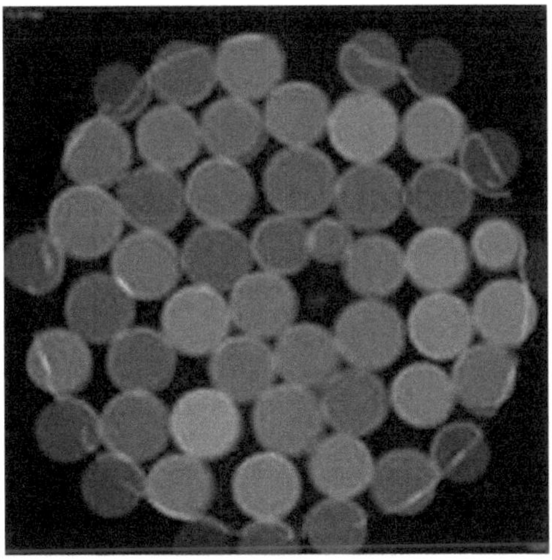

Fig. 5 X-ray image with extracted spheres

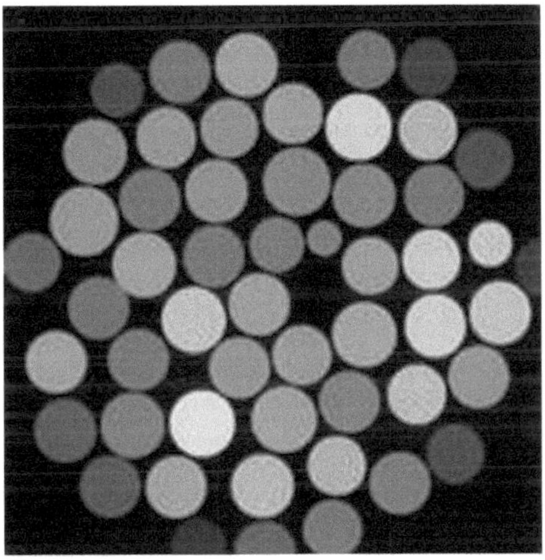

Fig. 6 Localized spheres

provides further robustness against disturbances in the spherical structure. Local maxima in the accumulator correspond to sphere centers in the original 3D image domain. The sphere and radius detection of an example slice of the 3D image is depicted in Figs. 4, 5 and 6. In Fig. 4 the superimposition of the image slice with the accumulator map is shown. The local maxima in the sphere centers are clearly visible. The spheres extracted from these local maxima are shown in Fig. 5 as an overlay with the original image. In Fig. 6 the resulting spheres are shown. These results underline that the proposed algorithm is well suited to localize the spheres reliably and estimate their radii correctly. In our example 3D images, an average 99% of the spheres have been localized correctly with 0.5% false positives.

4.3 FE Model Based on Real Structure

Depending on the sample-size the FE model based on the real structure – proper model resolution was chosen to 0.17 mm/voxel. The 3D image of the whole MHSS plate is 255 × 709 × 1818 voxels. Because the diameter of the spheres varries one voxel plus or minus, the resolution to detect the locations of the hollow spheres is sufficient. The architecture of the MHSS plate was reconstructed by the above mentioned sphere detection method based on the X-ray 3D image. Figure 7 shows the reconstructed CAD volume model of the MHSS plate with over 40,000 detected hollow spheres. In the next step, the mass point – beam model was prepared based on the center coordinates of each hollow sphere. The stiffnesses of sphere and adhesively joint were substituted by a Timoshenko beam with a quadratic cross section. In our calculation we assumed an averaged (homogenized) Young's modulus of about 60,000 MPa for sphere and joint. The rebuilt process from real sample to the FE model can be seen in Fig. 8. All simulations were carried out with the software system ANSYS® (Ansys Inc., Canonsburg, USA). The computation time for a single run took a few minutes on a standard office machine (DualCore AMD Opteron™, 2 × 2 CPU 2.4 GHz, RAM 8 GB). Beams are represented as BEAM4 elements, mass points as MASS21 elements. MASS21 is a point element having up to six degrees of freedom: translations in the local x, y, and z directions and rotations about the local x, y and z axes. A different mass and rotary inertia may be assigned to each coordinate direction. The mass element is defined by a single node. BEAM4 is an uniaxial element with tension, compression, torsion, and bending capabilities. The element is defined by two or three nodes, the cross-sectional area, two area moments of inertia (IZZ and IYY), two thicknesses (TKY and TKZ), an angle of

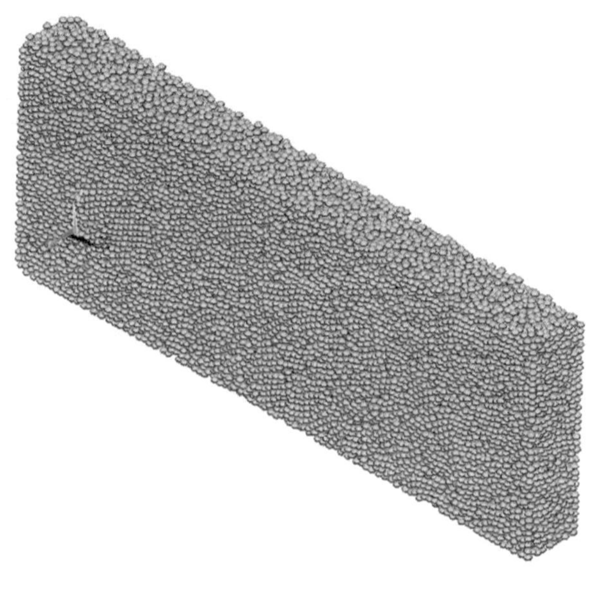

Fig. 7 Reconstructed CAD model of the MHSS plate

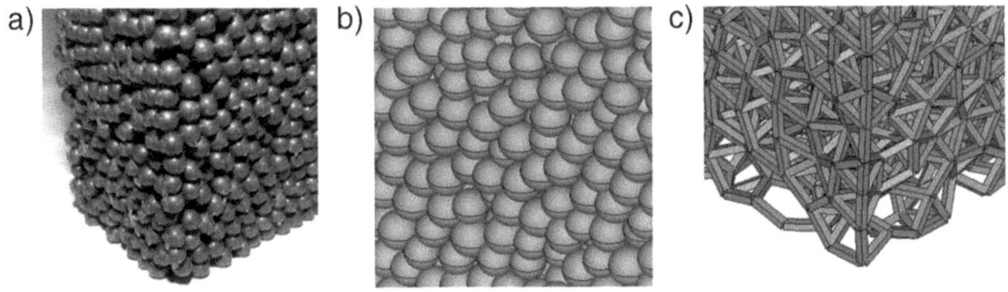

Fig. 8 Computed aided rebuilt process (**a**) real MHSS (**b**) scan data (**c**) FE model

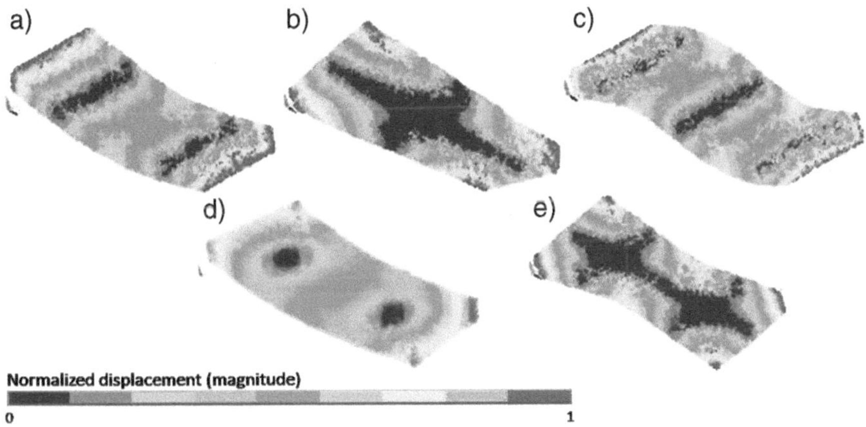

Fig. 9 (**a**) 1st, (**b**) 2nd, (**c**) 3rd, (**d**) 4th and (**e**) 5th mode shape calculated by FE proper model

orientation (Φ) about the element x-axis, the torsional moment of inertia (IXX), and the material properties. The element has six degrees of freedom at each node. In Figs. 9a–e the simulated mode shapes are plotted.

5 Experimental Vibration Analysis

There are different methods to measure the vibration of plates, e.g. with shaker and accelerometers. In our investigations for free vibrating plates, we used a laser scanning vibrometer with impulse hammer as excitation. The vibration behaviour of adhesively bonded hollow sphere plates was experimentally analysed. The pattern of the structure is not unique. It is neither regular, nor body or face centred nor hexagonal closest-packing. The specimens lay on a weak ground made of a foamed plastic to approximate the free–free (F–F) condition of vibration analysis. The vibrations are introduced by impulse hammer. The transverse displacement of

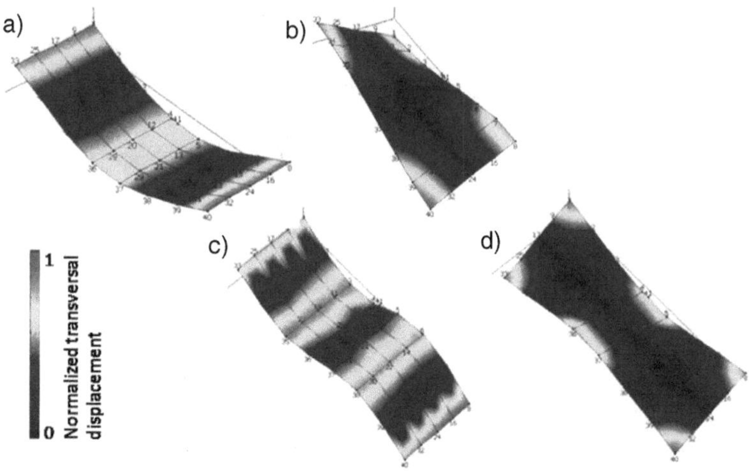

Fig. 10 (**a**) 1st, (**b**) 2nd, (**c**) 3rd and (**d**) 5th measured mode shape of adhesively bonded specimens

the specimen is measured by a laser scanning vibrometer. The post processing of the measurements is done with the Me'ScopeVES® (Vibrant Technology, Scotts Valley, United States) software. Figures 10a–d show the mode shapes of adhesively bonded MHSS specimen. In the first and fifth mode shape we see a bending mode, the second and third mode shape are torsional modes. The fourth mode cannot be detected by this test bench, because the displacements are in plane. Higher mode shapes than the fifth eigenmode cannot be activated because the required energy is too high and spheres will be deformed.

6 Validation

The validation process has to consider the mode shapes as well as the frequencies. The FE model of a proper model with mass points and beams shows equivalent mode shapes. The mode shapes from the FE simulations (Figs. 9a–e) correspond exactly to those gained by experiment and already presented in Figs. 10a–d. In Fig. 11 the eigenfrequencies calculated by the different FE models are plotted. The homogenised FE model are modeled as homogenised mass with constant density, which was defined by experiment. The FE model consists of 3D solid elements [26]. The Young's modulus of this model was assigned 1.100 Pa and Poisson's ratio as 0.15. The FE proper model represents the beam-mass point model with different architecture structure, simple cubic and rebuilt by sphere detection based on X-ray scans. The quadratic beam cross section of the simple cubic model is 0.85^2 mm^2, the cross section for the rebuilt model is 0.6^2 mm^2. So it can been seen by the dimension of the beam cross section that the rebuilt structure shows a stiffer behaviour than the simple cubic structure.Regarding the eigenfrequencies, the results of numerical and

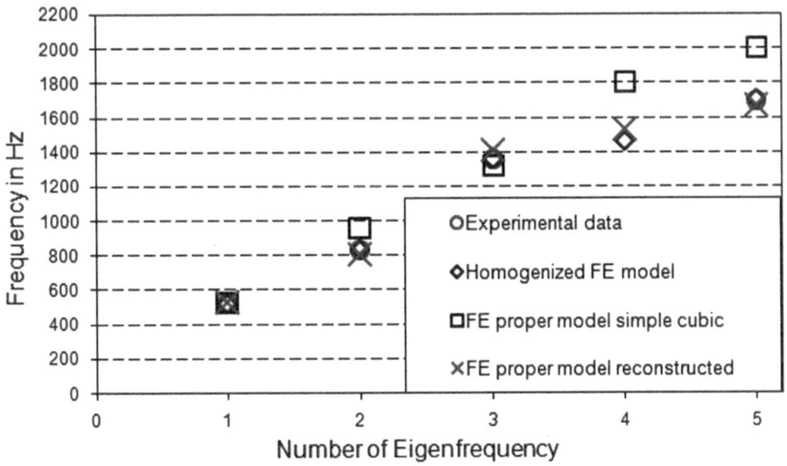

Fig. 11 Measured and simulated eigenfrequencies of an adhesively bonded MHSS plate

experimental analyses fit well. The beam mass point model with the reconstruction structure shows a very good correlation with the measured eigenfrequencies from the experiment.

7 Modeling Improvements

Despite the fact that a simple FE model with constant stiffness along the beam axis delivers quite good results, the need for a more detailed model is obvious when additional components will be added to the hollow sphere structure. These are expected to influence the vibration behaviour. Therefore, we will define an extended FE model, where the bending stiffness as well as the geometrical moment of inertia is formulated as variables along the principal beam axis. In Fig. 12a, the basic configuration is depicted. Two halves of spheres are connected by a joining element. In Fig. 12b, the resulting bending and torsional moment of inertia is plotted along the beam axis. The exact geometrical specifications of hollow sphere structures are described in detail in Öchsner et al. [27]. This non-constant curve for the moment in combination with different Young's moduli results in a non-constant bending stiffness. The FE model has to reflect this by several subsections. Another idea is to split the beam between the mass points into shorter sub-beams or use tapered beams (Fig. 13). Also, the quadratic cross section of the beams can be replaced by other geometries, e.g. circle or ellipse. Additionally, we will investigate different kind of structures with other packing density for the FE model, e.g. body center cubic (bcc), face center cubic (fcc) or hexagonal closest packing (hcp) (Fig. 14).

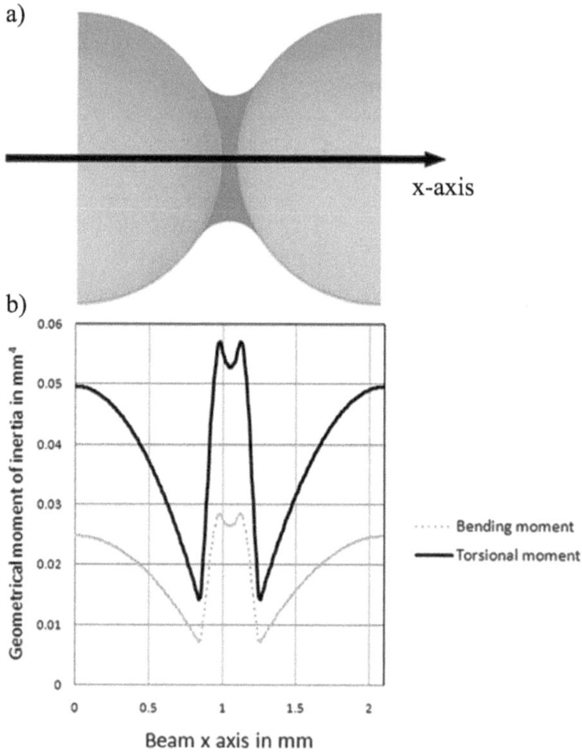

Fig. 12 Extended simulation model. (**a**) Spheres with joint (**b**) geometrical moment of inertia along the beam axis

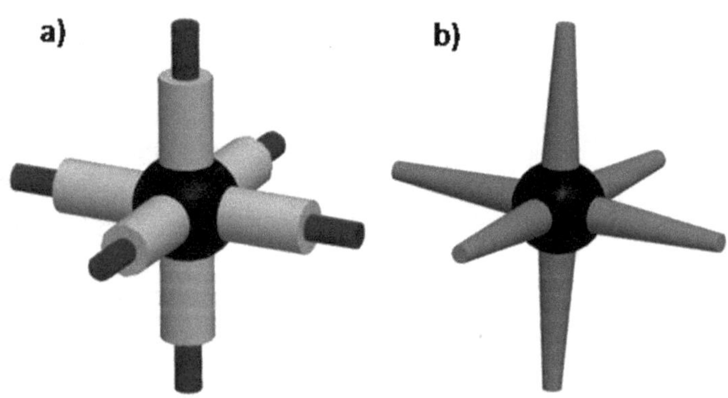

Fig. 13 Extended model with (**a**) sub-beams and (**b**) tapered beams

Fig. 14 (**a**) Bcc, (**b**) Fcc and (**c**) Hcp lattice model

8 Conclusion

In this study, we investigated the vibration behavior of adhesively bonded MHSS plates. Eigenfrequencies and mode shapes were determined experimentally by a vibro laser scanning bench. For numerical analysis there are different modeling methods, multi-scale respectively homogenization method, detailed micro-scale modeling or the presented beam-mass point model. A beam-mass point model approximates a real structure properly, because this model accounts for non-homogenized effects such as structure defects and non regular structures. The computer tomography technology enables to model on basis of real architecture structures. A new method was presented to localize the exact position of the hollow spheres in the 3D X-ray images. The computed results confirm the good quality of the proposed models and computational time is reasonable low.

Acknowledgements We acknowledged the kind help of Timo Bernthaler, Ralf Löffler (Materials Engineering, University Aalen, Germany) and Walter Leis (GTA Foundry Technology Aalen, Germany) for X-ray scanning our MHSS Samples. We thank Glatt Company for providing material. Janina Schulz was with the Chair of Pattern Recognition and Image Processing of the Albert-Ludwigs-University of Freiburg, Germany, while contributing to this work.

References

1. N.N., Metal foam – News. Composites **2**, 11 (1971)
2. H. Bray, Design opportunities with metal foam. Eng. Mater. Des. **16**, 16–16 (1972)
3. R.R. Barton, F.W.S. Carter, T.A. Roberts, Use of reticulated metal foam as flash-back arrestor elements. Chem. Eng. **291**, 708–708 (1974)
4. J. Banhart, Manufacture, characterisation and application of cellular metals and metal foams. Prog. Mater. Sci. **46**, 559–632 (2001)
5. J. Baumeister, überblick – Verfahren zur Herstellung von Metallschäumen. Technische Mitteilungen. **92**, 94–99 (1999)
6. A.G. Evans, J.W. Hutchinson, M.F. Ashby, Multifunctionality of cellular metal systems. Prog. Mater. Sci. **43**, 171–221 (1999)

7. S.L. Lopatnikov, B.A. Gama, J.W. Gillespie, Modeling the progressive collapse behaviour of metal foams. Int. J. Impact Eng. **34**, 587–595 (2007)
8. I.S. Golovin, H.R. Sinning, Damping in some cellular metallic materials. J. Alloys Compd. **355**, 2–9 (2003)
9. I.S. Golovin, H.R. Sinning, Internal friction and damping behaviour of metallic foams and some related structures. Mat. Sci. Eng. **370**, 504–511 (2004)
10. R. Neugebauer, T. Hipke, J. Hohlfeld, R. Thümmler, Metal foam as a combination of lightweight engineering and damping. In *Cellular Metals and Polymers*, ed. by R.F. Singer, C. Krner, V. Altstdt, H. Mnstedt (Trans Tech Publications, Zuerich, 2005), pp. 13–18
11. J. Hübelt, G. Bingel, Excellent sound absorption by metal hollow sphere structures. Cellmet News **1**, 1–2 (2006)
12. T.J. Lu, C. Chen, Thermal transport and fire retardance properties of cellular aluminium alloys. Acta Mater. **47**, 1469–1485 (1999)
13. J.W. Paek, B.H. Kang, S.Y. Kim, J.M. Hyun, Effective thermal conductivity and permeability of aluminum foam materials. Int. J. Thermophys. **21**, 453–464 (2000)
14. M.F. Ashby, A. Evans, N.A. Fleck, L.J. Gibson, J.W. Hutchinson, H.N.G. Wadley, *Metal Foams: A Design Guide* (Butterworth-Heinemann, Boston, MA, 2000)
15. T. Fiedler, A. Öchsner, J. Grácio, G. Kuhn, Modelling the mechanical behaviour of adhesively bonded and sintered hollow-sphere structure. Mech. Compos. Mater. **41**, 405–422 (2005)
16. Alm GmbH, Big AFS test structure for Ariane rocket V booster. Cellmet News **1**, 4–4 (2006)
17. J. Zhou, P. Shrotriya, W.O. Soboyejo, On the deformation of aluminum lattice block structures: From struts to structures. Mech. Mater. **36**, 723–737 (2004)
18. A. Rousset, J.P. Bonino, Y. Blottiere, C. Rossignol, Process for the Production of Porous Metal Bodies. French Patent 8707440, 1987
19. H.P. Degischer, B. Kriszt, *Handbook of Cellular Metals* (Wiley-VCH, Germany, 2002)
20. T. Fiedler, A. Öchsner, On the thermal conductivity of adhesively bonded and sintered hollow sphere structures (HSS). Mater. Sci. Forum **553**, 39–44 (2007)
21. H.J. Böhm, D.H. Pahr, T. Daxner, *Computional and Experimental Mechanics of Advanded Materials, Chapter Analaytical and Numerical Methods for Modeling the Thermomechanical and Thermophysical Behavior ofMicrostructuredMaterial* (Springer, Vienna, 2009), pp. 167–223
22. H. Schulz-Mirbach, Invariant features for gray scale images. *Proceedings of DAGM Annual Pattern Recognition Symposium*, Bielefeld (1995)
23. J. Fehr, O. Ronneberger, J. Schulz, T. Schmidt, M. Reisert, H. Burkhardt, Invariance via group-integration: A feature framework for 3D biomedical image analysis. *Proceedings of Computer Graphics and Imaging (CGIM)*, Innsbruck, Austria (2008)
24. J. Schulz, T. Schmidt, O. Ronneberger, H. Burkhardt, T. Pasternak, A. Dovzhenko, K. Palme, Fast scalar and vectorial grayscale based invariant features for 3D cell nuclei localization and classification. *Proceedings of DAGM Annual Pattern Recognition Symposium*, New York, NY (2006)
25. D.H. Ballard, Generalizing the Hough transform to detect arbitrary shapes. Pattern Recognit. **13**, 111–122 (1981)
26. R. Winkler, M. Merkel, A. Öchsner, W. Günter, On the vibration analysis of adhesively bonded hollow sphere structure. Materialwiss. Werkst. **39**, 139–142 (2008)
27. A. Öchsner, C. Augustin, *Multifunctional Metallic Hollow Sphere Structure* (Springer, Berlin, 2009)

Computational Modelling and Experimental Characterisation of Heterogeneous Materials

A.J. Beveridge, M.A. Wheel, and D.H. Nash

1 Introduction

A loaded material can be classed as either homogeneous; material behaviour is length scale independent, or heterogeneous; local variations in structure produce length scale dependence. This work describes the deformation of materials where the heterogeneous nature becomes significant. Classical elasticity is a continuum model for describing the deformation of homogeneous materials but it is insufficient when the scale of the local structure becomes significant. Therefore an approach differing from classical elasticity is required.

1.1 Generalized Elastic Continuum Theories

One approach is to use a model for a generalized elastic continuum, which is one that takes into account the detail of the underlying structure but is still a continuum model similar to classical elasticity. A number of these theories exist but one of the simplest is that of micropolar elasticity [1]. It is applicable to heterogeneous materials with a matrix that is stiffer than the inclusions [2].

Size effects have been observed experimentally. Structural polyurethane foam beams have displayed, in bending tests, an increase in stiffness as the beam depth tends to zero [3]. Work presented in [4] describes an experimental investigation of the bulk material response of a polymer MEMS cantilever beam. The results show a marked difference between the beam stiffness predicted by classical Euler–Bernoulli beam theory and the observed response. A good correlation was found to a micropolar theory based upon previous work on micropolar plates [5]. The observed beam response was one of increasing flexural stiffness with decreasing thickness.

A.J. Beveridge (✉)
Department of Mechanical Engineering, University of Strathclyde, 75 Montrose Street, Glasgow, G1 1XJ, UK
e-mail: andrew.beveridge@strath.ac.uk

A. Öchsner et al. (eds.), *Materials with Complex Behaviour*, Advanced Structured Materials 6, DOI 10.1007/978-3-642-12667-3_16, © Springer-Verlag Berlin Heidelberg 2010

In order to capture these size effects numerically using micropolar elasticity, additional constitutive properties need to be determined. These are in addition to the classical elastic constants, Young's modulus and Poisson's ratio. Unfortunately identifying these addition constitutive properties is more involved.

1.2 Experimental Determination of Constitutive Properties

The first experimental work carried out on micropolar materials, to determine their constitutive properties, proved to be inconclusive [5]. Indeed the difficulty arose from the inability to find a material that exhibited micropolar material behaviour to a sufficient extent, but the methods outlined in [6] were the first treatment of the micropolar theory to separate and determine the additional constitutive properties.

A more recent review of experimental methods in generalised elastic continua is presented in [7]. The experimental determination of the constitutive properties can be divided into three methods; size effect methods, field methods and wave methods.

The size effects method, used in [8], makes use of the dependency of stiffness upon size of sample. A method that is capable of determining all six micropolar constitutive properties by use of electromagnetic torque generation and interferometric determination of angular displacements is presented in [7]. The stiffness of circular rod specimens of decreasing size were tested in bending and torsion, and these results were compared against analytical solutions to extract the elastic constants. The electromagnetic torque generator was used so as to minimize local loading errors that can obscure size effects in smaller samples.

Field methods can be used to determine the continuum theory that the material is exhibiting. In [9] an analysis of the strain field on the surface of a rectangular section under torsion is presented. It was found that at the edge of the rectangular section the shear strain was none zero, which is not predicted in classical elasticity. A screening method was presented in [10] that used a holographic image to detect the motion of a small corner crack. This motion would be present in a micropolar material but not in a classical continuum material.

Wave methods use the propagation of stress waves to determine the constitutive properties. Micropolar materials exhibit dispersion of plane waves, although this can present difficulties as this dispersion can also be attributed to a viscoelastic response [7]. The advantage of the wave method is that they can be used for large scale materials, e.g. rock formations, which cannot be treated practically using the size effects method.

1.3 Numerical Determination of Constitutive Properties

The use of finite element modelling of the discrete micro structure to determine the micropolar in plane shear and rotational moduli of unidirectional fiber composites with fiber-matrix interfacial de-bonding is suggested in [11]. Subsequently in [12] a finite element homogenisation method to determine micropolar constitutive properties is presented in the context of composite laminates with high stress gradients.

1.4 Numerical Modelling

Computer methods, in particular the finite element method (FEM) have shown limited success in modelling micropolar elasticity. An alternative, the control volume finite element method (CVFEM), has been developed which has shown increased accuracy over the FEM [13]. This is assumed to be due to a condition of local equilibrium being imposed rather than just the global equilibrium that is enforced in the FEM.

1.5 Objective

The objective of this work is to predict an experimentally observed size effect, in a beam of a model micropolar material under a bending load, using numerical models incorporating micropolar behaviour. In order to achieve this an additional material constant is needed over and above the classical engineering constants. This material constant is gained in two ways. The first is experimentally using the size effects method [3] then secondly using a fully detailed finite element model, in ANSYS, once again using the size effect method.

2 Micropolar Elasticity

Micropolar elasticity is capable of describing size effects due to the introduction of a length scale dependent coupled stress, m, and an additional degree of freedom, a micro-rotation, ϕ. For a linear elastic isotropic micropolar material the force stress tensor and couple stress tensor respectively are,

$$\tau_{kl} = \lambda \varepsilon_{mm} \delta_{kl} + (2\mu + \kappa)\varepsilon_{kl} + \kappa e_{klm}(\theta_m - \phi_m) \quad (1)$$

$$m_{kl} = \alpha \phi_{m,m} \delta_{kl} + \beta \phi_{k,l} + \gamma \phi_{l,k} \quad (2)$$

where τ is the force stress tensor, m is the couple stress tensor, ε is the strain tensor, θ is the macro rotation. The macro rotation is usually kinematically distinct from the microrotation but in this formulation a special case where they are equal will be used to derive a simple bending equation for a slender micropolar beam [14]. This special case is usually referred to as coupled stress theory in the literature.

From the moment curvature relationship, see Appendix, the maximum displacement, v_{max} for a micropolar beam under 3 point bending is,

$$v_{max} = \frac{WL^3}{48(E_m I + \gamma A)} \tag{3}$$

where W (N) is the central applied load, L (m) is the length of the beam, E_m (Nm^{-2}) is the micropolar Young's Modulus, I (m^4) is the second moment of area, γ (N) is a length scale dependent micropolar constant and A (m^2) is the crossectional area. This can be rearranged to express the stiffness, K (Nm^{-1}) in terms of the beam depth d.

For a rectangular cross section the substitutions for the second moment of area, I, and area, A, are,

$$I = \frac{bd^3}{12} \tag{4}$$

$$A = bd \tag{5}$$

where b is the breadth and d is the depth, see Fig. 1. The stiffness K is,

$$K = 4E_m b\left(\frac{d}{L}\right)^3\left(1 + \left[\frac{l_r}{d}\right]^2\right) \tag{6}$$

where l_r (m) is the characteristic length in bending for a rectangular cross section,

$$l_r = \sqrt{\frac{12\gamma}{E_m}} \tag{7}$$

For a classically elastic beam the equation for determining the maximum deflection of a beam under three point bending load is

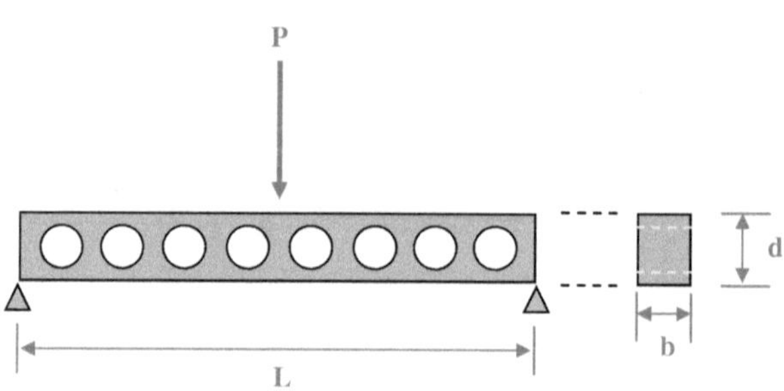

Fig. 1 Heterogeneous beam in 3 point bending. Applied load P, length L, breadth, b and depth d

$$v_{max} = \frac{WL^3}{48E_m I} \tag{8}$$

therefore the stiffness is

$$K = 4E_m b \left(\frac{d}{L}\right)^3 \tag{9}$$

from this it can be seen that in Eq. (6) the expression outside the bracket is that of the classical beam equation and inside is the term associated with the micropolar stiffening. It can also be seen that as the depth of the beam increases the significance of the characteristic length reduces and the solution converges to the equation for a classically elastic beam.

3 Micropolar Beam Element

A four degree of freedom straight micropolar 1D beam element was developed to capture the size effect numerically. The four degrees of freedom describe a cubic lateral displacement field (w_n, θ_n) $(n = 1,2$ cycling for the number of nodes). The derivation follows that of a standard classical beam element [15] but the constitutive relationship has been altered to take account of the micropolar elasticity. The stiffness matrix, **K**, is

$$\mathbf{K} = \frac{(E_m I) + (\gamma A)}{l^3} \begin{bmatrix} 12 & 6l & -12 & 6l \\ 6l & 4l^2 & -6l & 2l^2 \\ -12 & -6l & 12 & -6l \\ 6l & 2l^2 & -6l & 4l^2 \end{bmatrix} \tag{10}$$

where the symbols have the meaning already stated and l is the element length.

4 Experimental and Numerical Results

A model material was manufactured from aluminium bar, $E = 70.0$ MNm^{-2}, $\nu = 0.3$, with a regular pattern of holes, where the holes pass through the axis of bending (Fig. 1). The bar was tested in 3 point bending for various beam depths, d, while maintaining a constant ratio of length, L, to depth (Fig. 2). If the material were classical, the stiffness would remain constant, as the L/d ratio is constant, see Eq. (9).

The hole pattern is arranged in a hexagonal lattice (Fig. 3) with horizontal pitch $P_1 = 0.016$ m and vertical pitch $P_2 = 0.0127$ m. Four test samples were made, varying from one hole to depth to four holes to depth (See Table 1). Once the load deflection data had been gained the stiffness of each sample can be determined.

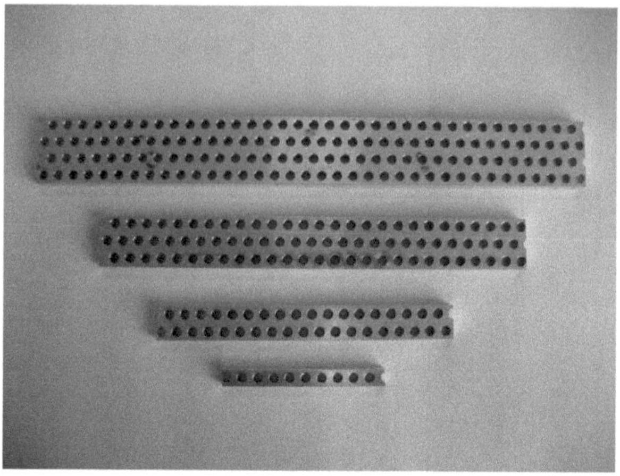

Fig. 2 Model heterogeneous beam test samples, constant L/d

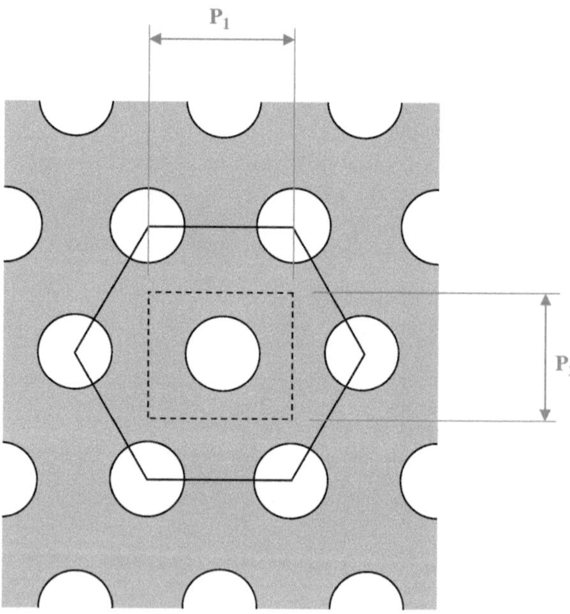

Fig. 3 Hexagonal hole pattern of model heterogeneous material with horizontal pitch, P_1 and vertical pitch P_2

Experimental results are shown in Fig. 4 where stiffness is plotted as a function of the reciprocal of beam depth squared. In addition an FEA analysis of the beam, modelling all the discrete detail was carried out for the model heterogeneous beams. These FEA results are superimposed on Fig. 4. There is a distinct size effect present in both the physical test and the FEA results. Fitting a straight line to the data to gain E_m and γ for both experiment and FEA analysis give rise to the results shown in Table 2. Although Eq. (6) implies that the variation in stiffness should be linear,

Table 1 Size effect test sample. Dimensions and stiffness, K, results

Beam	Depth (mm)	Length (mm)	L/d	K_{EXP} (N/m)	K_{ANSYS} (N/m)
1	12.7	128	10.08	2.832e6	2.903e6
2	25.4	256	10.08	2.147e6	2.195e6
3	38.1	384	10.08	2.038e6	2.043e6
4	50.8	512	10.08	1.964e6	1.982e6

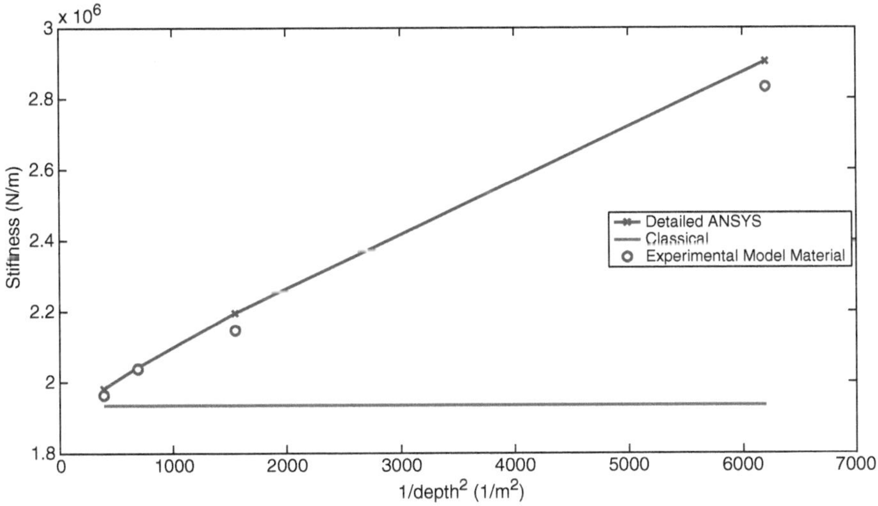

Fig. 4 Stiffness data for experimental and detailed ANSYS results for the model material

Table 2 Constitutive properties from size effect experiments. Physical test, EXP and detailed ANSYS model, ANSYS

	E_m (N/m^2)	γ (N)	l_r (m)
EXP	3.871e10	2.469e5	0.00875
ANSYS	3.900e10	2.629e5	0.00899

both sets of data actually exhibit a slight deviation from linearity. This suggests that the model material is exhibiting truly micropolar material behaviour rather than couple stress behaviour. However, this deviation is difficult to quantify because the shear deformation in the relatively slender beam samples is small in comparison to the bending deformation.

Now that the constitutive properties have been found experimentally and by FEA they can be used in both the numerical models. Figures 5 and 6 show the results from a micropolar beam element while Figs. 7 and 8 depict the results for the micropolar plane stress CVFEM element. The difference between the numerical and experimental results is due to approximations in the trend line fit.

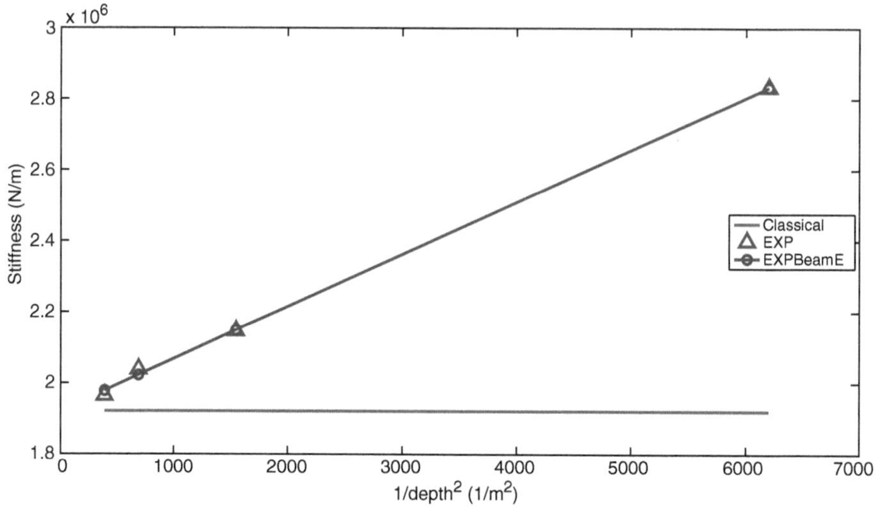

Fig. 5 Stiffness result from Micropolar Beam element (EXPBeamE) using constitutive properties from experimental 3 point bending test of model micropolar material, for a given load (100 *N*). Plotted against stiffness from the experimental procedure (EXP)

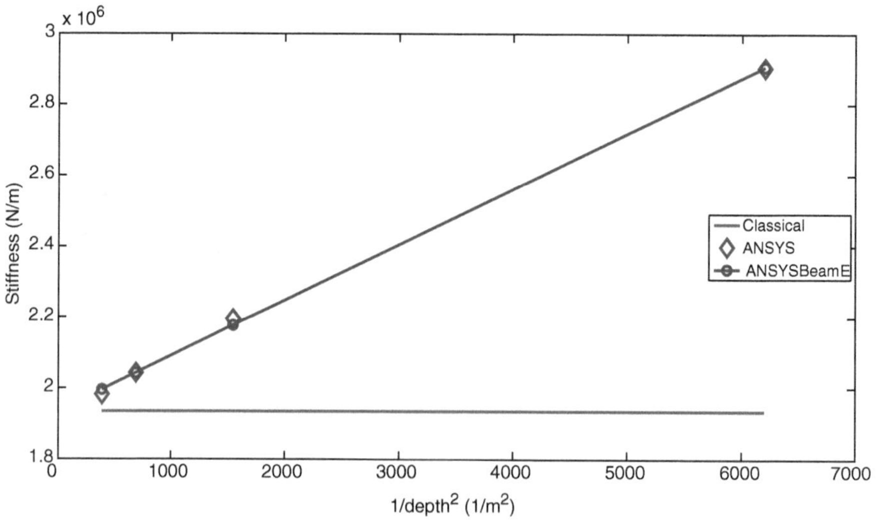

Fig. 6 Stiffness result from Micropolar Beam element (ANSYSBeamE) using constitutive properties from detailed ANSYS model material in 3 point bending, for a given load (100 *N*). Plotted against displacements gained from the ANSYS procedure (ANSYS)

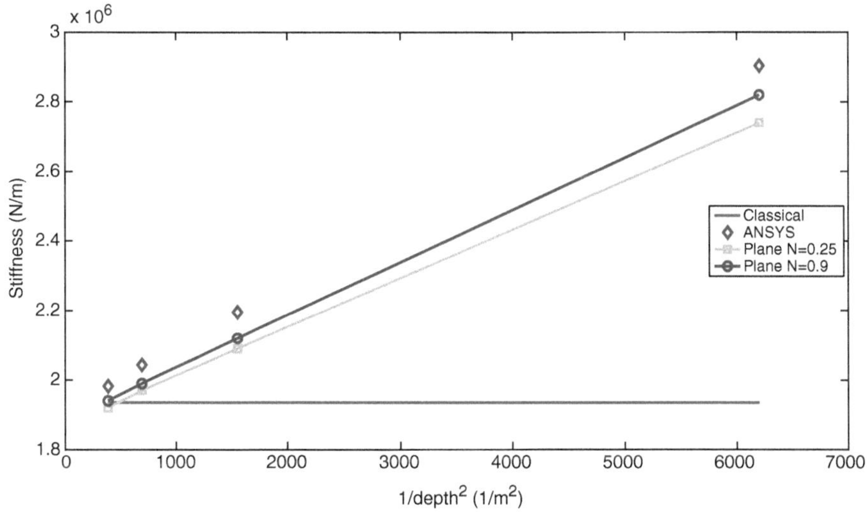

Fig. 7 Stiffness results from micropolar plane stress CVFEM element (plane), for various coupling numbers *N*, using constitutive properties from detailed ANSYS model material in 3 point bending, for a given load (100 *N*). Plotted against stiffness gained from the ANSYS procedure (ANSYS)

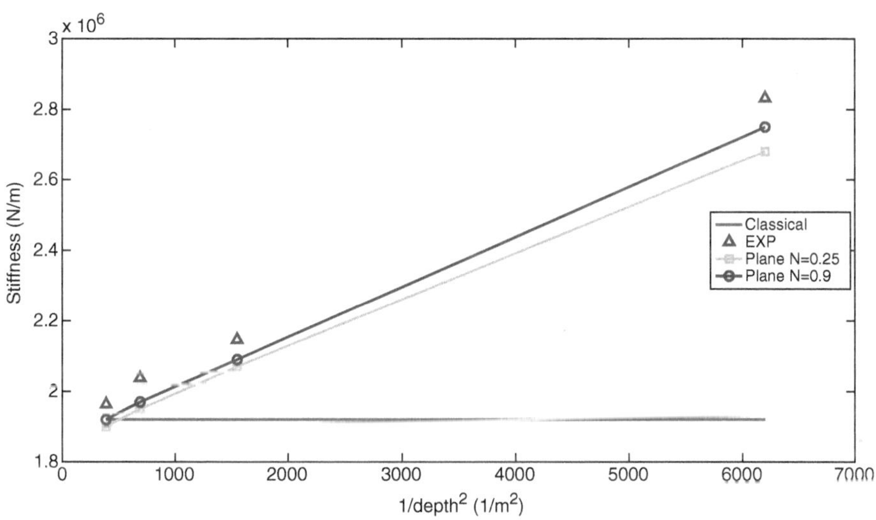

Fig. 8 Stiffness results from micropolar plane stress CVFEM element (plane), for various coupling numbers *N*, using constitutive properties from experimental 3 point bending test of model micropolar material, for a given load (100 *N*). Plotted against stiffness gained from the experimental procedure (EXP)

There is good agreement between the experimental data, the detailed FEA and the numerical procedures. Clearly the size effect has been captured in the numerical procedures incorporating micropolar continuum model. Both provide stiffness predictions close to the detailed ANSYS FEA model albeit at a significantly reduced computational cost.

5 Discussion

As already noted the experimental and detailed FEA data show some deviation from linearity suggesting the presence of some shear deformation in the otherwise relatively slender sample. In full micropolar elasticity shear deformation is governed by an additional parameter, the coupling number, which the beam analysis given in the appendix ignores. However micropolar plane element results, Figs. 7 and 8, indicate that predicted deformation is relatively insensitive to coupling number, $0 \leq N \leq 1$ (although the deviation is noted for $N = 0.25$), therefore the role of shear deformation is secondary for the L/d ratio considered. However, while at present experiments to determine constitutive properties are limited to procedures where analytical solutions exist, the coupling number, usually found from torsion tests, could be extracted using the micropolar plane stress CVFEM element by introducing shear into the same beam sample used in this work. Shear can be introduced by decreasing the L/d ratio. From these results the plane stress CVFEM element could then be used in an inverse procedure to determine the coupling number. This has the distinct advantage over previous methods that only one sample geometry is required.

The voids in the matrix material are a sizeable fraction of the beam depth to reduce the influence of the systematic error and emphasis the size effect. Problems have been identified in past work indicating that the micropolar behaviour is often masked by the error in the experimental procedure [3]. If the void size were sufficiently small the increase in bending stiffness could be within the systematic error of the test procedure. However, the disadvantage of large voids is that they increase the local loading effects. It is therefore suggested that there is a region in which the number of voids is large enough to average out any local loading variation but not so great that the size effect is masked by the testing procedure itself.

If a fully 3D problem were to be addressed then a further two constitutive properties require to be determine; the characteristic length of torsion, dictating torsional size effects and the polar ratio, similar to Poisson's ratio but applicable to the micro-rotation. A 3D analysis is an unrealised challenged. Limited work has been done to determine all the required constitutive properties and an accurate 3D finite element procedures remain unrealised [14].

It is hoped that the presentation of this simple case of beam bending can show the potential for the use of micropolar elasticity in the computational modelling and experimental characterisation of heterogeneous materials. Work is on going to develop further techniques to determine all constitutive properties and to expand the types of elements available.

6 Conclusion

It has been shown that a size effect can be identified in a model heterogeneous material which can be described by micropolar elasticity. The size effect can also be identified in an FE model of the discrete detail of the material which opens the possibility of determining the correct constitutive relationships without lengthy physical testing. Furthermore analytical solutions and numerical methods for solving the micropolar beam problem have been developed that are able to correctly determine the stiffness of the micropolar beam with the size effect present.

7 Appendix: Micropolar Beam Derivation

Once the microrotation is no longer kinematically distinct this simplifies the formulation and the curvature, R of a beam under pure bending, being bent through a small angle is,

$$\frac{1}{R} = \frac{d\theta}{dx} = \frac{d\phi}{dx} = -\frac{d^2v}{dx^2} \tag{11}$$

Considering only the out of plane couple stress, m_z and direct stress σ_x

$$m_z = \gamma \frac{d\phi}{dx} \tag{12}$$

$$\sigma_x = \frac{E_m y}{R} \tag{13}$$

Taking the internal resisting moment equal to any externally applied moment, M,

$$M = \int\limits_A (y\sigma_x + m_z)dA \tag{14}$$

Substituting for $m_z = \frac{\gamma}{R}$ and $\sigma_x = \frac{E_m y}{R}$,

$$M = \frac{1}{R}\int\limits_A (y^2 E_m + \gamma)dA \tag{15}$$

Completing the integration where the second moment of area, I is,

$$I = \int\limits_A y^2 dA \tag{16}$$

and area, A is

$$A = \int_A dA \tag{17}$$

And substituting for $\frac{1}{R} = -\frac{d^2v}{dx^2}$ the curvature relationship is,

$$\frac{d^2v}{dx^2} = -\frac{M}{E_m I + \gamma A} \tag{18}$$

References

1. D. Bigoni, W.J. Drugan, Analytical derivation of cosserat moduli via homogenization of heterogeneous elastic materials. J. Appl. Mech. **74**(4), 741–753 (2007)
2. R.D. Cook, D.S. Malkus, M.E. Plesha, *Concepts and Applications of Finite Element Analysis*, 3rd edn. (Wiley, New York, NY, 1989)
3. A.C. Eringen, *Microcontinuum Field Theories I: Foundations and Solids* (Springer, New York, NY, 1999)
4. R.D. Gauthier, *Experimental Investigations of Micropolar Media* (World Scientific, Singapore, 1981)
5. R.D. Gauthier, W.E. Jahsman, A quest for micropolar elastic constants. J. Appl. Mech. **42**, 369–374 (1975)
6. R.D. Gauthier, W.E. Jahsman, Bending of a curved bar of micropolar elastic material. J. Appl. Mech. **43**, 502–503 (1976)
7. F.Y. Huang, B.H. Yan, J.L. Yan, D.U. Yang, Bending analysis of micropolar elastic beam using a 3-d finite element method. Int. J. Eng. Sci. **38**(3), 275–286 (2000)
8. P. Hutapea, P. Qiao, Micropolar in-plane shear and rotation moduli of unidirectional fiber composites with fiber-matrix interfacial. J. Compos. Mater. **36**(11), 1381–1399 (2002)
9. P. Hutapea, F.G. Yuan, N.J. Pagano, Micro-stress prediction in composite laminates with high stress gradients. Int. J. Solids Struct. **40**(9), 2215–2248 (2003)
10. R. Lakes, Experimental microelasticity of two porous solids. Int. J. Solids Struct. **22**(1), 55–63 (1986)
11. R.S. Lakes, Experimental methods for study of cosserat elastic solids and other generalized elastic continua. In *Continuum Models for Materials with Micro-structure*, ed. by H. Muhlhaus (Wiley, New York, NY, 1995), pp. 1–22
12. R.S. Lakes, D. Gorman, W. Bonfield, Holographic screening method for microelastic solids. J. Mater. Sci. **20**(8), 2882–2888 (1985)
13. A.W. Mcfarland, J.S. Colton, Role of material microstructure in plate stiffness with relevance to microcantilever sensors. J. Micromech. Microeng. **15**(5), 1060–1067 (2005)
14. H.C. Park, R.S. Lakes, Torsion of a micropolar elastic prism of square cross-section. Int. J. Solids Struct. **23**(4), 485–503 (1987)
15. W.A. Wheel, A control volume-based finite element method for plane micropolar elasticity. Int. J. Numer. Methods Eng. **75**(8), 992–1006 (2008)

Model Experiment and Numerical Modelling of Dynamic Soil-Structure Interaction

Noriko Kodama and Kazuhito Komiya

1 Introduction

Earth force or earth pressure during earthquake have been causing numerous ill effects on many buildings and utility infrastructures. In order to overcome the earth force, many seismic design methods have been developed in engineering since early time, for example Mononobe–Okabe earth pressure theoryMononobe-Okabe earth pressure theory [1]. In modern time, seismic technologies like seismic isolation systems have been developed in the building and civil engineering.

Because of the complex boundary conditions of a seismic problem, the use of the Finite Element Method is one of the popular methods to investigate the seismic response of structures. In general, the Finite Element analysis results are closely linked to the dynamic interaction between the soil and the structure.

In a soil-structure dynamic interaction problem, earth pressure applied to the structure and displacement of the structure and soil are the main concerns. The soil-structure interaction is often modelled by soil-springs to apply external forces, as shown in Fig. 1 [2], to introduce traction and to force displacements at the boundaries. However, when the soil deforms largely or fractures, its behaviour depends on soil properties and the stress history. Therefore, in order to understand the soil-structure interaction mechanism, the pressure-displacement behaviour caused by the motion needs to be investigated associated with soil properties.

In this study, the behaviour of an underground structure under seismic loading is simulated numerically. Generated earth pressure and displacement of an underground structure due to earthquake calculated by Finite Element analysis is compared to those of laboratory model shaking experiments. Features of different constitutive models in numerical simulation are discussed.

N. Kodama (✉)
Waseda Institute for Advanced Study, Waseda University, 3-4-1 Okubo, Shinjuku, Tokyo 169-8555, Japan
e-mail: kodama@aoni.waseda.jp

A. Öchsner et al. (eds.), *Materials with Complex Behaviour*, Advanced Structured Materials 6, DOI 10.1007/978-3-642-12667-3_17, © Springer-Verlag Berlin Heidelberg 2010

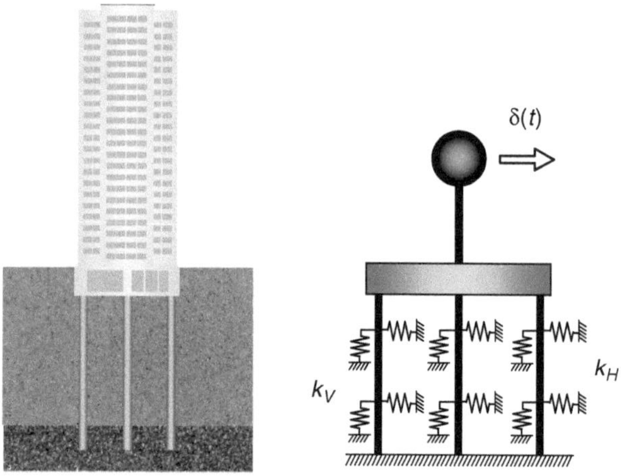

Fig. 1 Typical example of numerical model for dynamic seismic verification

2 Model Shaking Test

Laboratory shaking tests were carried out to investigate a dynamic interaction between soil and a structure. A schematic diagram of the laboratory shaking test apparatus is shown in Fig. 2 [3–5]. The shaking of the tank was allowed in only one horizontal direction as shown in Fig. 2.

For the tests, rectangular parallelepiped model soil grounds of 400 mm width, 300 mm depth and 450 mm height were made at both ends in the shaking tank as shown in Fig. 3. Before the soil grounds were made, a hollow rectangular structure model made of steel with 50 kg mass, 400 mm width, 300 mm depth and 800 mm height was placed at the centre of the tank. On the interface between the base of the structure model and the tank, ball bearings were installed as shown in the side view of Fig. 3, and lateral force transmitted through the interface was cut off by them. The soil grounds were then filled like bridging a gap between the structure model and the tank. When the tank was shaking, therefore, only the lateral forces coming from the soil grounds acted upon the structure model except for bearing vertical support forces.

The earth force applying to the structure model was measured by two pressure transducers located on the structure model as shown in Fig. 3. The relative displacement between the structure model and the tank was measured by a laser distance sensor attached to the tank. The shaking waveform was 6 Hz sinusoidal wave with amplitude 6.0 mm.

Soil ground used in the test was made of Toyoura Sand. The water content of the sand was approximately 9.4%. The soil ground was made by tamping the sand. The mass density of the ground was measured to be 1.10 g/cm^3.

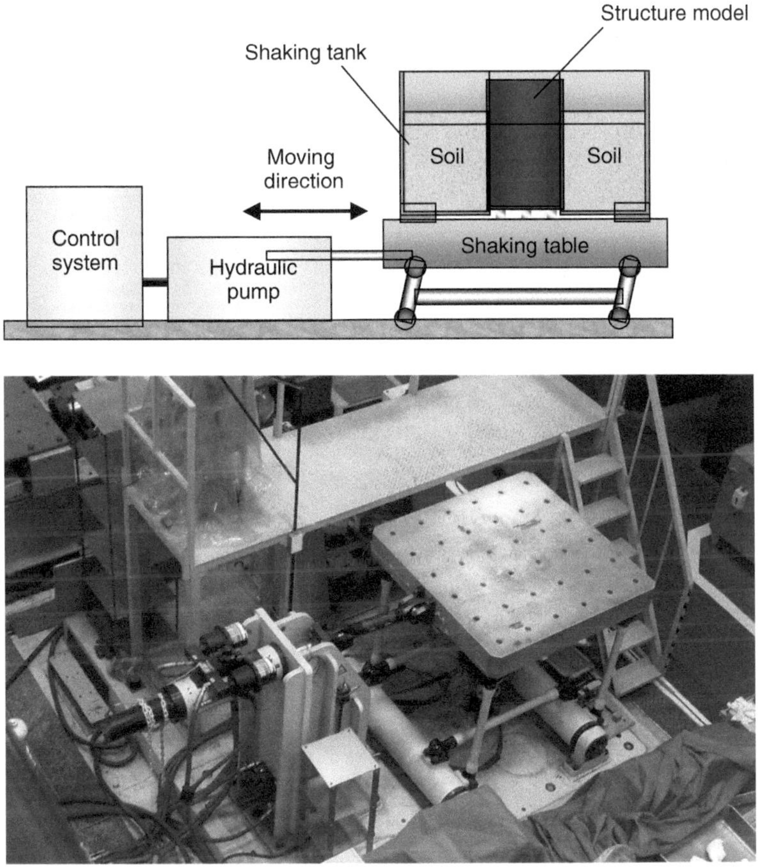

Fig. 2 Laboratory shaking test apparatus

3 Numerical Simulation

3.1 Finite Element Modelling

Two-dimensional Finite Element analyses were carried out to simulate the model shaking test described in the previous section. The finite element model used in the calculation is shown in Fig. 4. The soil and the structure are modelled by two-dimensional plane-strain elements. Sinusoidal acceleration boundary condition was applied to the leftmost and the rightmost nodes of the soil elements. For comparison, the elements for soil were modelled by elasto-viscoplastic materials with five different yield criteria: elastic, von Mises, Tresca, Drucker-Prager and

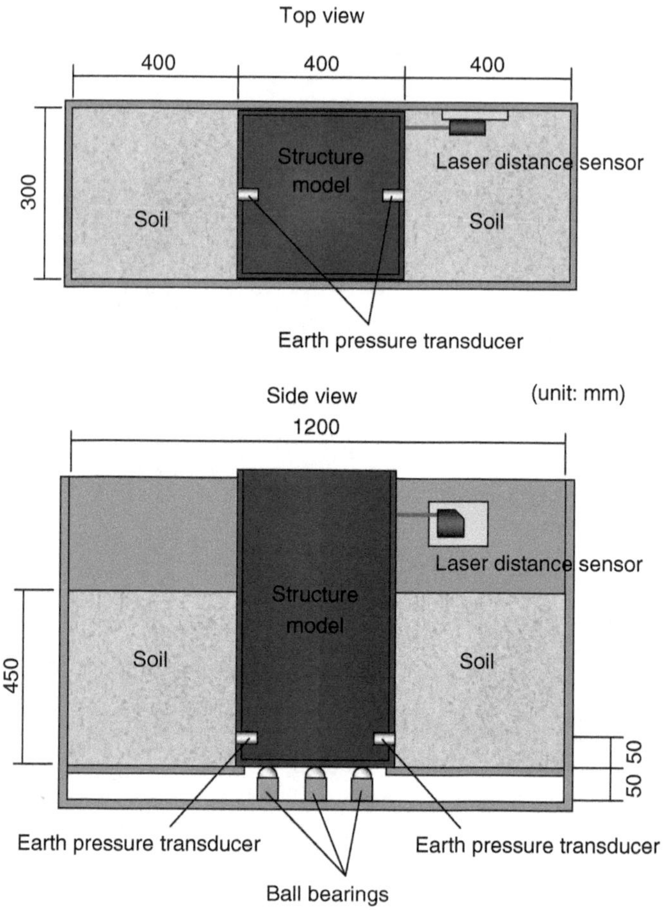

Fig. 3 Shaking tank details

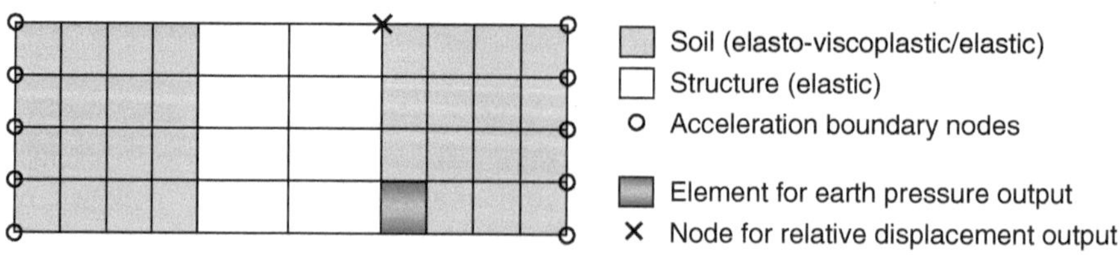

Fig. 4 Finite element model of the shaking tank

Mohr-Coulomb. The latter two criterias (Drucker-Prager and Mohr-Coulomb) can consider the internal friction, which is often observed in soil materials. The structure elements were modelled by elastic material behaviour. Viscoplastic strain velocity $\dot{\varepsilon}_{vp}$ is described as Eq. (1) [6].

$$\dot{\varepsilon}_{vp} = \gamma \frac{F - F_0}{F_0} \frac{\partial \mathbf{F}}{\partial \boldsymbol{\sigma}} \tag{1}$$

where γ is the fluidity parameter, F is the equivalent stress, $F_0 = \sigma_Y$ or $c \cos \phi'$ is the uniaxial yield stress, $\mathbf{F}$ is the yield surface function and $\boldsymbol{\sigma}$ is the stress vector. Material parameters are shown in Table 1. Equation of motion was solved by an explicit direct time integration with a time step 1.0E–6 s and a damping coefficient 4.0 Ns/m.

Table 1 Material parameters in finite element analysis

	Soil					Structure
Parameter	Elastic	von Mises	Tresca	Drucker-Prager	Mohr-Coulomb	(Elastic)
Elastic modulus E (kPa)	2,000	2,000	2,000	2,000	2,000	2.1E+08
Poisson's ratio ν	0.35	0.35	0.35	0.35	0.35	0.30
Mass density ρ (kg/m^3)	1,100	1,100	1,100	1,100	1,100	925.9[a]
Cohesion c (kPa)	–	–	–	1.0E–04	1.0E–04	–
Uniaxial yield stress σ_Y (kPa)	–	0.1	0.1	–	–	–
Friction angle ϕ' (deg)	–	–	–	35.0	35.0	–
Fluidity parameter γ	–	0.005	0.005	0.005	0.005	–

[a] Equals to (structure mass)/(structure elements volume).

3.2 Simulated Earth Pressure and Displacement of Underground Structure

The experimental and the numerically simulated earth pressure and displacement of the structure model are compared in Fig. 5. The calculated hysteresis curve of the Mohr-Coulomb soil model was found to be in good agreement with the experimental results (Fig. 5e). The hysteresis curves of the elastic, the von Mises and the Tresca soil models differed form the experimental results and they gave excessively small values of earth pressure and displacement (Fig. 5a–c). Time histories

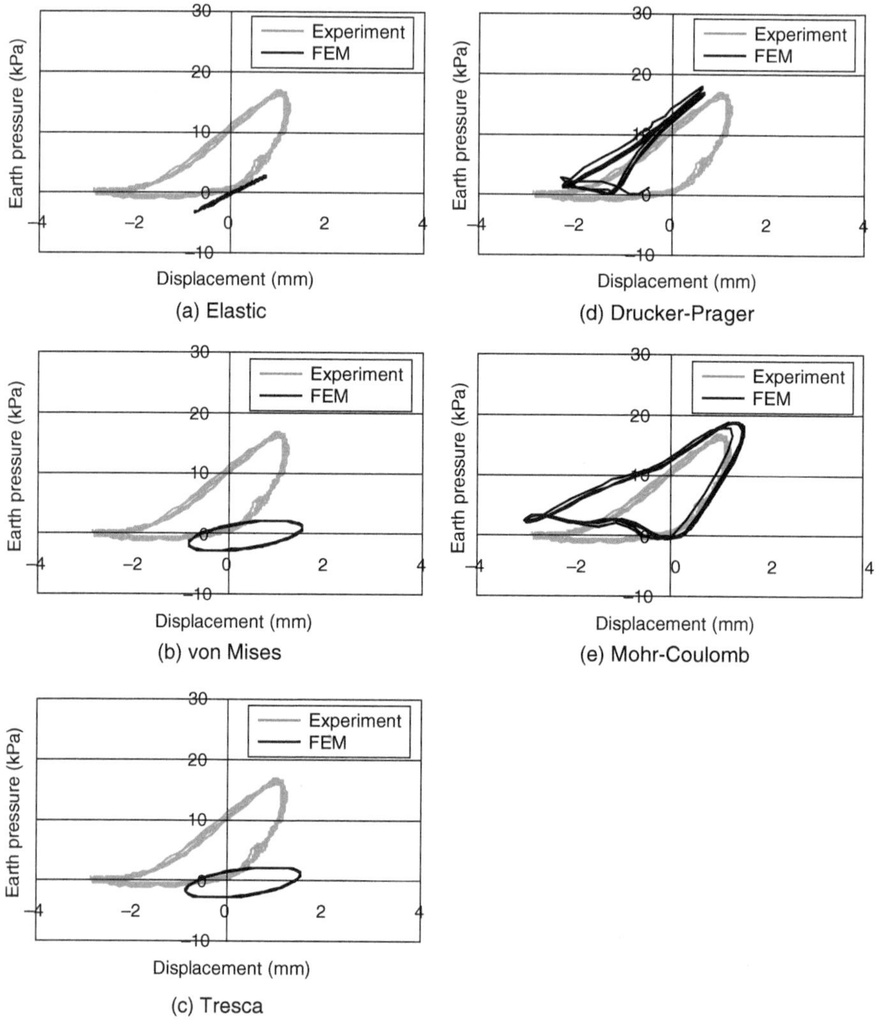

Fig. 5 Earth pressure-displacement relations of the structure model calculated with different constitutive laws for soil elements

of the earth pressure and the displacement obtained experimentally and numerically are shown in Fig. 6. The difference between the time histories seems to be a result of inadequate constitutive modelling during a soil element undergoes tensile deformation.

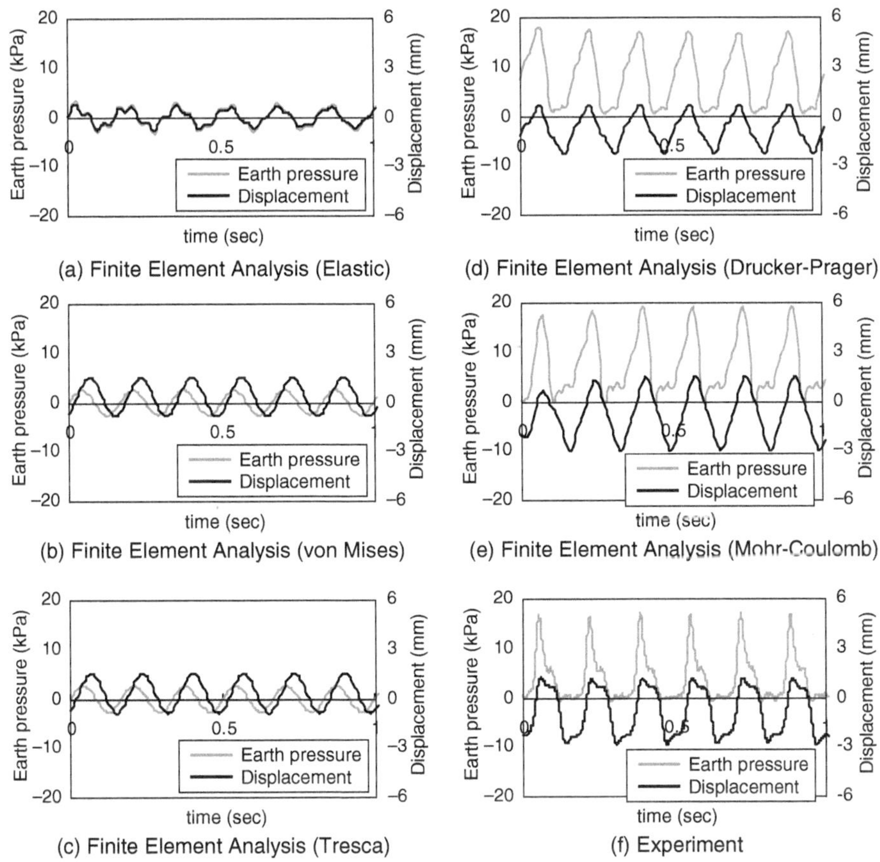

Fig. 6 Experimental and numerically simulated time histories of earth pressure and displacement of the structure model

4 Conclusions

Earth pressure and displacement of a structure model in sand were measured in laboratory shaking tests, and compared to numerically simulated results of two-dimensional Finite Element analysis.

It is concluded that the elasto-viscoplastic model with the Mohr-Coulomb yield criteria is adequate for simulations of soil behaviours under seismic loading. To obtain more precise prediction of earth pressure and displacement time histories, some modification of constitutive model for soil is still needed. Elastic material is used conventionally for modelling of soil in seismic designs, but it may predict excessively low earth pressure which acts on an underground structure.

Acknowledgments The authors would like to acknowledge the financial support form the Grant-in-Aid for Scientific Research (C) (No.19560548) of the Japan's Ministry of Education, Culture, Sports, Science and Technology.

References

1. N. Mononobe, H. Matsuo, *On the Determination of Earth Pressure During Earthquakes*. Proceedings of the World Engineering Conference, vol. 9, pp. 177–185 (1929)
2. Japan Road Association, *Specifications for Highway Bridges, Part V Seismic Design* (Tokyo, Japan Road Association, 2002)
3. K. Komiya, N. Kodama, K. Shikata *Model Experiments on Dynamic Soil-Structure Interaction*. Proceedings of the International Geotechnical Conference on Development of Urban Areas and Geotechnical Engineering, Saint Petersburg, Russia, pp. 245–248 (2008)
4. K. Komiya, N. Kodama, *Model Tests on Dynamic Soil-Structure Interaction*. Proceedings of the 3rd International Symposium on Advanced Fluid/Solid Science and Technology in Experimental Mechanics, Tainan, S1-2-14 (2008)
5. N. Kodama, K. Komiya, T. Morozumi, *Numerical Simulation of Soil-Tunnel Dynamic Interaction*, Proceedings of the 2nd International Conference on Computational Methods in Tunnelling, Bochum, pp. 71–76 (2009)
6. D.R.J. Owen, E. Hinton, *Finite Elements in Plasticity* (Swansea, Pineridge Press, 1980)

Part III
New Technologies

The Laser Butt Welding Simulation of the Thin Sheet Metal

Takeji Arai

1 Introduction

Basically laser welding is a thermal processing, but there are few papers and documents available on distortion caused by laser welding [1–4]. If any, most of the distortion described is entirely-focused on qualitative explanation and few papers provide concrete values of distortion and strain. Furthermore, they deal with bead-on-plate with a continuously-moving heat source on a thin metal plate, which does not accurately reflect an actual welding. However, thermal distortion of the material is extremely important in the joining technology and it cannot be ignored, especially for thin plate welding and precision welding where accuracy is required. Unlike bead-on-plate, there is always a gap between two materials in an actual welding such as butt welding. Therefore, it is necessary to clarify the distortion caused by laser welding performed under the condition where a gap exists.

In this research, an experiment with a butt welding was performed. The shape of the melting zone and the effect of the gap on the distortion were studied. In butt welding with a gap, the relative position between the focal point of laser beam and the material, as well as the rate of laser energy that affects each part of the material are the important factors. Based on this analysis, simulations of the butt welding were performed and the basic distortion of a plate was analyzed.

As a result, it was proved that the overall distortion is smaller in butt welding with a gap compared with bead-on-plate, and the wider the gap is, the smaller the thermal distortion becomes. The distortion behaviors such as thermal expansion and constriction at the molten cross section were studied for small gaps and large gaps that were large enough compared to the plate thickness.

T. Arai (✉)
Chuo University, Tokyo 112-8551, Japan
e-mail: tarai@tamacc.chuo-u.ac.jp

A. Öchsner et al. (eds.), *Materials with Complex Behaviour*, Advanced Structured Materials 6, DOI 10.1007/978-3-642-12667-3_18, © Springer-Verlag Berlin Heidelberg 2010

2 Simulation of Bead on Plate

2.1 Calculation Method

This section describes the calculation model for laser welding. First of all, Fig. 1 shows the experimental set-up of the welding with an actual high power laser.

To clarify the distortion mechanisms of the thin plate and the thermal stress caused in the materials, the finite element method was used and 3-D nonstationary elastoplasticity analysis was performed in which the laser heat source was emitted moving at a constant rate. In this calculation, radiation and convection were taken into account for contact heat transfer at the nodal points of the element [5].

Figure 2 illustrates the flow chart of the simulation calculation. This simulation is basically a thermal analysis simulation using the finite element method but it obtains the temperature field by using a heat transfer equation which takes eradiation, convection and contact heat transfer into account. Then elastic-plastic stress distortion analysis was carried out by giving the thermal strain obtained from the temperature. In this simulation, the laser heat source is replaced with a heat flux which was converted to heat quantity, and the temperature was calculated considering the ratio of beam absorption and the thermal property of the material.

Moreover, stress, thermal strain, transforming strains, plastic strain and elastic strain were determined by using the elastic constants of the materials such as the Young's modulus and the Poisson's ratio. Finally the distortion amount of the entire plate was calculated. Also, temperature dependence was taken into account for the physical properties of the materials and the latent heat was considered as well in the molten state.

In the temperature field, nonstationary heat conduction equation, temperature dependence physical properties, contact heat transmission, joule generation of heat

Fig. 1 Experiment set-up of the laser welding

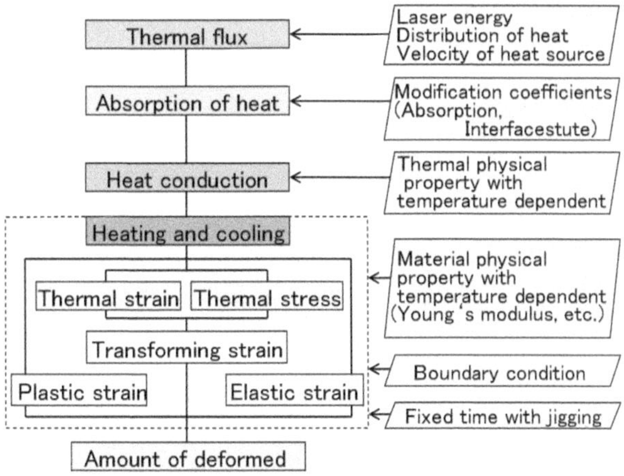

Fig. 2 Flow chart of the calculation

and cooling condition were used as the main equation in the temperature place. At the same time, the equilibrium equation, the temperature dependence physical properties, elastoplasticity, and the thermal stress were used to the calculation in the displacement field. In the calculation, the radiation and convection were also considered.

On the other hand, the equation of continuity, the equation of energy, the motion equation, and the turbulent flow motion equation were used in the calculation of the behavior of the molten metal. In the calculation, gravity, the viscosity, and the surface tension are considered. The calculation concerning the behavior of the molten metal was done by the thermo-fluid analysis.

2.2 Calculation Model of Bead on Plate

For the calculation model, stainless steel (SUS304) of 1 mm thickness was adopted and for the bead-on-plate experiment, a $100 \times 100\,\mathrm{mm}^2$ plate was used. To adjust to the actual power level of the laser, the power was set to 3.4 kW at the material surface with a spot diameter of φ=0.6 mm. To fit into the actual experiment, both sides of the plate were assumed to be held with a jig from the upper side and to be released in a specific time (10 s) after laser irradiation in the calculation model.

Figure 3 shows the analysis model of bead-on-plate which was used for calculation. In this calculation process, the volume expansion and the mass increase caused by oxidization are calculated, which occur in the melting process caused by irradiation of the laser heat source. This clarifies the total distortion in time-series and then the remaining stress and the distortion are obtained as a final state. In bead-on-plate simulation, the calculation result becomes symmetric across the heat source, so that half side was calculated after excluding the angular distortion in the model and then synthesized the whole. Under these conditions, the amount of distortion of the plate, which is supposed to vary depending on the welding speed, was simulated and analyzed.

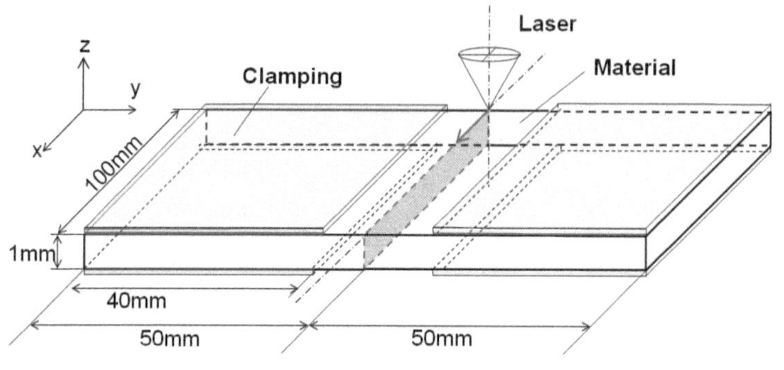

Fig. 3 Calculation model of bead-on-plate

2.3 Calculation and Experimental Results of Bead on Plate

The shape of the heat input was adjusted to the shape of the molten metal cross section created in an actual processing, and the convergent laser beam moved on the material surface at a constant speed. In the calculation, several different focus spot diameters were used. The result with a focus spot diameter $\varphi=0.6$ mm is shown as an example. The beam mode of the laser which is transferred through fiber was assumed to be a leveling top hat type.

As a calculation result, Fig. 4 shows the stress distribution in the direction of the welding line and in the direction of perpendicular to the welding line after the heat source passed. It is recognized that the stress distribution differs between the start and the middle point of the material. The start and end points are subjected to compressive stresses, and the middle part is subjected to tensile stresses.

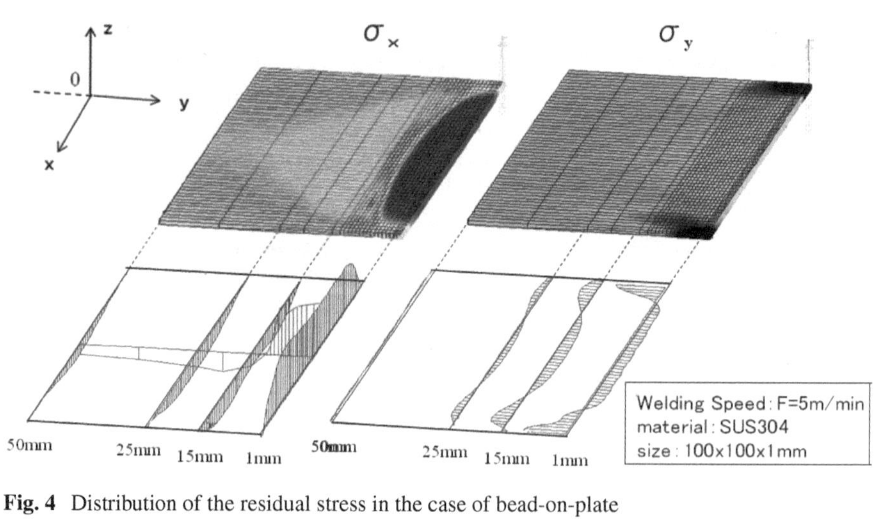

Fig. 4 Distribution of the residual stress in the case of bead-on-plate

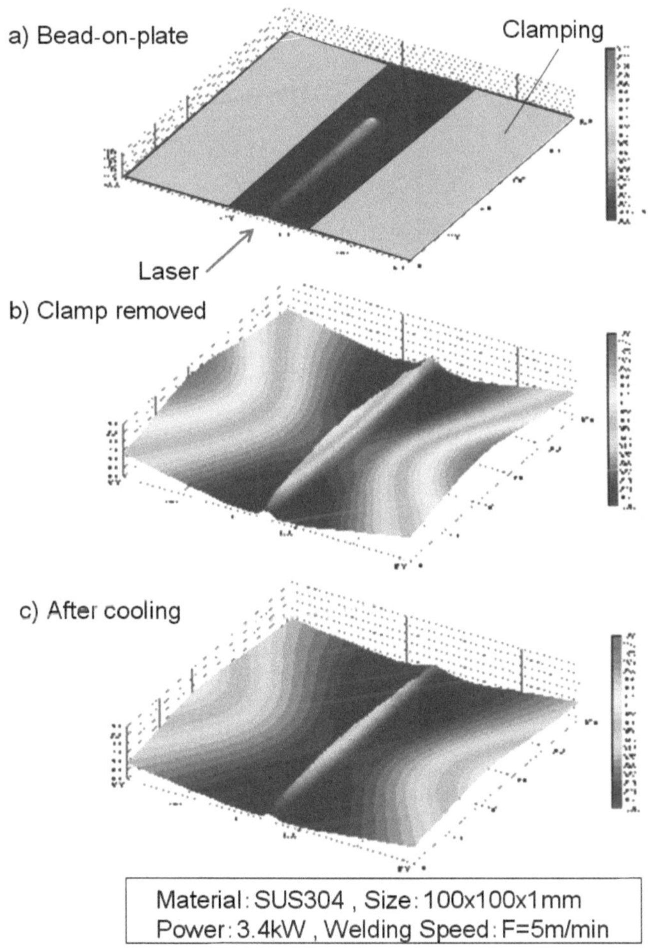

Fig. 5 Moving picture simulation of the distortion in the case of bead-on-plate

The moving picture simulation of bead-on-plate is shown in Fig. 5 This animation was based on the simulation, and showed the excerpted main screen from animation here. Figure 5a shows the welding material fixed from the upper side by clamping. Figure 5b shows that the material *is a rebound* and is greatly deformed when fixation from the upper side is removed. And Fig. 5c shows the material shrinks after cooling and is saturated to a constant deformation value. The amount of deformation depends at time of clamping by jig of the material.

To confirm the accuracy of the simulation, the actual amount of distortion was obtained through an experiment. In the experiment, the distortion amount was measured by a laser displacement gauge. A CCD laser displacement gauge (Keyence LK-G) was used. After welding, the sample was left to stand for sufficient time and was put on the measuring stand in a distorted condition. 50 lines were set every 2 mm

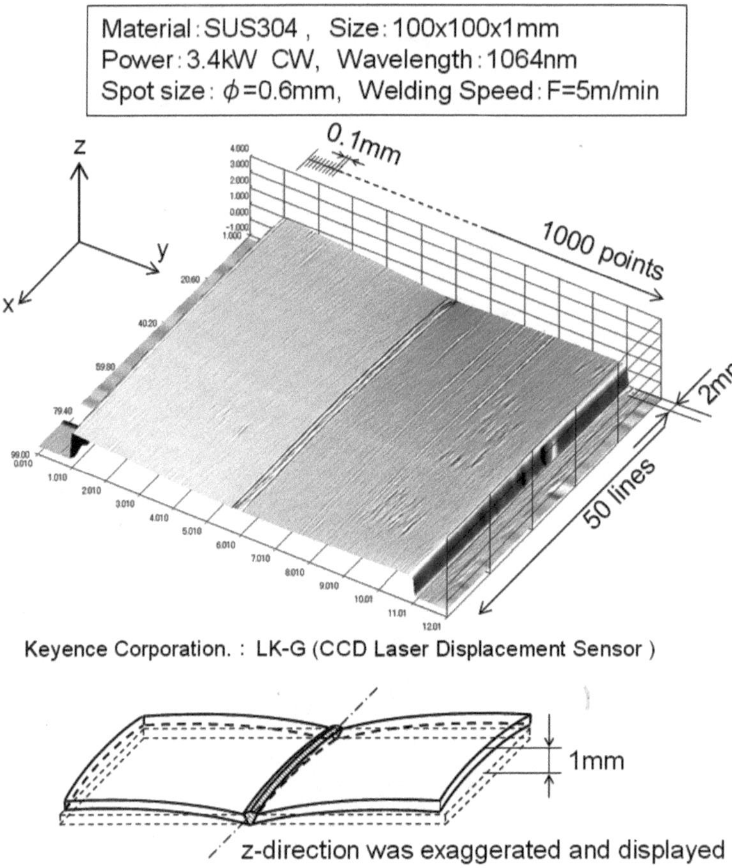

Fig. 6 Measurement result of distortion in the case of bead-on-plate

along the welding line in an orthogonal direction. Each line was divided by 0.1 mm distance and 1,000 points in total were measured. Figure 6 shows one example of the result of a measurement that uses this measuring instrument. The simulation result of the distortion was examined, in which the plate was assumed to be cooled down sufficiently after the heat source passed over.

Figure 7 shows the distortion close to the welding line calculated for bead-on-plate. In this calculation, a plate with 1 mm thickness was processed with the same power level as the above, setting the welding speed in the range of 3–5 m/min. In laser welding, the slower the welding speed gets, the larger the swelling becomes near the center upward (z direction in Fig. 3).

Furthermore, the amount of distortion differs between penetration welding and non-penetration welding. The result shows that the slower the welding speed gets, the larger the distortion amount becomes. In addition, the simulation and the actual

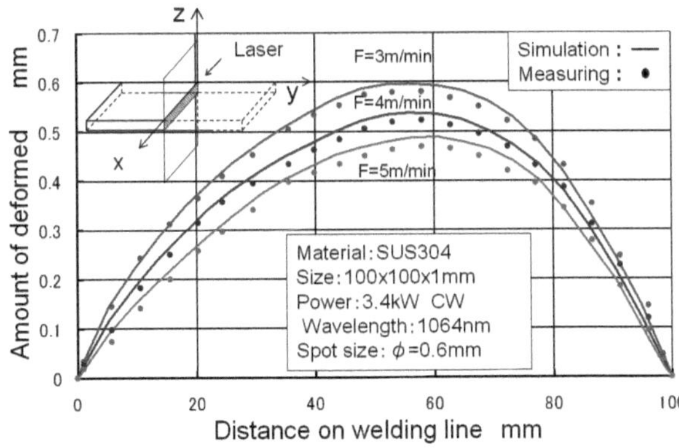

Fig. 7 Result of deformation in bead-on-plate

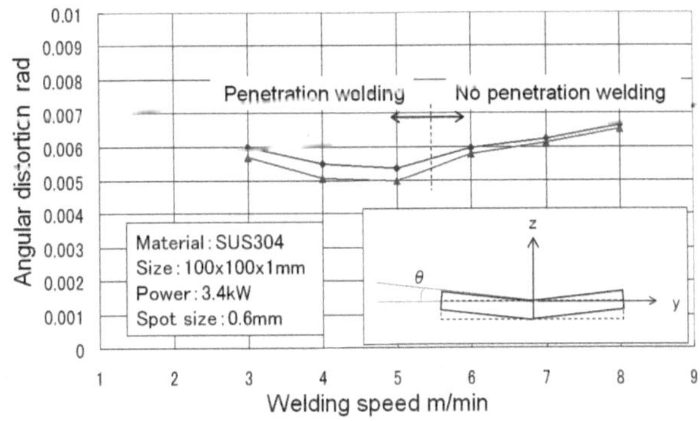

Fig. 8 Angular vs. welding speed

experiment compare reasonably well. The solid line corresponds to the simulation, and the points refer to the measurement result in the experiment (in Fig. 7).

The angular distortion corresponding to welding speed was computed based on the calculation model. The angular distortion in penetration welding is smaller if the welding speed is fast, and the angular distortion in non-penetration welding is bigger if the welding speed is fast, as shows in Fig. 8.

When the laser heat source passes over and the welding process is completed, and after the plate fixed by a jig is released, the following distortion behavior is seen (in Fig. 9). After the laser heat source moves and the welding process ends, the fixation by jig is released and the plate was deformed greatly. When the heat source passes over, the welding portion starts to release the heat, and the distortion amount goes down to a certain amount. As for the duration of the fixation in the jig, the shorter the duration is, the greater the initial rebound and the final distortion amount are. On the contrary, if the fixation duration is long, the initial rebound and the distortion amount are small. By taking enough time to hold the material, no rebound is observed and the distortion amount converges to a certain small amount.

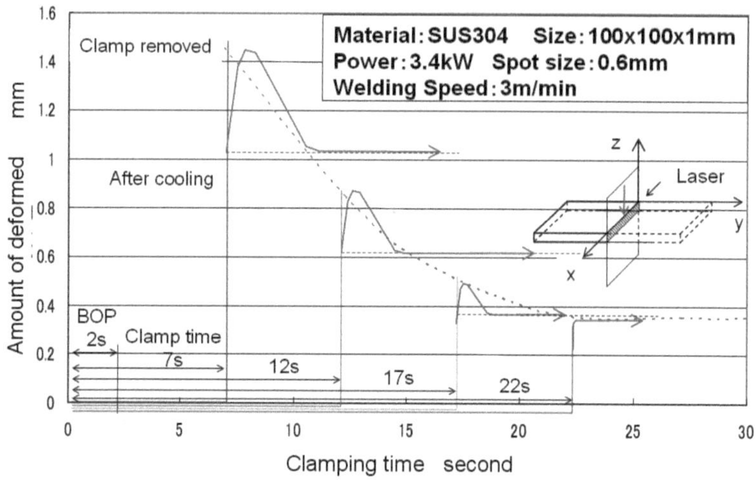

Fig. 9 Clamping time vs. Thermal deformation in the case of bead-on-plate

3 Simulation of Butt Welding

3.1 Calculation Model of Butt Welding

Stainless steel of 1 mm thickness with $100 \times 50\,\text{mm}^2$ size was used. Calculations for butt welding were made with changing the gap by 0.05 mm from 0.05 to 0.25 mm. The calculation model is shown in Fig. 10. The testing piece for butt welding was precisely pre-processed by electro-discharge to get 5 μm roughness at the welding surface. As a result, the material was stuck tight enough so that any gap at the welding surface with visual observation was not recognizable.

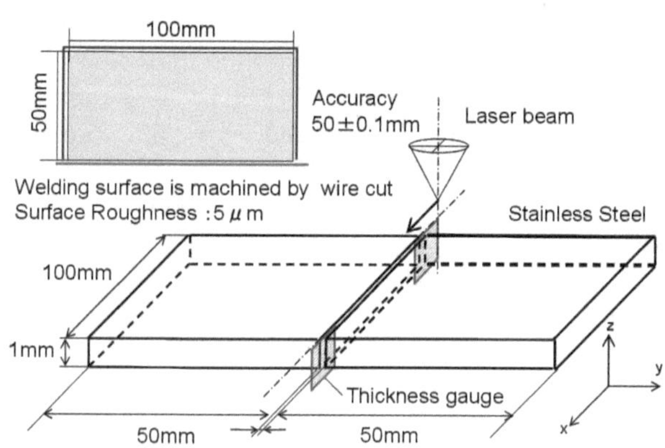

Fig. 10 Actual butt welding model

Through optical measurement, a 5 μm clearance was observed so that the minimum gap at welding was assumed to be 0.005 mm.

3.2 Calculation and Experimental Results of Butt Welding

In the same manor, a laser beam with a focus spot diameter φ=0.6 mm was applied to butt welding with gap. When the beam was concentrated with a lens, there was some aberration near the focal point. Therefore, in the calculation for butt welding with gap, ray tracing method was used at first to calculate the point where in the target material the concentrated beam makes thermal effects [6]. For calculation of ray tracing, 200 lines in the diameter direction were used. Although the place of the heat effect that the concentrating beam gives is different according to the relative position between focuses of a material surface and the beam, Fig. 11 shows an example with "just on focus (focal position is just on the material surface)" where the focal point is right on the material surface. In butt welding, the laser beam is dispersed being. A part of it goes to the material surface or welding surface and some energy passes through without involved in welding. Figure 12 shows how the gap size affects the energy distribution in butt welding. Based on this study, the distortion in butt welding with gap was calculated.

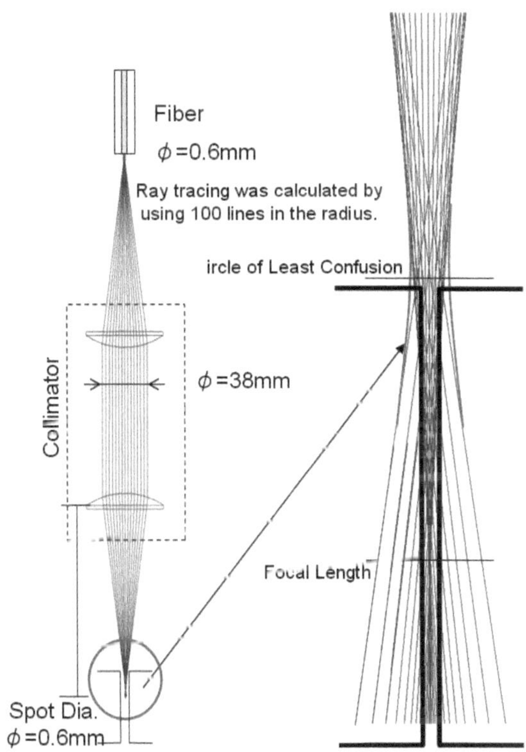

Fig. 11 Optical system of fiber transmission in the case of butt welding

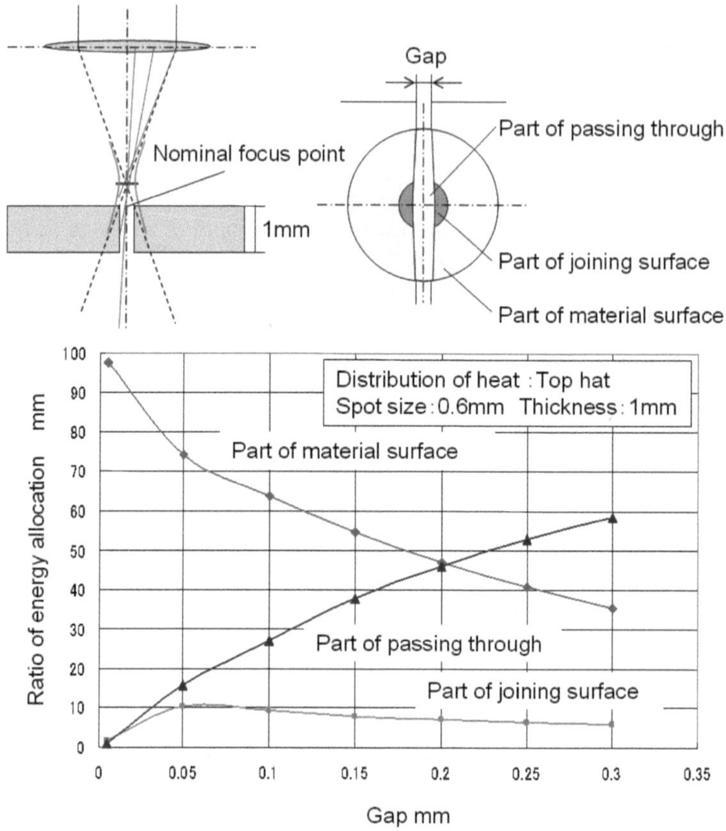

Fig. 12 Laser energy distributions on material and joining gap (butt welding)

The difference of the temperature distribution by the simulation was confirmed by three patterns as shown in Fig. 13. Figure 13a corresponds to bead-on-plate. This is a phenomenon of only the thermal conduction's happening when the laser light beam is irradiated to the surface. Figure 13b is an example of the weld in a butt joint. In this case, the laser beam is given only to the surface of the material. The laser beam goes along a gap and goes through it. And the heat spreads slightly below by the effect of the surface heat transfer of happening in the upper part of the gap of the material. However, the fact is not expressible only by this. All of real phenomena of laser butt welding are not explained. The real cross section by experiment of welding shows deeper penetration depth. A correct calculation of an actual butt welding is obtained by considering that the laser beam irradiates to the joint surface in the gap. Figure 13c shows the calculation example of the temperature distribution considering all laser energy allocations. Therefore, the allocation of laser energy in butt welding must be considered by the computer calculation.

In butt welding, not only the simple thermal stress but also the thermal expansion and oxidization as well as the constriction due to cooling of molten metal

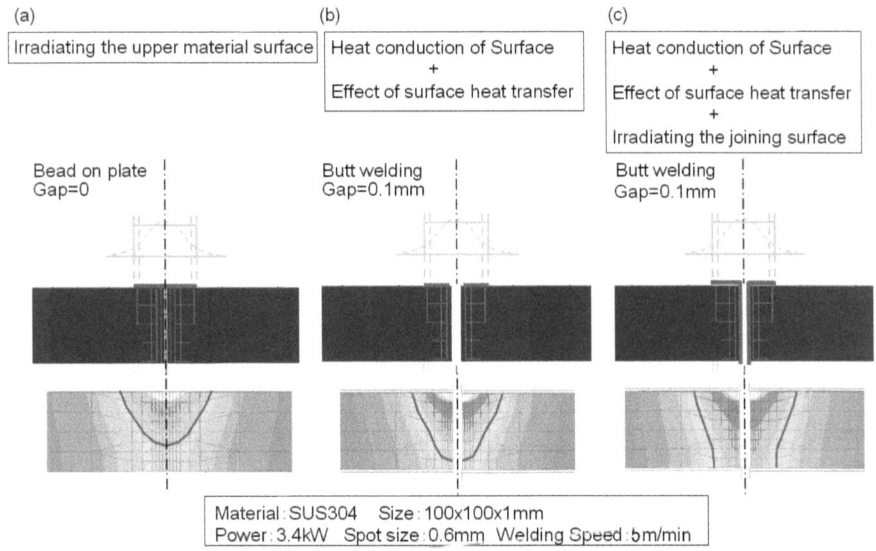

Fig. 13 Temperature distribution by difference of computing model comes from difference of irradiation place of laser beam

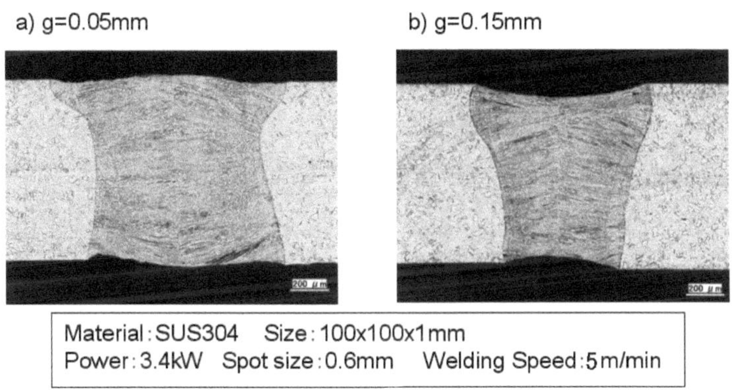

Fig. 14 Cross sections in butt welding

are deeply involved in the distortion, especially at the molten layer where welding characteristics are largely linked with the distortion. Those factors were taken into consideration in the calculation. The distortion in the molten part changes as the laser beam moves. The greatest expansion is observed right after the beam center passes the target. When the beam has passed over, cooling down starts and constriction occurs. Figure 14 shows the cross section of a) with small gap (g=0.05 mm) and b) with big gap (g=0.15 mm).

Expansion of material is seen when the material heated up after laser beam reaches the welding part. However, when the gap is large, the volume of the molten metal is insufficient to fill the entire gap. The shape of the molten metal changes into the state of "under-fill (dented state)" by insufficient amount of molten metal to bury the entire gap. Gravity and the surface tension act on there.

The model specially designed was used to calculate the butt welding to fit into the actual experiment, In this computing model, after the laser heat source passes the welding point, the gap of two plates on the welding surface is buried with the molten metal. Figure 15 shows a new simulation model.

When half the part of the heat source passes the welding point, the same shape of cross section of molten metal buried in a blank ditch where the gap was assumed is set in the computing model of the butt joint. In this case, the ditch is buried with molten metal in melting temperature (temperature is 1,400). Afterwards, the thermal expansion of the molten metal is caused by laser irradiation. The molten metal cooling starts though it reaches the maximum temperature due to the heat source's passing. And molten metal gradually returns to its former volume. Both plates are deformed by this expanding and cooling. With this model, the distortion amount and the angular distortions were obtained for various gaps by setting the welding speed as a parameter.

A moving picture simulation (Fig. 16) shows the time-series distortion in the butt welding. If the gap is small, expansion is observed as swelling at the top and bottom but if the gap is large, the molten part has dent which is called under fill.

In butt welding, if the gap is small, the butt wall surface expands more and when the molten metal join at the center, it is pushed out both upward and downward and it forms a large swelling (Fig. 17).

However, when the beam passes over and cooling starts, the swelling slightly shrinks. If the gap between materials is large (Fig. 18), the molten metal expands at the butt end face and rises slightly both upward and downward but the volume of the molten metal is insufficient to fill the gap when the expanded molten metal comes to the center of the joint so that under fill is seen on the upper and lower sides of the butt surface.

Figure 19 includes the result obtained from bead-on-plate as a reference. The wider the gap, the smaller the distortion becomes as a whole (Fig. 19). In addition,

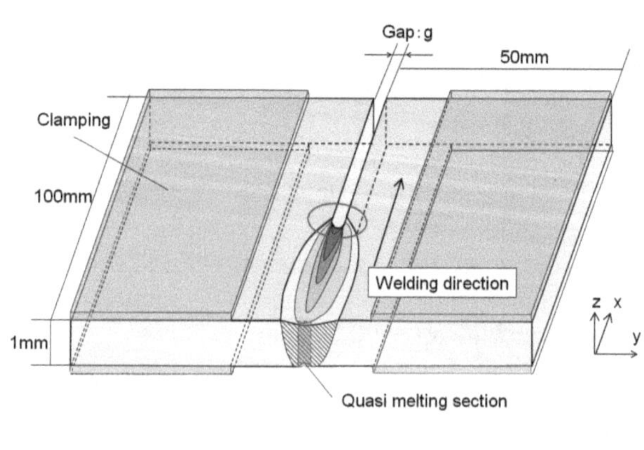

Fig. 15 Calculation model of butt welding

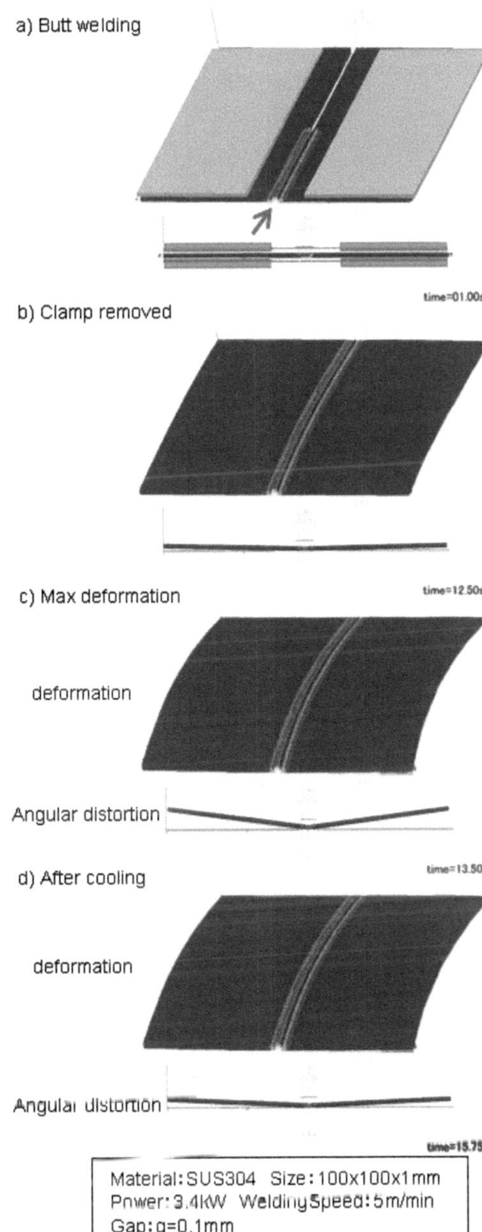

Fig. 16 Moving picture simulation of butt welding

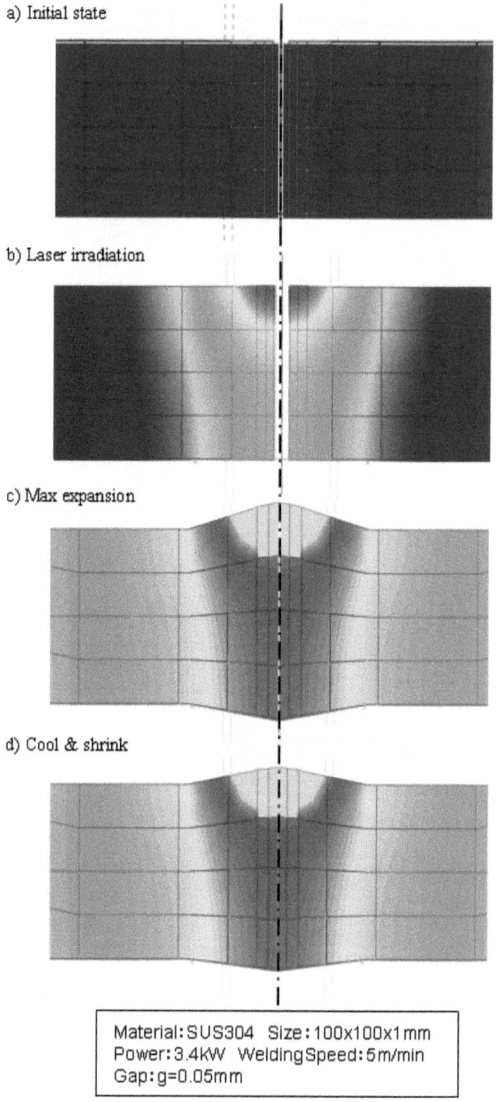

Fig. 17 Simulation model in the case of a narrow gap in butt welding

the angular distortion corresponding to gap variation was computed based on the calculation model (Fig. 20). The angular distortion is a bending angle caused by the heat source passing over the welding line. Both the right and the left sides of the plate bend upwards, symmetric about the welding line. The angular distortion is smaller if the gap is bigger. Moreover, in butt welding, the greater the gap is, the smaller the distortion amount along the welding line is. This is because the amount of laser beam that passes through the gap without involving increases. If the amount of laser beam increases, the angular distortion gets smaller.

Fig. 18 Simulation model in the case of wide gap in butt welding

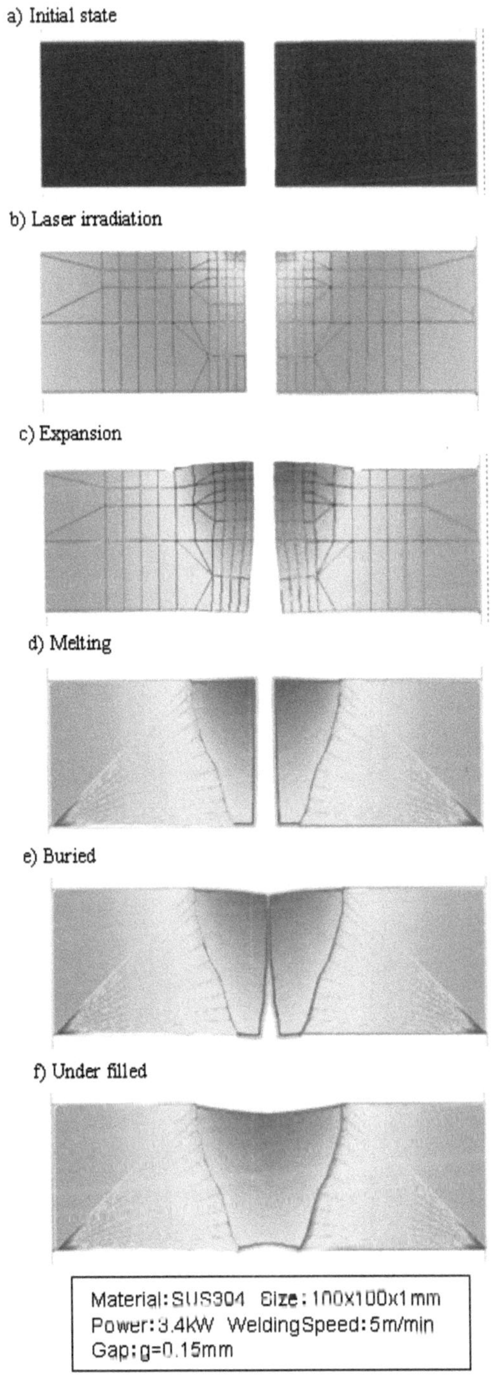

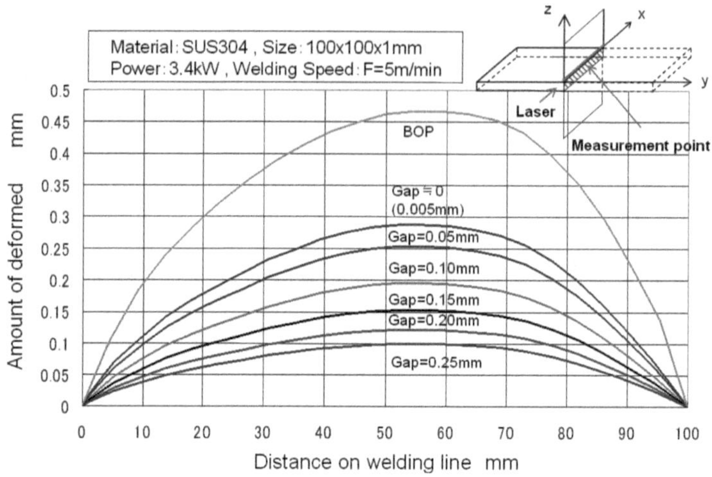

Fig. 19 Amount of deformation function of gap

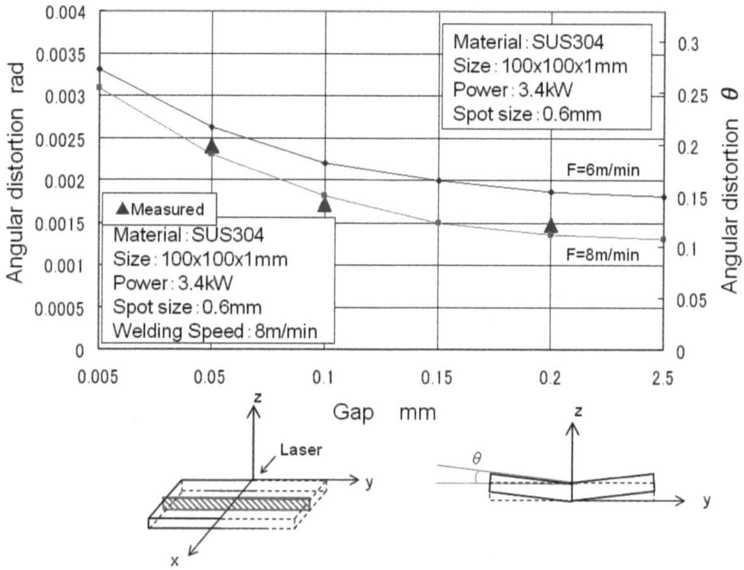

Fig. 20 Angular distortion as function of gap

4 Discussion of the Results

The distortion obtained through a simulation which targets the material sufficiently cooled down, is almost identical to the result from the actual measurement. In the laser welding simulation for bead-on-plate that assumes the ideal state where no gap exists, swelling distortion was observed along with the welding line when the heat source moves at a constant rate in the middle of the sheet metal plate. The same trend was found at both edges of the plate with a larger swelling compared with the middle of the plate.

This distortion shape almost corresponds to the thermal stress distribution (cf. Fig. 4). The distortion amount represented by swelling along the welding line is inversely proportional to the welding speed. The higher the welding speed, the smaller the distortion amount is. If the speed gets slower, more distortion is observed. This means that the distortion amount is proportional to the amount of heat energy input.

In the penetration welding where a molten layer is generated on the lower surface of a thin plate, the tension stress on the upper and lower surfaces get balanced out so that the angular distortion which shows the bending amount is smaller compared with the non-penetration welding.

To decide energy distribution in case of the butt welding with the gap, it was assumed that the laser beam passed through a lens obeyed the ray tracing method. This method shows that the amount of the laser energy on the butt surface is changing as the focal position shifts.

5 Conclusions

From the set of calculations, the followings can be concluded. In bead-on-plate, penetration welding is performed by generating a keyhole but in butt welding where gap exists, keyhole is not generated. However, as for the butt welding, welding performance (welding depth etc.) can be obtained similar to penetration welding due to the surface heat transfer along the welding surface of the gap and due to the laser beam emitted inside the gap wall.

It is necessary for the swelling volume to exceed the gap volume to fill up the gap at the butt surface. However, if the swelling volume falls below the gap volume, an under fill state is observed. This means that the melting and swelling metal does not completely fill up the gap while welding.

Since laser welding is basically thermal processing, the distortion due to thermal stresses is inevitable. According to the simulation results, the distortion amount varies depending on the plate thickness, size and power of laser. With regard to the welding speed, the faster the speed gets the smaller the distortion. The reason is that the exposure time that influences the material decides the amount of the distortion of the welding thin plate.

In the welding process, the material expands and exceeds the center of the welding point, so that it is presumed that the gap is filled at that time. Therefore, the following calculation can be drawn. If there is sufficient swelling volume of molten metal to exceed the center of the welding point when heated, it forms swelling on the upper and lower sides of the butt surface. On the other hand, if the gap is too large to get sufficient swelling volume of molten metal to exceed the center of the welding point, an under fill state is observed or joining cannot be done.

References

1. J.-U. Park, H.-W. Lee, H.-S. Bang, Effects of mechanical constraints on angular distortion of welding joint. Sci. Technol. Weld. Join. **7**(4), 232 (2002)
2. A. Kitagawa, et al., A Study on Laser Welding Deformation of 304 Stainless Steel, Q. J. Jpn. Weld. Soc. **20**(2), 295–300 (2002) (in Japanese)
3. G. Yu, K. Masubuchi, T. Maekawa, N.M. Patrikalakis, FEM Simulation of Laser Forming of Metal Plates. J. Manuf. Sci. Eng. **123**(3), 405 (2001)
4. Y. Dain, P.D. Kapadia, J.M. Dowden, The distortion gap width and stresses in laser welding of thin elastic plates. J. Phys. D Appl. Phys. **32**, 164–175 (1999)
5. T. Arai, N. Asano, Studies on deformation of the thin metal sheet laser welding (1st Report) Proceedings of JSPE Autumn Meeting, Japan (2007) (in Japanese)
6. T. Arai, N. Asano, Studies on deformation of the thin metal sheet laser welding (4th Report) Proceedings of JSPE Spring Meeting, Japan (2009) (in Japanese)

Laser Drilling Simulation Considering Multiple Reflection of Laser, Evaporation and Melt Flow

Etsuji Ohmura and Satoru Noguchi

1 Introduction

A keyhole is formed in the melting region of the base metal during laser drilling with high energy density. When a keyhole is formed, drilling of a hole with a large aspect ratio becomes possible. On the other hand, the formation of a keyhole becomes the cause of defects such as porosities. Therefore, it is important that the keyhole behavior in laser drilling is understood. There are some studies [1, 2] in which the keyhole behavior and the molten metal flow were observed by X-rays. However, generally the observation of the keyhole is difficult, because the keyhole occurs inside the material. Computer simulation is effective as a means to understand the keyhole behavior and predict hole shape.

In conventional simulations of laser welding, heat conduction calculations [3, 4] where a keyhole wall surface was fixed were performed. Thermohydrodynamic calculations [5] where a cylindrical keyhole was assumed were also conducted. When a keyhole is formed, many events occur such as the multiple reflection of laser, evaporation, a surface shape variation by the evaporation recoil pressure and flow of the molten metal. There is a welding simulation described in [6] in which the multiple reflection of laser at the free surface was considered, but the calculation method of the multiple reflection was not described. The purpose of the present study is the unsteady thermohydrodynamic analysis of the melting / evaporation of metal in laser drilling considering the multiple reflection of laser, evaporation of the material and evaporation recoil pressure.

The multiple reflection calculation of the laser is essential in the laser drilling simulation. The present authors [7] have already proposed methods to calculate the multiple reflection of laser by the ray tracing technique and calculate the power distribution in the wall surface of the hole shape expressed by the Volume of Fluid (hereafter VOF) method. In this paper, an analytical model is constructed for laser

E. Ohmura (✉)
Department of Management of Industry and Technology, Graduate School of Engineering, Osaka University, Osaka 565-0871, Japan
e-mail: ohmura@mit.eng.osaka-u.ac.jp

A. Öchsner et al. (eds.), *Materials with Complex Behaviour*, Advanced Structured Materials 6, DOI 10.1007/978-3-642-12667-3_19, © Springer-Verlag Berlin Heidelberg 2010

drilling of metal and the governing equations are described. The multiple reflection of the laser, evaporation recoil pressure, mass loss by the evaporation and the variation of the free surface are considered. A thermohydrodynamic analysis is conducted on the basis of the constructed model. The variation of the laser power distribution absorbed at the wall surface of the keyhole, the variation of the velocity distribution of the molten metal and the difference of the hole shape dependent on the material are investigated.

2 Analysis Method

2.1 Governing Equations of the Thermohydrodynamics [7]

The molten metal is supposed to be an incompressible Newtonian fluid. The governing equations are the equation of continuity

$$\nabla \cdot \boldsymbol{v} = 0. \tag{1}$$

The Navier-Stokes equation

$$\rho\left[\frac{\partial \boldsymbol{v}}{\partial t} + (\boldsymbol{v}\cdot\nabla)\boldsymbol{v}\right] = -\nabla p + \mu\nabla^2 \boldsymbol{v} + F, \tag{2}$$

and the equation of energy

$$\rho\left[\frac{\partial H}{\partial t} + (\boldsymbol{v}\cdot\nabla)H\right] = \nabla\cdot\left(\frac{K}{C_p}\nabla H\right) + w, \tag{3}$$

where $\boldsymbol{v}$ is the flow velocity vector, p the pressure, F the body force vector, ρ the density, μ the coefficient of viscosity, K the thermal conductivity, C_p the specific heat at constant pressure, H the enthalpy and w the internal heat generation. The internal heat generation is obtained by the multiple reflection calculation of the laser by the ray tracing, which is reported in Ref. [8].

2.2 Calculation of Free Surface

2.2.1 VOF Method and Advection Calculation

In this study, the VOF method [9] was adopted as the expression method of the surface shape. In the VOF method, when the domain is divided into elements, the shape of the free surface is expressed by the fluid volume proportion F for the element region V.F takes a value of $0 \leq F \leq 1$ by definition. A position of $F = 0.5$ was considered here to be the interface of the liquid and vapor phases.

It is supposed that the physical properties in each element can be expressed by the next equation with F:

$$\phi = \phi_l F + \phi_g (1 - F), \tag{4}$$

where ϕ_l and ϕ_g are the physical properties of the fluid and the gas, respectively. In this simulation, the gas is assumed to be air.

The advection equation of the fluid volume proportion F, which represents the free surface flow, can be expressed by the next equation:

$$\frac{\partial F}{\partial t} + \nabla \cdot (F\boldsymbol{v}) = 0. \tag{5}$$

The MARS (Multi-interfaces Advection and Reconstruction Solver) method [10] was used for calculation of the advection quantity.

2.2.2 Treatment of Surface Tension

Consideration of surface tension is necessary for the treatment of a fluid having a free surface. In the conventional method where the free surface is the boundary of the calculation region, the treatment of the boundary condition is complicated. Therefore, the CSF (Continuum Surface Force) method [11] was used in this study, where the surface tension, that is a surface force, is converted into a body force, and then introduced into the Navier-Stokes equation. The interfacial normal vector $\boldsymbol{v}(\mathbf{x})$ and the body force vector $\boldsymbol{F}_{SV}(\boldsymbol{x})$ are expressed in the following equations, respectively:

$$\boldsymbol{n}(\boldsymbol{x}) = \frac{\nabla \rho(\boldsymbol{x})}{[\rho]}, \tag{6}$$

$$\boldsymbol{F}_{SV}(\boldsymbol{x}) = \sigma \kappa(\boldsymbol{x}) \boldsymbol{n}(\boldsymbol{x}) \frac{\rho(\boldsymbol{x})}{\langle \rho \rangle}, \tag{7}$$

where σ is the surface tension coefficient, κ the interfacial curvature, $[\rho] = \rho_l - \rho_g$ and $\langle \rho \rangle = (\rho_l + \rho_g)/2$.

2.3 *Vaporization Model and Evaporation Recoil Pressure*

The laser energy absorbed at the surface region S by multiple reflection is diffused into the material and is also used for evaporation after evaporation start. The evaporation mass m_v of the material during the time interval Δt is evaluated by

$$m_v = \frac{\int_t^{t+\Delta t} Q\,dt + \int_t^{t+\Delta t} \int_{S_{in}} \frac{K}{C_p} \frac{\partial H}{\partial n}\Big|_{S_{in}} dS\,dt}{L_V}, \tag{8}$$

where Q is the laser power absorbed at the surface area S calculated by the multiple reflection, L_V the evaporative latent heat, and S_{in} is the inside boundary surface of the material region whose temperature reached the boiling point. The evaporation recoil pressure p_r can be expressed by the product of the gaseous flow velocity v_T at the Knudsen layer termination and the evaporated mass per unit time and per unit area $\dot{m}_v$:

$$p_r = \dot{m}_v v_T = \frac{m_v}{S\Delta t} v_T, \tag{9}$$

$$v_T = \frac{1}{4}\sqrt{\frac{8kT_s}{\pi m_a}}, \tag{10}$$

where k is the Boltzmann constant, T_s the temperature at the material surface and m_a the mass of a single atom [12].

The evaporation recoil pressure is also converted into a body force by the above CSF method, and it is introduced into the Navier-Stokes equation. The body force vector $\boldsymbol{F}_{\mathrm{PV}}(\boldsymbol{x})$ is expressed in the next equation.

$$\boldsymbol{F}_{\mathrm{PV}}(\boldsymbol{x}) = p_r\, \boldsymbol{n}(\boldsymbol{x}) \frac{\rho(\boldsymbol{x})}{\langle \rho \rangle} \tag{11}$$

2.4 Computational Algorithm

The SMAC (Simplified Marker and Cell) method [13] is a velocity correction method that is used as a computational algorithm for the flow velocity and the pressure. The discrete equation of the Navier-Stokes Eq. (2) is as follows:

$$\frac{\boldsymbol{v}^{n+1} - \boldsymbol{v}^n}{\Delta t} + (\boldsymbol{v}^n \cdot \nabla)\, \boldsymbol{v}^n = -\frac{1}{\rho}\nabla p^{n+1} + \frac{\mu}{\rho}\nabla^2 \boldsymbol{v}^n + \frac{1}{\rho}\boldsymbol{F}^n, \tag{12}$$

where n is the time step and Δt the time division. Because the unknown p^{n+1} is included in the right-hand side of Eq. (12), the predicted value of the flow velocity at the next time $\tilde{\boldsymbol{v}}$ is calculated by

$$\tilde{\boldsymbol{v}} = \boldsymbol{v}^n - \left[(\boldsymbol{v}^n \cdot \nabla)\, \boldsymbol{v}^n - \frac{\mu}{\rho}\nabla^2 \boldsymbol{v}^n - \frac{1}{\rho}\boldsymbol{F}^n \right] \Delta t. \tag{13}$$

The true flow velocity $\boldsymbol{v}^{n+1}$ is expressed by the sum of the predicted value of the flow velocity $\tilde{\boldsymbol{v}}$ and the corrected quantity of the velocity $\delta \boldsymbol{v}$, as shown in the following equation.

$$\boldsymbol{v}^{n+1} = \tilde{\boldsymbol{v}} + \delta \boldsymbol{v} \tag{14}$$

On the other hand, Eq. (15) is obtained from Eq. (12) and Eq. (13).

$$\frac{\delta \boldsymbol{v}}{\Delta t} = -\frac{1}{\rho}\nabla p^{n+1} \tag{15}$$

By taking the divergence of both sides of Eq. (15) and considering Eq. (1), Poisson's Eq. (16) about pressure p^{n+1} is obtained as follows:

$$\nabla \cdot \left(\frac{1}{\rho} \nabla p^{n+1} \right) = \frac{1}{\Delta t} \nabla \cdot \tilde{\boldsymbol{v}}. \tag{16}$$

The pressure p^{n+1} can be calculated by solving Eq. (16). The flow velocity can be revised by Eq. (17), introduced from Eq. (14) and Eq. (15).

$$\boldsymbol{v}^{n+1} = \tilde{\boldsymbol{v}} - \frac{\Delta t}{\rho} \nabla p^{n+1} \tag{17}$$

The flow chart of this simulation is shown in Fig. 1. First of all, the surface normal vector is calculated from Eq. (6). The surface tension and the evaporation recoil pressure are converted into body forces $\boldsymbol{F}_{SV}$ and $\boldsymbol{F}_{PV}$ with the obtained normal vector, and they are substituted into the Navier-Stokes equation. Then, the velocity field $\boldsymbol{v}^{n+1}$ and the pressure field p^{n+1} are calculated by the SMAC method. With the obtained velocity field, a new F value of each element is calculated by the advection Eq. (5).

Afterwards, the multiple reflection of the laser beam is calculated by ray tracing at the new surface, and the absorbed laser power distribution in the surface

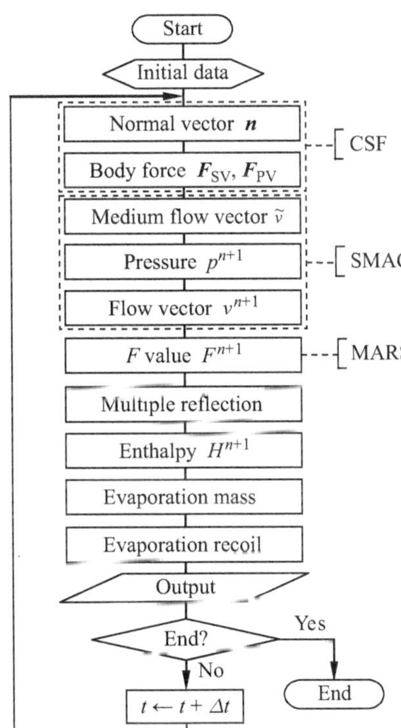

Fig. 1 Flowchart of the laser drilling simulation considering multiple reflection

is calculated. Using the obtained laser power distribution, the energy equation is solved, and the enthalpy of the material H is calculated. When there is an element whose temperature is beyond the boiling point, the evaporation mass is calculated by Eq. (8), and the evaporation recoil pressure p_r is calculated by Eq. (9).

When a calculation step is performed, the normal vector at the surface is calculated again.

3 Analysis Results

3.1 Analysis Condition

The finite element method was used for the numerical calculation. The analysis area was 141 μm × 141 μm × 135 μm (see Fig. 2). The region of 105 μm high was taken in the negative side and the region of 30 μm high was taken in the positive side, when the initial surface is $z = 0$. The central axis of the laser beam was taken along the z axis. The analysis area was divided into 33 × 33 × 40 with a hexahedron linear element. The analysis area divided by elements is shown in Fig. 2. The material in the negative side is pure iron, whose physical properties are shown in Table 1. The boundary condition was heat insulation and the flow velocity at the surroundings of the analysis area was zero. The flow velocity of the nodes whose temperature was lower than the melting point and F was not less than 0.5 was set also to zero.

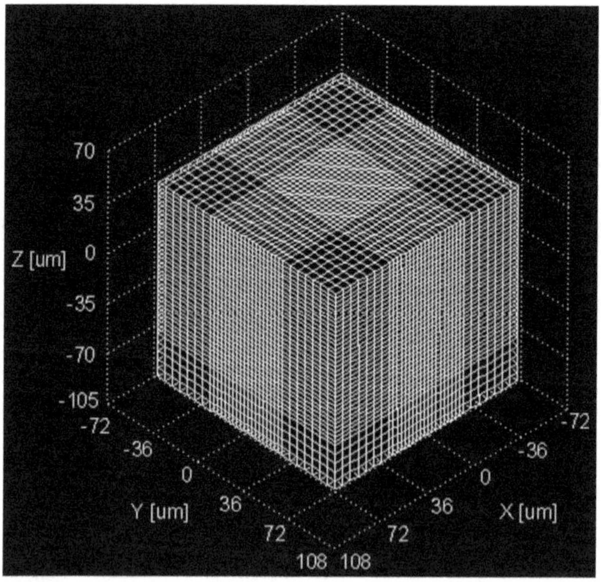

Fig. 2 Analysis area divided by elements

Table 1 Thermal and Physical properties of pure iron

Property	Value
Melting point [K]	1,809
Boiling point [K]	3,160
Density [kg/m^3]	7,870
Specific heat [J/kg]	460
Thermal conductivity [W/m K]	$83.5 - 4.57 \times 10^{-2}T$ $(293 \leq T \leq 1273)$
	29.7 $(1273 < T)$
Latent heat of melting [J/kg]	2.47×10^5
Latent heat of evaporation [J/kg]	6.29×10^6
Viscosity [N s/m^2]	3.7×10^{-3} exp$(4.14 \times 10^4 / 8.3144T)$
Surface tension coefficient [N/m]	$1.872 - 4.9 \times 10^{-2}(T - 1536)$

A laser whose wavelength is about 1 μm such as a Nd:YAG laser and fiber laser was assumed. The influence of the plasma can be ignored for such wavelengths [7]. The laser intensity was a Gaussian distribution, and a collimated beam was irradiated perpendicularly to the initial surface. The beam radius was 20 μm and the laser power was 250 W. A reflectivity of 0.65 was used, which is the reflectivity of iron for a wavelength of 1 μm. The inside of a circle with a radius of 26.5 μm of the present laser beam includes 97% of the whole power. Therefore, the part of beam which was included in a square of 26.5 μm × 26.5 μm was divided in a bunch of 210 × 210 rays, and power was given to each ray by considering a Gaussian distribution. The power of the rays outside the circle with a radius of 26.5 μm was set to zero, because it is small enough compared with the paracentral power. The irradiation time was assumed to be 25 μs.

3.2 Hole Formation Process

The time variation of the surface shape during laser irradiation is shown in Fig. 3. Formation of a hollow begins after about 1 μs from the start of laser irradiation. A large hollow is formed at time of 5 μs, as shown in Fig. 3b. The hole depth increases at 10 μs and 15 μs, and the hollow grows as a hole (Fig. 3c, d).

At 20 μs, the central part of the hole begins to fall down conspicuously, and the formation of the keyhole begins (Fig. 3e). The keyhole grows up increasingly after that, and the keyhole depth becomes about 60 μm at 25 μs, as shown in shown in Fig. 3f.

3.3 Time Variation of Laser Power Distribution Absorbed at Wall Surface

Figure 4 shows a part of the 44,100 rays displayed and the laser power distribution absorbed at the wall surface in the $x-z$ cross section of the hole are shown in Fig. 4.

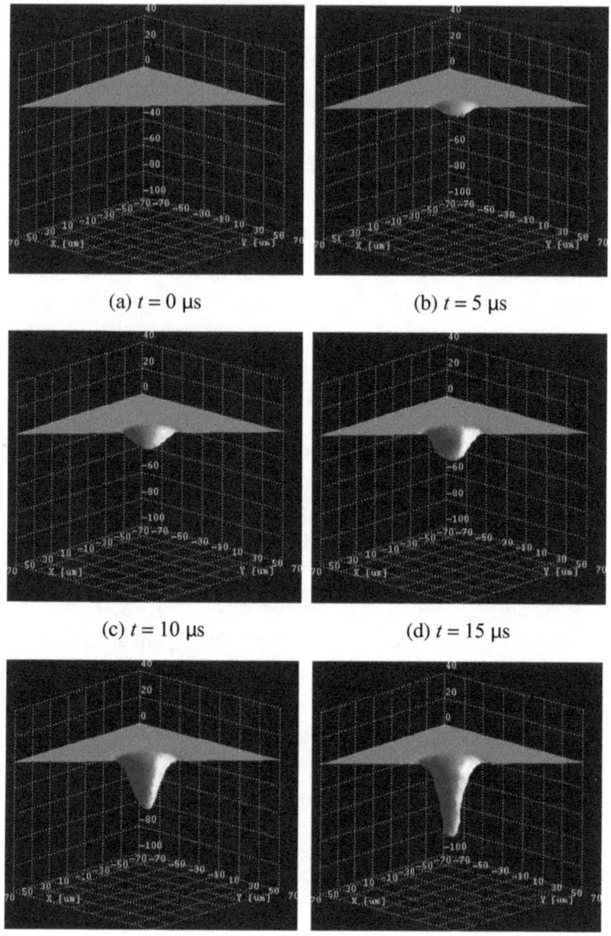

(a) $t = 0$ μs (b) $t = 5$ μs

(c) $t = 10$ μs (d) $t = 15$ μs

(e) $t = 20$ μs (f) $t = 25$ μs

Fig. 3 Surface variation in laser drilling (laser power: 250 W, $1/e^2$ radius: 20 μm)

As shown in the left-hand side of Fig. 4b, the laser is reflected only once at 5 μs because the hole depth is extremely shallow. Therefore, the maximal value of the absorbed laser power in Fig. 4b is the same as in Fig. 4a.

At 10 μs, the laser is reflected twice. However, the maximal value of the absorbed laser power does not change (left-hand side of Fig. 4c). Because the hole is shallow, rays are reflected at the hole wall of the opposite side after the first reflection and go out of the hole.

However, the maximal value of the laser power absorbed at the hole bottom increases at 15 μs, as shown in Fig. 4d, because the gradient of the hole wall becomes steep and the rays reach the hole bottom by multiple reflection.

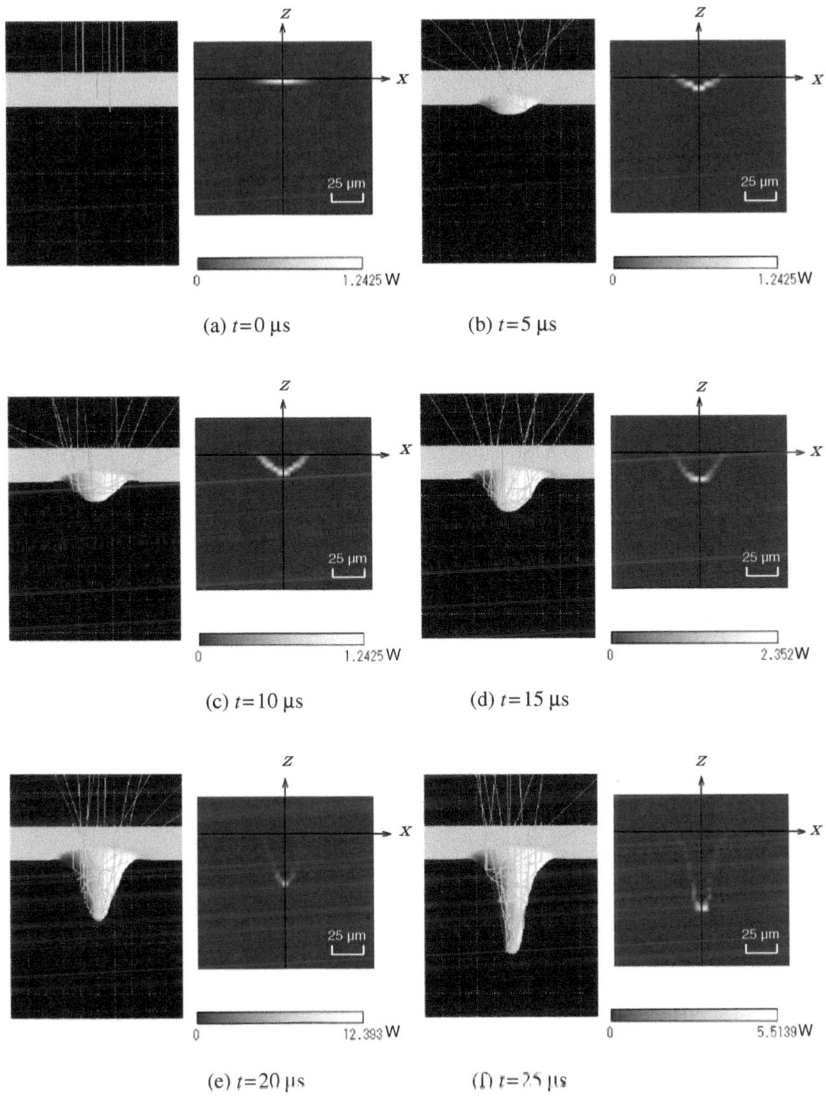

Fig. 4 Relationship between hole shape and laser power distribution

At 20 μs, the increase of the power at the hole bottom by the multiple reflection is remarkable (Fig. 4e). The hole becomes deeper at 25 μs, as shown in Fig. 4f, and the number of reflections till the rays reach the hole bottom increases. Therefore, the maximal value of absorbed laser power decreases compared with Fig. 4e, however, it is about 4.4 times compared with Fig. 4a.

The relationship between the laser irradiation time and the hole depth in this simulation is shown in Fig. 5. The drilling velocity is about 1.3 m/s until 15 μs,

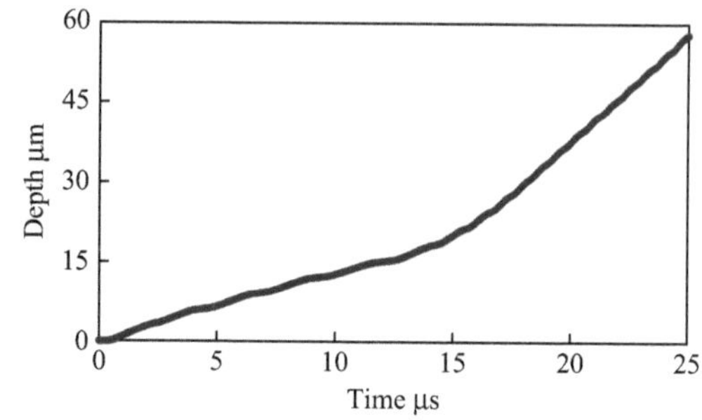

Fig. 5 Time variation of keyhole depth

but it is accelerated to about 3.8 m/s when over 15 μs. 15 μs is the time when the laser power absorbed at the hole bottom begins to increase by multiple reflection, as mentioned above.

It is understood from Fig. 4 that the power at the hole bottom increases by multiple reflection even if the depth of the hole is comparatively shallow. This is a factor to grow a shallow hole to a keyhole. On the other hand, multiple reflection becomes too remarkable when a hole becomes much deeper. As a result, it is estimated that the laser power at the hole bottom decreases and the growth rate of keyhole falls.

3.4 Relationship between Hole Depth and Absorptance

The time variation of the hole depth and the absorptance of laser energy in laser drilling was investigated. Here, the absorptance ε was defined as

$$\varepsilon = \frac{\text{Absorbed laser power by material}}{\text{Incident laser power}}. \tag{18}$$

The relationship of the hole depth and the absorptance ε is shown in Fig. 6. Absorptance begins to increase when the hole depth exceeds about 10 μm, and abruptly increases after that. It is understood that the multiple reflection is effective from the stage when the hole depth is comparatively shallow. The increase of

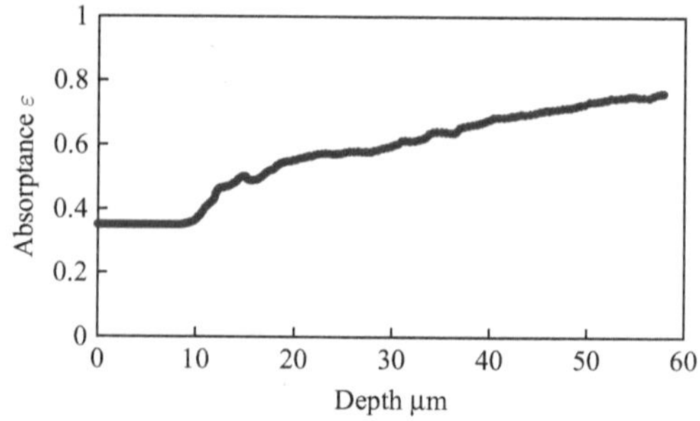

Fig. 6 Relationship between absorptance and keyhole depth

the absorptance slows down when the hole depth exceeds 20 μm. The hole depth reaches about 58 μm at 25 μs when the laser irradiation is finished, and the absorptance rises to about 0.75. This indicates that multiple reflection becomes remarkable and the absorption efficiency of the laser power becomes sufficiently high when the keyhole becomes much deeper.

3.5 Time Variation of Velocity Distribution in the Process of Hole Formation

The surface shape and flow velocity vectors in the $x-z$ cross section at 5 μs, 15 μs, 20 μs and 25 μs are shown in Fig. 7a, b, c, d, respectively.

Figure 7 was taken from a depression angle of about 6 degrees. At 5 μs, little time passes since the surface begins to be pushed down by evaporation recoil pressure. Therefore, flow pushing down the surface at the hole central part is generated, but the flow velocity is about 0.04 m/s, which is not so large (Fig. 7a). Compared with Fig. 7a, at 15 μs, the flow pushing down the surface at the hole central part is accelerated to about 0.33 m/s by evaporation recoil pressure, as shown in Fig. 7b.

At 20 μs, the eddy which moves from the surface toward the inside at the upper part of the molten pool is generated, as shown in Fig. 7c. The flow toward the surface of the molten metal is changed to the flow moving toward the inside of the molten pool by surface tension. As a result, an eddy is formed.

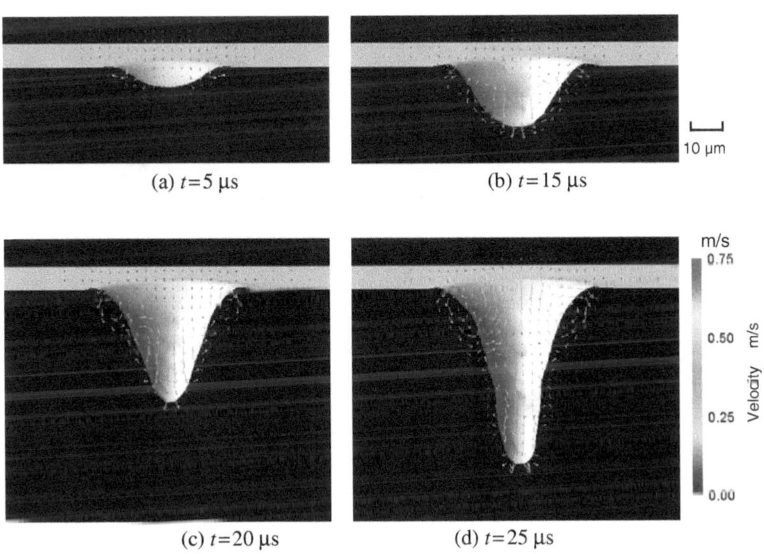

Fig. 7 Velocity vectors and hole shape during laser irradiation (material: iron, laser power: 250 W, $1/e^2$ radius: 20 μm)

At 25 μs, the eddy becomes more remarkable (Fig. 7d). It is estimated that the shape of the molten pool becomes a nail head type by this eddy.

3.6 Effect of Material on Hole Shape

The effect of material on the hole shape was examined using aluminum. As well as in the case of the simulation for iron, a 1 μm laser was assumed. Reflectance of aluminum for a wave length of 1 μm is about 0.93 [14, 15]. The laser power used in the drilling of aluminum was set to 1.25 kW so that the laser power absorbed at the initial surface corresponds to the case of iron. The reason is that the purpose of the present simulation has the purpose to investigate the effect of the physical properties of the material on the hole shape. The irradiation time was 25 μs which is the same as that of iron.

The surface shape and flow velocity vector in the $x - z$ cross section at 25 μs are shown in Fig. 8. This figure is also taken from depression angle of about 6 degrees. Compared with Fig. 7d, the hole shape of the aluminum is narrower than that of iron. Furthermore, it is understood that compared with iron, the flow velocity vectors in the inside of aluminum are distributed more widely. This means that the melting layer of aluminum is larger compared with that of iron. The thermal diffusivity of aluminum is larger than that of iron. Therefore, the energy lost by thermal diffusivity increases, and the energy used for evaporation decreases. On the other hand, in comparison with the melting point of iron which is 1,809 K, the melting point of aluminum is 933 K, which is lower. Therefore, the hole of aluminum is narrower, and its molten pool becomes wider.

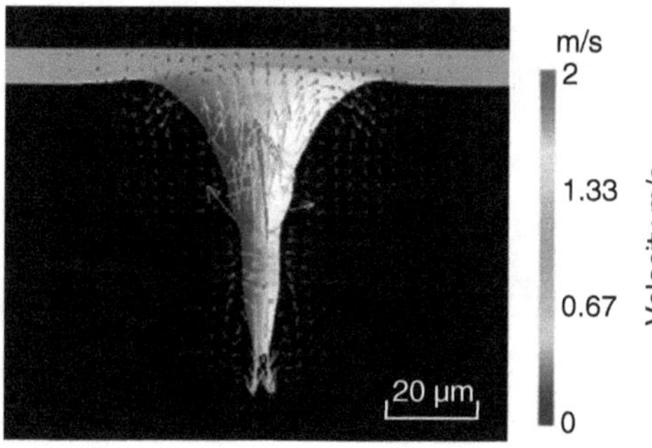

Fig. 8 Velocity vectors and hole shape at $t = 25\,\mu s$ (material: aluminum, laser power 1.25 kW, $1/e^2$ radius 20 μm, reflectivity 0.93)

4 Conclusion

A thermohydrodynamic analysis method for laser drilling of metal was proposed by considering the multiple reflection of laser, evaporation of the material, and evaporation recoil pressure. Time variations of laser power distribution absorbed at the hole wall and the velocity distribution of the molten metal in the keyhole formation process and the effect of material on the hole shape were investigated. The following conclusions can be drawn:

(1) At the stage where the depth of the hole is comparatively shallow, the power at the bottom of the hole increases by the effect of multiple reflection. This becomes a factor in the growth of the shallow hole into a keyhole.
(2) When the hole becomes deep, the effect of multiple reflection becomes remarkable and the absorptance increases. On the other hand, the laser power absorbed at the hole bottom decreases. In this simulation, the absorptance increases to a value of about 0.75.
(3) When the hole becomes deep, the flow field shows circulation of molten metal that rises near the walls of the keyhole and by surface tension, moves away from the keyhole at the surface, forming an eddy.
(4) Because the thermal diffusivity of aluminum is larger than that of iron, the energy lost by thermal diffusivity increases, and the energy used for evaporation decreases. On the other hand, in comparison with the melting point of iron which is 1,809 K, that of aluminum is 933 K, which is lower. Therefore, in case of aluminum, the molten pool broadens and the hole becomes narrow.

This paper shows a simulation of laser drilling by considering the multiple reflection of laser and evaporation of material. The usefulness of this simulation method will be investigated in future by comparing it with a laser drilling experiment.

Acknowledgement Part of this study was supported by Grants-in-Aid for Scientific Research from the Japan Society for the Promotion of Science in 2001, 2002, 2004, 2005 and 2006.

References

1. Y. Arata, N. Abe, T. Oda, Fundamental phenomena in high power CO_2 laser (report II). Trans. JWRI **14**, 5–11 (1985)
2. S. Fujinaga, H. Takenaka, T. Narikiyo, S. Katayama, A. Matsunawa, Direct observation of keyhole behavior during pulse modulated high-power Nd:YAG laser irradiation. J. Phys. D Appl. Phys. **33**, 492–497 (2000)
3. K.N. Lankalapalli, J.F. Tu, K.H. Leong, M. Gartner, Laser weld penetration estimation using temperature measurements. J. Manuf. Sci. Eng. **121**, 179–188 (1999)
4. G. Simon, U. Gratzke, J. Kroos, Analysis of heat conduction in deep penetration welding with a time modulated laser beam. J. Phys. D Appl. Phys. **26**, 862–869 (1993)
5. S. Rabier, M. Medale, R. Fabbro, 3-D numerical modeling laser welding. *Proceedings of ICALEO 2001*, Miami, FL (2001)
6. H. Ki, P.S. Mohanty, J. Mazumder, Multiple reflection and its influence on keyhole evolution. Proceedings of the ICALEO2001, Orlando, FL (2001), pp. 933–942
7. S.H. Lamb, *Hydrodynamics* (Cambridge University Press, Cambridge, 1993).
8. S. Noguchi, E. Ohmura, Thermohydrodynamics analysis of laser drilling considering multiple reflection of laser and evaporation – 1st report, multiple reflection simulation of laser using ray tracing technique. J. Jpn. Laser Proc. Soc. **14**, 33–38 (2007) (in Japanese)
9. C.W. Hirt, B.D. Nichols, Volume of fluid (VOF) method for the dynamics of free boundaries. J. Comput. Phys. **39**, 201–225 (1981)

10. T. Kunugi, Direct numerical algorithm for multiphase flow with free surfaces and interfaces. Trans. Jpn. Soc. Mech. Eng. B **63**, 1576–1584 (1997) (in Japanese)
11. J.Blackbill, D.B. Kothe, C. Zemach, A continuum method for modeling surface tension. J. Comput. Phys. **100**, 335–343 (1992)
12. V.V. Semak, J.A. Hopkins, M.H. McCay, T.D. McCay, A concept for a hydrodynamic model of keyhole formation and support during laser welding. Proceedings of the ICALEO'94, Orlando, FL (1994), pp. 641–650
13. A.A. Amsden, F.H. Harlow, A simplified MAC technique for incompressible fluid flow calculations. J. Comput. Phys. **6**, 322–325 (1970)
14. E.D. Palik (ed.), *Handbook of Optical Constants of Solids II* (Academic Press, San Diego, CA, 1991), pp. 385–387
15. T. Araya, Answer your question – From technical service Q&A laser process (4) welding technology. J. Jpn. Weld. Eng. Soc. **53**, 117–127 (2005) (in Japanese)

Effect of Flight Spectrum Loads on the Damage Tolerance Evaluation of a Helicopter Frame

Marco Giglio and Andrea Manes

1 Introduction

The structural safety of helicopters is guaranteed with a deeply fatigue analysis in the design phase [1–3] and a clear schedule of inspection during service life. However, the design and maintenance of helicopters are particularly important and complex with respect to aircrafts. The peculiarity lays in two ways: the load spectrum that is composed by a high number of low-amplitude cycles, which result from the mechanical rotation of the rotor blades (severe vibratory loads), and the low velocity impact damage. These types of loads can lead to high fatigue damage accrued in short time or rapid crack propagation from accidental flaws or damages. Due to this, the condition based maintenance for fatigue is not the most typical approach for helicopter fatigue, since short inspection intervals are frequently required. The FAR (Federal Aviation Regulations) and also the new CS (Certification Specification) standards, promoted by the European Aviation Safety Agency for the global aviation scenario, indeed points out how the determination of the real operative usage is a fundamental issue. It should be for the design of every aircraft, but even more critical in the case of helicopters. Actually, adoption of redundancy and low stress level is recommended but cannot be always implemented. Furthermore, failures need to be monitored also due to not directly predicted damage like accidental damage, environmental damage, life extension [4].

Considering these issues, the development of Health and Usage Monitoring Systems (HUMS) has received considerable attention from the helicopter community in recent years [5, 6] with the declared aim to increase flight safety, increase mission reliability, extend duration of life limited components and, of course, reduce maintenance costs [7].

Structural Health Monitoring (SHM) seems capable to help in reducing the maintenance cost, which is about 25% of the direct operating cost of the helicopter [5] and plays an important role especially in the case of the ageing helicopters.

M. Giglio (✉)
Dipartimento di Meccanica, Politecnico di Milano, Milano, Italy
e-mail: marco.giglio@polimi.it

A. Öchsner et al. (eds.), *Materials with Complex Behaviour*, Advanced Structured Materials 6, DOI 10.1007/978-3-642-12667-3_20, © Springer-Verlag Berlin Heidelberg 2010

With the aim to reduce the direct maintenance costs of the airframe without compromising any safety or reliability issues, in the last years some vibration monitoring systems were integrated into the gears and rotating shafts of the helicopter transmission, and now the HUMS is an integral part of the new generation helicopters. However, it is important to underline that HUMS is actually mainly based on monitoring vibrations generated in components critical for the flight performance. Data are continuously recorded using accelerometers, processed and compared to a threshold value and models that describes the allowable and accumulated damage. HUMS is installed on various helicopters such as the Sikorsky SH-60B Black Hawk, S-61, and S-76, the AgustaWestland EH-101 and AW139, the Boeing CH-47D and MH-47E Chinook, WAH-64 and AH-64D Apache and UH-60A Black Hawk, but are based only on the principle of usage monitoring acquiring loads and vibration without any direct indication of incipient failure. An exception is the on-board diagnostic systems like the structural life monitoring of the Bell-Boeing V22.

Thus regarding the structural helicopter fuselages, only partial attempt to apply reliable methods to monitoring directly on-line the damage nucleation, accumulation and propagation during life were carried out. In this field, an integrated and reliable system for monitoring the damage in the fuselage (diagnosis) and for evaluating the time inspections and remaining life (prognosis) is missing.

For what concerns the aicraft scenario, both Boeing and Airbus are starting to use HUMS, in particular, approximately 150 Comparative Vacuum Monitoring CVM sensors were used in the qualification of GLARE on the A380 full scale fatigue test rig [8].

According to the main aim of the sensor network (the degradation mechanism is the main phenomena directly dedicated to the health monitoring), the candidate sensors would identify when a fatigue crack has initiated or when an existing crack grows, and monitor crack growth in the most stressed hotspot. Considering also the future requirement to be embedded in real operating frames, sensor reliability becomes important when measurements are required over a long period of time, as long as costs, environmental resistance and weight are concerned.

Specifically, for the SHM purpose of aerospace frames, the potential sensor types are crack gage, comparative vacuum monitoring (CVM), Acousto-Ultrasonic (AU), Acoustic emission (AE), Lamb waves, fibre-optic strain (in particular, Fibre Bragg Gratings (FBG)), Eddy Current, Microelectromechanical system (MEMS) etc. A change in the material local behavior (and hence a damage) can be picked up and localized by an array of such sensors.

The crack gage is a well known sensor for crack monitoring and consists of a thin coupon that can be bonded to a structural member in the vicinity of a known stress raiser. The gage utilizes the indirect potential drop method for measuring the crack growth. Comparative Vacuum Monitoring (CVM) [8, 9] technique provides a novel and interesting method for crack initiation detection and long term monitoring of fatigue cracks in aircraft structures. Open cracks generate leaks in a series of galleries bonded to the structures. Pressure changes in a system of hair-fine capillaries provide an indication of structural defects (cracks, corrosion and loss of bonding contact), tracked with a remote monitoring device. CVM has the ability to monitor external surfaces of materials for crack initiation, propagation and corrosion. In addition, CVM sensors can also be embedded between components (e.g. lap joints) or within material compounds such as fibre composite. Airbus and Boeing are using CVM technology for both laboratory and structural tests. In particular, Airbus has developed sensors for early detection of fatigue cracks within riveted lap and butt joints. This is achieved by placing the sensor between the lap/butt joint components. The sensors are inert and may be left in-situ (in the structure) for real time or periodic monitoring. Furthermore, the sensors do not suffer the restriction of wire crack gauge, which have a significant probability of failure if a bending moment is repeatedly applied to the sensor. CVM sensors can be placed in fatigue critical hotspots and are sensitive enough to pick up cracks as they initiate Tests with CVM on helicopter have been executed for Sea King helicopters being operated by the British Royal Navy and Royal Air Force.

AE/AU technology can detect structural defects long before possible catastrophic failures [8]. This is possible because discontinuities will produce detectable emissions, long before structural integrity is compromised and structural failure occurs. Acoustic Emission (AE) is based on elastic radiation generated by the rapid release of energy from sources within a material (impacts, crack initiation, crack growth, delamination). AE sensors are passive small piezoelectric sensors mounted to a convenient surface of the material. The sensor response and front end filters can remove frequencies below about 100 kHz, which includes most audible noise. The result is that acoustic emission can be used to monitor a structure for active damage even when ambient noise levels are extremely high. Acoustic emission is sensitive enough to detect newly formed cracks. AE helicopter HUMS has been used for detecting damage in SH-60 helicopter drive trains. Acousto-Ultrasonic (AU) is a technique that sends acoustic waves into the structure and intercepts them when they emerge on the other side. Deviations from the expected wave pattern indicate the presence of discontinuity like cracks. The European project AISHA (Aircraft Integrated Structural Health Assessment) aimed to contribute in realizing an aircraft monitoring technology by using ultrasonic Lamb waves as the basic sensing principle [10]. The special potential of Lamb waves for damage detection arises from their propagation capabilities. Lamb waves are guided acoustic waves propagating in plate-like structures. In the case of damage, the propagation of an ultrasonic Lamb wave will be disturbed resulting in a characteristic reflection and attenuation pattern. Experiments performed within AISHA on lab-scale and on selected full-scale parts showed the ability of Lamb waves or other guided waves to give information on correlations between acoustic parameters and damage in structural parts.

Fibre Bragg Gratings (FBG) are fibre-optic sensors with elastic properties that mirror those of the tested material, and can be used to monitor temperature, thermal and mechanical stress, damage caused by collision or impact, and delaminations. These sensors have the advantage of being light weight, having all passive configurations, low power utilization, immunity to electromagnetic interference, and bandwidth, compatibility with optical data transmission and processing, long lifetimes, low cost and high sensitivity, comparing with typical electrical strain gauges, such as resistive type, piezoelectric, semiconductor, and capacitance gauges. These sensors in fact generally have a small dynamic range, gauge factors of less than 5, and are affected by environmental conditions such as moisture and temperature.

The use of Microelectromechanical systems MEMS is also an interesting option. MEMS are miniature electromechanical sensor and actuator systems. Advances in MEMS technologies have led to dramatic reductions in size, power consumption, and cost for wireless communications. Their small size allows them to be used in applications where conventional sensors and actuators would be intrusive. Because of the economies of scale achievable from the conventional chip manufacturing processes, they can be mass produced and copiously applied in a cost-effective manner. These types of sensors have been used on aircraft structures due to their minimal aerodynamic disruption [11]. MEMS sensing technologies are appropriate for local SHM applications, such as those that identify crack initiation, propagation and corrosion. However deploying a large number of MEMS devices over a large area in a cost-effective manner is a difficult problem. Moreover reliability and measurement accuracy are still problems that must be addressed for successful implementation of MEMS technologies together with wireless data transmission and connection to a power source.

Traditionally, cracks must be monitored by conducting visual and non destructive testing on a large area of the aircraft during the operating life. Often, this can require significant downtime to get access to such an area with a handheld testing machine. The advantage of SHM is that once the sensors network is installed and set up, inspection is possible not only without disassembling parts but also in continuum using a real time strategy to detect failures (existence, location, type, extent). Thus is possible to make a prognosis of their effect on the overall reliability of the structure. In case of damage, the system directly identifies the location and follows-up actions that can be taken. The application of low-power sensors will bring these benefits with low costs in terms of overall energy consumption of the helicopter.

The data elaboration during the whole structural health monitoring process involves selecting the quantities to be measured, the types of sensors to be used, the locations where the sensors should be placed, the number of sensors, the sensor resolution, bandwidth and the data acquisition/storage/transmittal hardware. The data fusion process is extremely important: the integration of the data from a multitude of sensors with the objective of making a more robust and confident decision than is possible with any one sensor alone.

It is important to state that the base of this advanced prognostic program is the availability of finite element models (FEM) of the structure with and without damage. This is the key step for a complete knowledge of the structure in the damaged and undamaged condition. Thus, the information obtained from the FE analysis allow for the extraction of reliable information from sensor data, i.e. the identification of the damage-sensitive properties, derived from the measured dynamic response, which allows to distinguish between the undamaged and damaged structures.

Actually, the finite element method represents a useful tool in the analytical evaluation of local stress and strain. In particular, the finite element method makes it also possible to study the fatigue and damage tolerant behaviour of complex structures using fracture mechanics approach [12–14].

In spite of the versatility and power of the FE method, experimental tests cannot be completely removed from the design process. Currently, a complete fatigue assessment approach (including damage tolerant behaviour) include of both the experimental test phase and the numerical model development [15]. Moreover, due to the complexity of the aerospace structure, FE submodeling technique is widely used [16–18], in particular for the calculation of the fracture mechanics parameter [14].

The passage of the crack through a rivet hole in a typical aerospace frame construction (skin, stringers and ribs), connecting the skin with a stringer, has been considered in [17]. An analytical model from literature and a FE model of the whole panel specimen, with a submodel of the rivet hole passing through, have been carried out to investigate the crack parameter, the Stress Intensity Factor (SIF). The analytical and FE models give coincident results when the same test configuration is modelled. Otherwise, the great advantage of the FE model is the capability to describe complicated and non usual states of stress and failure conditions. Moreover, the FE model allows to know the complete state of stress of the panel during the propagation of the crack and thus permits a more complete fatigue and damage tolerant assessment of the component.

In this work, a finite element model of the whole rear fuselage of an actual helicopter is presented. The first purpose is to obtain the curve of the stress intensity factor K_I versus crack length for a crack introduced in the most stressed zone. Therefore, the crack propagation vs. real contingent loads (represented as a combination different spectrum loads) can be calculated in order to asses a dedicated danger and/or performance index of the components with respect to the damage phenomena in the real usage scenario.

2 The FE Model

The structure under exam was a typical aerospace frame with panel skin, stringers and ribs made in Al-Li 8090 T 81, Fig. 1. A FE Model was developed using MSC Patran as preprocessor and ABAQUS/Standard as solver and post processing, Fig. 2. The model started from the modular joint (the joint between the rear and the central part of the fuselage) and included the rear fuselage and the tail unit. The panels and the ribs' cores were modelled using shell elements S4R5 with four nodes and with the appropriate thickness. The ribs were strengthened with beam elements to give out the plane stiffness in the same way used for the stringers. The stringers were modelled using beam elements. Considering that the tail unit was built with a folding system, the folding beams were modelled using rigid beams. These components are not under examination in this assessment because they were previously analyzed in detail [18].

With the aim to obtain a refined mesh in the areas where there were stress concentrations, the final size was obtained through an iterative process.

The skin panels of Rear Fuselage and the stringers were made in Al-Li 8090-T81 alloy (Young modulus E = 79,000 [MPa]).

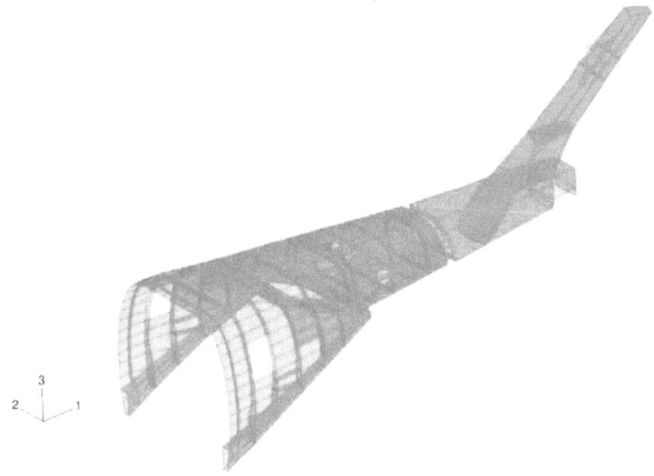

Fig. 1 Translucent view of the finite element model of the rear fuselage. The stringers and ribs are visible

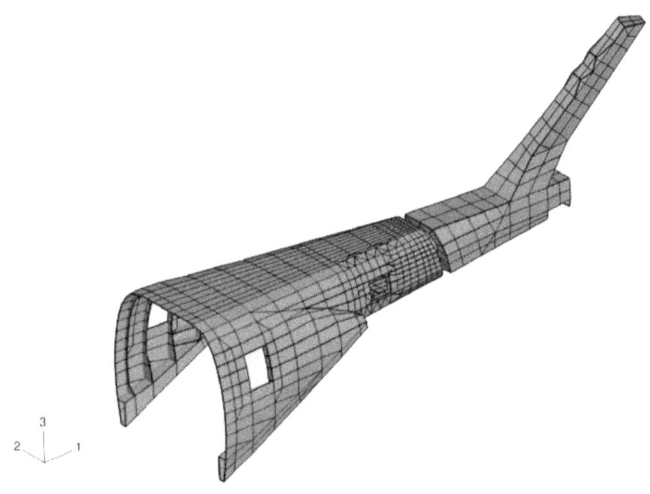

Fig. 2 Finite element model of the rear fuselage

The model was constrained fixing all the degrees of freedom of the nodes highlighted in Fig. 3. Three different loads were applied to the model: F_y (tail rotor traction), F_z (tail plane lift) and M_y (reaction torque from the tail rotor), Fig. 4. The load spectrum applied to the helicopter is described next and it was build using different combinations of these three loads.

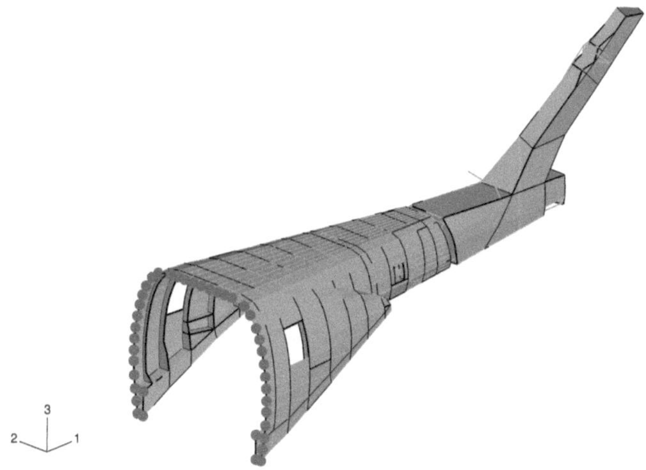

Fig. 3 Constraints applied to the rear fuselage finite element model. The nodes highlighted have been fixed

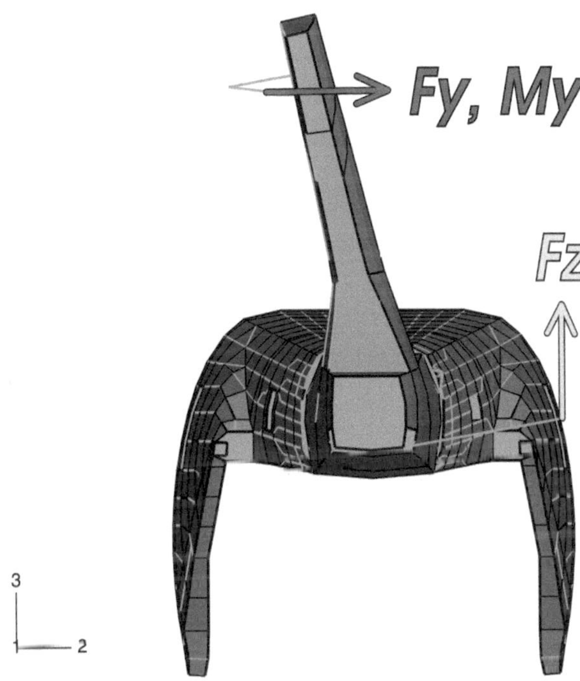

Fig. 4 Forces F_y and F_z and torque M_y applied to the rear fuselage acting in their positive direction

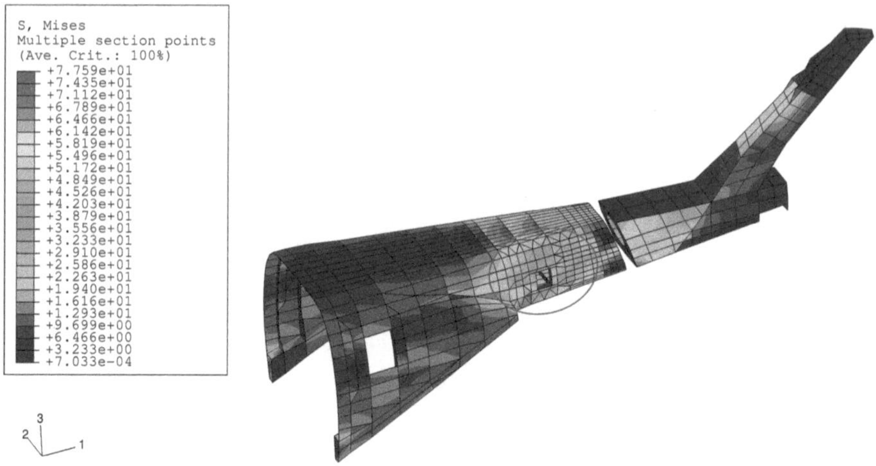

Fig. 5 Mises equivalent stresses. The most stressed zone is highlighted

According to the maximum force applied and considering that the model was linear (due to negligible amount of non linear phenomena to be modelled), an analysis of the most stressed zone was carried on separately for the three load. The results shown that F_y is the most stressing load and the most stressed zone of the fuselage is illustrated in Fig. 5.

In order to calculate carefully the stress distribution, and thus the stress intensity factor SIF versus crack length curve (in the further step), a detailed modelling of the most stressed area was carried on as shown on Figs. 6 and 7. The SIF calculation

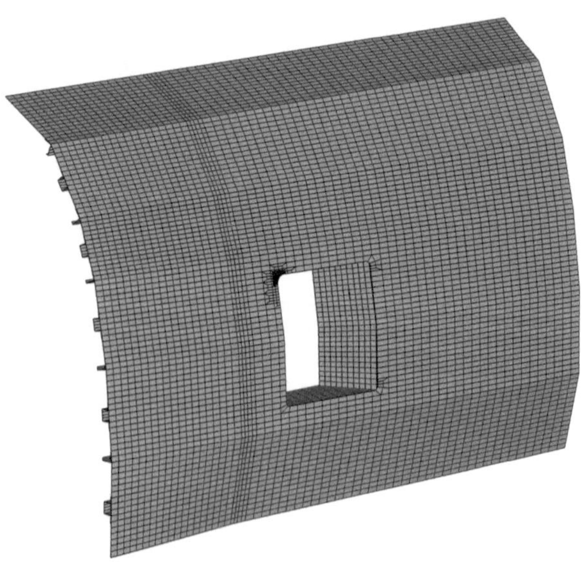

Fig. 6 External view of the submodel of the window area of the rear fuselage

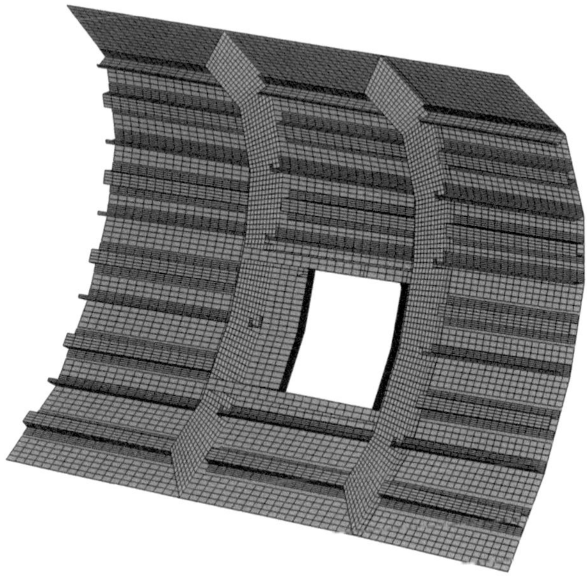

Fig. 7 Internal view of the submodel of the window area of the rear fuselage

in fact could not be carried out using the model of the whole fuselage due to the extreme difference between the dimensions of the Rear Fuselage (8,700 mm) and the crack (few millimetres, at least at the beginning of the propagation). The elements used in the model were eight nodes shells. No contact interaction was imposed between the different parts of the model. Gap elements were used instead. This means that the model was non linear at all, however the non linearity was checked to be negligible. Moreover, beam elements were used to simulate the rivets. The dimension of the elements in the submodel was between 0.7 and 23 mm.

A detailed reproduction of the thickness of the panels (including the reinforcement) of the rear fuselage was used for the submodel.

2.1 Experimental Validation of the Model

The FE model was validated comparing the numerical results obtained with the experimental data available from fatigue tests on the rear fuselage.

Four fatigue tests were executed. Three tests were conducted applying only one component of the load and in the fourth test all the components of the load were applied at the same time. Deformation was recorded on the left side of the rear fuselage using four strain gauges. The position of the strain gauges is visible in Fig. 8, where numbers 1, 2, 5, 6 indicate the strain gauges rosettes. Also, displacement data was collected by means of six LVDTs positioned in different zones of the fuselage.

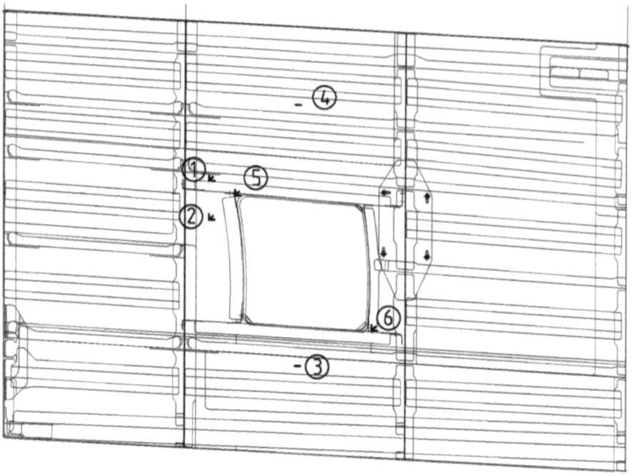

Fig. 8 Position of the strain gauges

According to the experimental data, in agreement with FE analysis, it could be stated that the worst type of load for the part of the rear fuselage analysed is the F_y load. The influence of force F_z and torque M_z was very limited. Thus, for the validation of the model, only the test data obtained from the F_y load was used.

Only LVDT 6, positioned in the rear zone of the fuselage, reported a significant value of displacement for test involved only F_y. Therefore only this LVDT was used to validate the FE model.

Comparing the experimental data and the numerical results no problem arise for strain gauges 3 and 4 because they are located in an area in which the stress and strain gradient was low. The FE value was calculated selecting the nodes closest to the position of the strain gauge and evaluating the mean value of the strain at these nodes.

On the other hand, all other strain gauges were located in areas where the stress and strain gradient was high. In this case the value of strain predicted by the FE model was strongly influenced by the position of the real strain gauge and by its orientation in such a manner that a rotation of a few degrees could modify the predicted value by 100 microstrains or more. However, being the strain gauges rosette, the maximum principal stress was used for this purpose. The influence of the positioning of the strain gauge was reduced selecting the nodes surrounding the location of the strain gauge on a bit larger area and evaluating the mean value of the maximum principal stresses at these nodes. The results of the comparison for the stresses obtained by the strain gauges values are reported in Table 1. The maximum principal stresses for the experimental data were calculated from the three experimental strains measured by each strain gauge rosette. In Table 2, the comparison of the global displacement as acquired from LVDT 6 is reported. As previous remarked, this comparison was executed only with the F_y load applied (experimental and FE model).

Table 1 Comparison between FE results and experimental data obtained from strain gauges (test with only F_y applied). The parameter used for the comparison is the maximum principal stress

Drawing position	FE value (MPa)	Experimental value (MPa)	Error (%)
1	45.8	47.5	−3.6
2	34.2	28.9	18.4
3	388.0	342.0	13.0
4	244.0	272.0	−10.0
5	141.0	82.2	71.5
6	81.6	81.5	0.2

Table 2 Comparison between experimental data and FE results for LVDT 6 (test with only F_y applied)

Drawing position	Direction	FE value (mm)	Experimental value (mm)	Error (%)
6	Y (aligned with F_y)	32.3	35.4	-8.7

With the exception of strain gauge 5, the error between FE and experiments for the strain gauges and the LVDT was less than 20%. This is an acceptable error. Strain gauge 5 was very close to the edge of the window in the Rear Fuselage in an area where the stress and strain gradient is very high. However, the results obtained with the model were conservative. Thus, it can be concluded that the results predicted by the FEM model were reliable and the model was validated.

2.2 Analysis of the Cracked Structure

Once the model was validated, a crack was introduced in the most stressed area of rear fuselage.

According to the previous analysis (only with F_y), most stressed area was carefully selected, see Figs. 9 and 10. A 1.27 mm initial crack was introduced in this area Fig. 10 in the direction orthogonal to the maximum principal stress. Discrete steps were used to simulate the propagation of the crack. For each step, the crack was extended in a straight direction by a fixed length and the direction of propagation was evaluated using the maximum energy release rate criterion. For the reason explained in the previous paragraph, only force F_y was used in the determination of the crack propagation direction even if the three analyses, reported in Table 3, were carried out for each step of the propagation.

The calculation of the stress intensity factor K_I was done for a crack length between 1.27 mm, Fig. 10 and 85 mm, Fig. 11. At about 60 mm, the crack reached a stringer. In order to obtain a conservative solution, the stringer was broken near the crack path for the remaining propagation, see Fig 12. The results of the simulation are reported in Table 3.

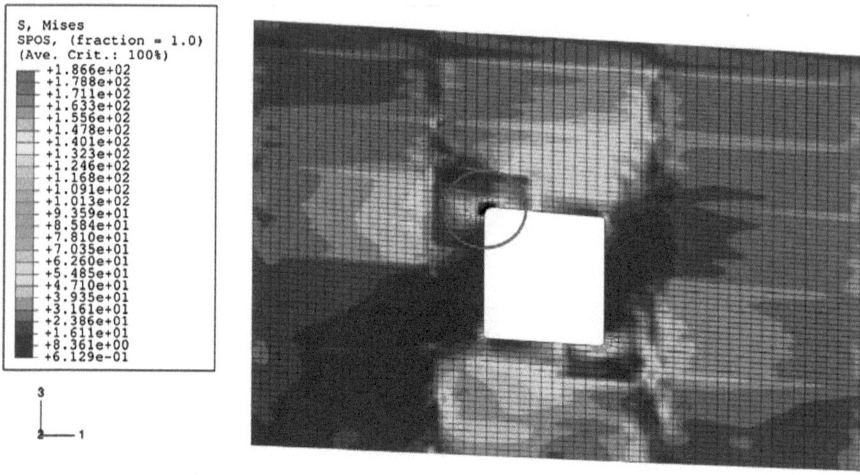

Fig. 9 Mises equivalent stresses. The most stressed area is highlighted

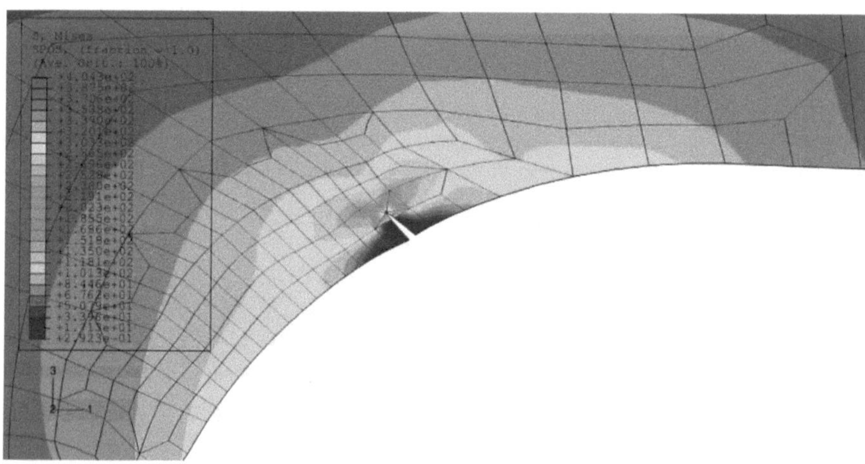

Fig. 10 Initial crack introduced in the submodel (external view), crack length = 1.27 mm, Mises equivalent stresses

Table 3 Values of K_I obtained from the FE analyses for different crack lengths

Crack length (mm)	$K_I - F_y$ (MPa·$\sqrt{mm}$)	$K_I - F_z$ (MPa·$\sqrt{mm}$)	$K_I - M_y$ (MPa·$\sqrt{mm}$)
1.27	199.9	0.4	0.1
1.60	264.2	2.3	0.5
3.20	392.9	3.5	0.8
6.40	473.1	4.3	1.0
12.80	521.7	5.2	1.2
25.60	568.3	6.8	1.6
41.40	573.8	8.4	1.9
85.00	476.5	4.5	1.0

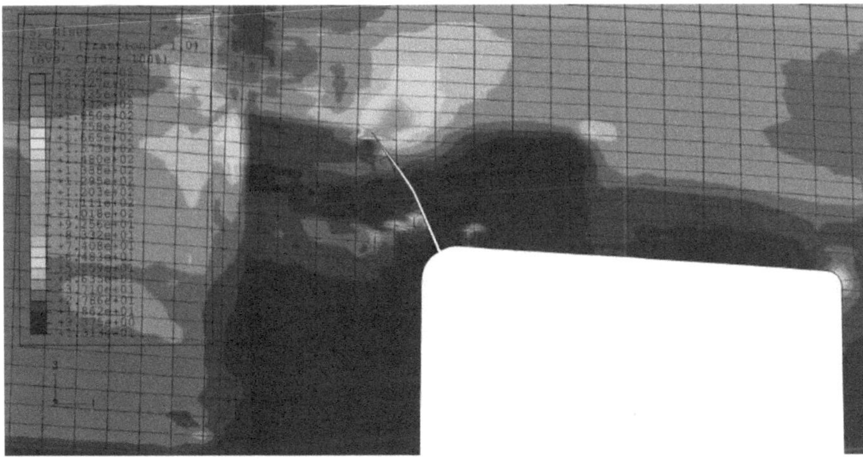

Fig. 11 Final configuration of the crack (external view), crack length = 85 mm, Mises equivalent stresses

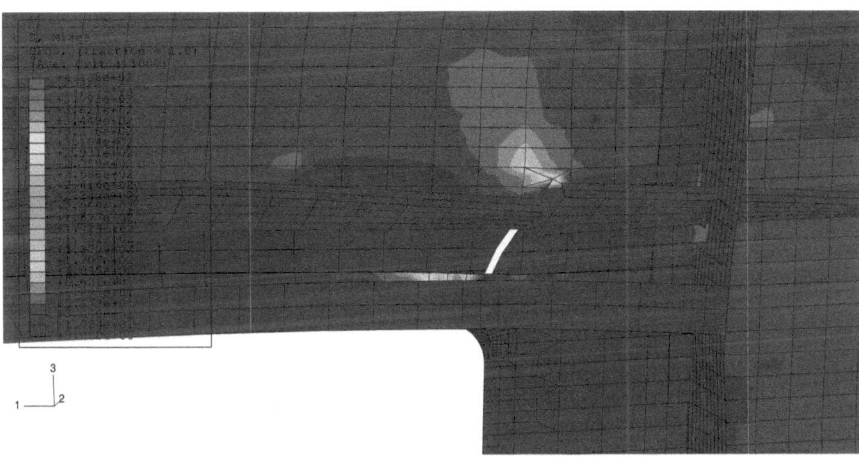

Fig. 12 Final configuration of the crack (*internal view*), crack length = 85 mm, Mises equivalent stresses. The stringer has been broken

3 Crack Propagation Analysis in Rear Fuselage

The loads used to simulate the crack propagation were defined on the basis of the loading spectra for the certification of the helicopter. The load history was based on the definition of three different flight types (high, medium and low loading level) reported in Table 4, each one corresponding to a different typical mission of the helicopter, and on the definition of vibratory loads (loading cycles not directly linkable

Table 4 Number of flights foreseen in a life of 10000 flight hours (FH) for the manoeuvre loads

Flight spectrum	Occurrences in 10,000 FH (number of flights)	Cycles in each flight
High loading factor	4,000	11
Medium loading factor	16,500	10
Low loading factor	21,000	10

to specific flight manoeuvres) but generally present because of the interaction between the rotor's downwash and the tail structure.

The different loading cycles, both the deterministic ones from manoeuvre spectra and the vibratory ones, were defined referring to the value of the two forces (the tail rotor traction F_y and the tail plane negative lift F_z) and the couple (the reaction torque from the tail rotor M_y) .

The vibratory loads acting on the structure were considered as constant amplitude cycles. Different kinds of vibratory loads were considered in the analysis: low frequency vibrations (0.05 and 4.10-4 Hz) and high frequency vibrations at 17.5 Hz. Each kind of vibratory load had an estimated percentage of application evaluated for 10,000 flight hours. Thus, it was possible to calculate the number of cycles expected for each vibratory load for 10,000 flight hours.

The different manoeuvre and vibratory loads were organized in a simple spectrum. The spectrum was defined to be as realistic as possible, maintaining the correct proportion between the manoeuvring spectra and the vibratory loads. Some trial simulations were carried out in order to study the effects of alternative solutions.

Thus, the load history was used in the simulation of the crack propagation using NASGRO [19] equation as the propagation law. The relationship was fitted using the data obtained from the experimental test at constant amplitude loading on the material Al-Li 8090-T81: fatigue crack propagation tests (ASTM E647/91 standard) with specimens of geometry M(T) with a central crack, and load ratio of R = 0.1, 0.3 and 0.7 were carried out.

Crack growth rate calculations in NASGRO use a relationship called the NASGRO equation. Forman and Newman of NASA, Shivakumar of Lockheed Martin, De Koning of NLR and Henriksen of ESA developed different elements of this equation and Forman and Mettu first published it. It is given by:

$$\frac{da}{dN} = C \cdot \left[\left(\frac{1-f}{1-R} \right) \cdot \Delta K \right]^n \cdot \frac{\left(1 - \frac{\Delta K_{th}}{\Delta K}\right)^p}{\left(1 - \frac{\Delta K_{max}}{K_c}\right)^q} \tag{1}$$

where N is the number of applied fatigue cycles, a is the crack length, R is the load ratio, ΔK is the stress intensity factor range, C, n, p, and q are constants which are empirically derived, ΔK_{th} is the threshold stress intensity factor, K_c is the critical stress intensity factor and f is crack opening function.

Some material, such as Al-Li 8090-T81 alloy, show only a marginal effect in respect to crack closure. Therefore it is assumed that $f = R$ in Eq. (1) and NASGRO equation becomes:

$$\frac{da}{dN} = C\Delta K^n \cdot \frac{\left(1 - \frac{\Delta K_{th}}{\Delta K}\right)^p}{\left(1 - \frac{\Delta K_{\max}}{K_c}\right)^q} \quad (2)$$

To eliminate the crack closure effect in the NASGRO program, the bypass parameter was assumed to be $\alpha = 5.845$ and $S_{max}/\sigma_0 = 1$.

Further elaborations were executed for the calculation of the threshold stress intensity factor range, ΔK_{th} and for critical stress intensity factor K_c. The experimental data resulting from the test carried out from Department of Mechanics were used to obtain the coefficient of Wheeler's equations, also looking for those coefficients that minimise the gap between the curve and the data test.

Two different simulations were carried out, one using Wheeler retardation model [20] (to account for the delay retardation due to applied overloads) and the other one without retardation.

The stress intensity factor K_I was evaluated by superposition for each single load. Even if the model is non-linear, according with the previous assertion of substantial linear behaviour, the superposition of the effects can be successfully applied. In fact an analysis was carried out applying the three loads at the same time. The value of the stress intensity factor evaluated in the analysis was compared to the one calculated by the superposition of the single loads, showing a negligible difference. The resulting curves showing the crack length vs. flights hours are shown in Figs. 13 and 14 and Table 5. The simulation of the propagation was carried out

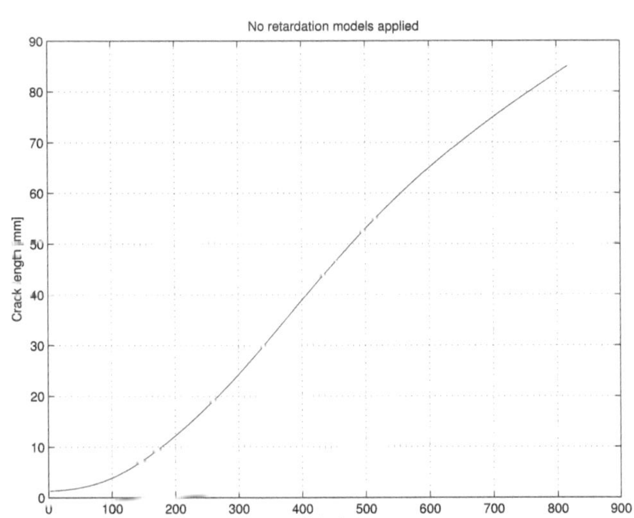

Fig. 13 Crack length vs flight hours curve describing the crack propagation. No retardation has been used. Initial crack length: 1.27 mm

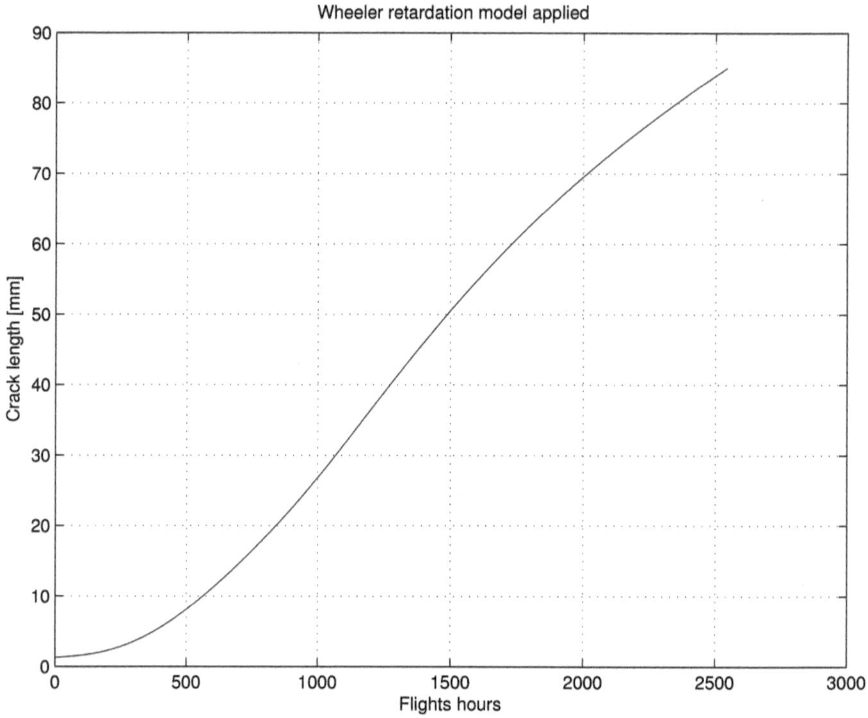

Fig. 14 Crack length vs flight hours curve describing the crack propagation. The Wheeler retardation model has been used. Initial crack length: 1.27 mm

for an initial crack length of 1.27 mm and a crack length up to 85 mm, that was the maximum length used in the finite element models. The value of the stress intensity factor K_I had never reached the critical value K_{Ic}, which is no failure was occurred.

Moreover comparing the experimental path obtained from the fatigue test with the one obtained from the FE simulations, Fig. 15, it is possible to see the good agreement and thus the capability of the numerical approach to describe the degradation mechanism.

Table 5 Flight hours needed to reach three particular events (crack length)

	Flight hours	
Event	No retardation model	Wheeler retardation model
a = 7.5 mm	151	480
a = 15 mm	227	715
a = 85 mm	817.5	2,542

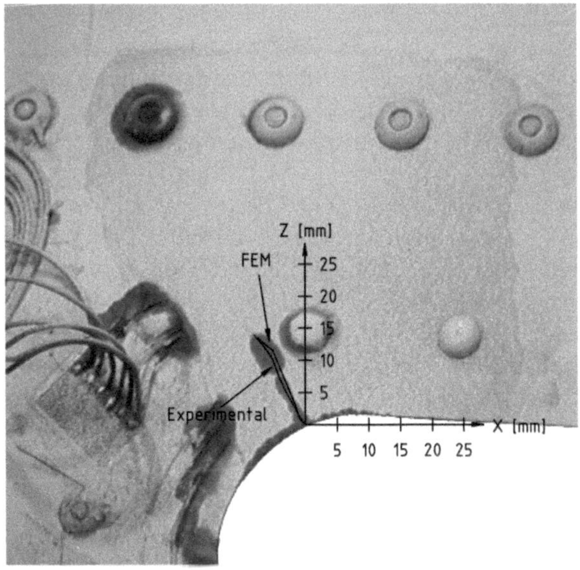

Fig. 15 A comparison between the crack path during the experimental test and in the FE simulations

4 Conclusions

The SHM is the next future key factor of all the cutting edge structures and in particular for what concern helicopters and (going ahead) tiltrotors. The raw instruments (sensor, algorithms and stress analysis) are ready for this task. They form basic elements of the SHM that should be integrated into one expert system. The sensor network together with data acquisition, processing and fusion units are key elements to obtain information about the current status of the system. Numerical models based on the FE analysis represent a powerful tool to develop a virtual image of the real system, as shown in this work. They are fundamental to study various scenarios that might occur under different conditions. Moreover, the numerical models results can be integrated with the data from the sensors and be used in situ to identify a degradation mechanism based on the current status of the system and its history. Thus they can be used to evaluate the impact of the degradation on the overall structure and, accordingly, to precisely predict its future development. The results processed by the adaptive prognostic assessment are the basis for advisory generation regarding the maintenance.

Considering that the primary damage in aircraft metallic structures results from fatigue cracks initiated at holes of joints and fasteners, a SHM system for aircraft, and in particular for helicopter subjected to severe vibratory loads, can reduce the repair and maintenance costs in two ways: first, the direct costs related to the repair can be reduced by detection damage at a very early stage; alternatively, the repair can be postponed until the next scheduled major overhauls to reduce indirect costs.

The fatigue damage related to a flight task is not exactly estimable / predictable: for similar flight tasks, the fatigue load and damage on a structure will vary due to variability in pilot performance and external conditions; the uncertainty related to the fatigue damage of a structure can be quantified by collecting data through sensors and making statistical inference on fatigue damage parameters of a data model.

Thus, a SHM integrated design can lead to a more optimized structure with the possibility to increase payload and performance and of course yet still provide required strength and safety level. Moreover, SHM is a powerful tool for the ground but also air crew, monitoring the health condition in order to make a correct real time prognosis for the achievement of the mission task. In addition, considering the cost of a new machine and thus the possibility to perform life extension program of existing machine, SHM can improve life of ageing airframe without reduce safety.

References

1. A. Ahmed, J.G. Bakuckas Jr., J. Awerbuch, A. CLau, T.M. Tan, Fatigue testing of a stiffened lap joint curved fuselage structure. 46th AIAA/ASME/ASCE/AHS/ASC Structures, Structural Dynamics & Materials Conference, Austin, TX 18–21 April (2005)
2. C. Boller, M. Buderath, Fatigue in aero structures where structural health monitoring can contribute to complex subject. Phil. Trans. R. Soc. A **365**, 561–587 (2007)
3. D. Colombo, M. Giglio, Valutazione del danno balistico su linea rotore di elicottero. Proceedings of the 34th AIAS Symposium, Milan, Italy (2005)
4. D. Colombo, M. Giglio, A methodology for automatic crack propagation in planar and shell FE models. Eng. Fract. Mech. **73**(4), 490–04 (2006)
5. D. Colombo, M. Giglio, A. Manes, 3D fatigue crack propagation analysis of a helicopter component. Int. J. Mater Prod. Technol. **30**(1/2/3),107–123 (2007)
6. Fatigue crack growth computer program NASGRO version 3.0.11. Lyndon Johnson Space Center, NASA, USA, October (2000)
7. M. Giglio, FEM submodelling fatigue analysis of a complex helicopter component. Int. J. Fatigue **21**, 445–55 (1999)
8. M. Giglio, A. Manes, Studio sulla propagazione di cricche in pannelli sandwich (Al-nomex). Proceedings of the 34th AIAS Symposium, Milan, Italy (2005)
9. M. Giglio, A. Manes, Crack propagation on helicopter panel: Experimental test and analysis. Eng. Fract. Mech. **75**, 866–879 (2008)
10. M. Giglio, A. Manes, U. Mariani, M. Vicario, Fatigue behaviour of Al-nomex sandwich panels. Proceeding of ICAF 2005 – International Committee on Aeronautical Fatigue, Hamburg, Germany (2005)
11. L. Lazzeri, U. Mariani, Application of damage tolerance principles to the design of helicopters. Int. J. Fatigue **31**, 1039–1045 (2009)
12. T.E. Munns, R.M. Kent, Structural health monitoring degradation mechanism and system requirements. Proceedings of the 19th Digital Avionics Systems Conference DASC, Philadelphia, PA (2000)
13. P.M. Pawar, R. Ganguli, Helicopter rotor health monitoring – A review. Proc. Inst. Mech, Eng. G J. Aerosp. Eng. **221**, 631–647 (2007)
14. H. Pfeiffer, M. Wevers, Aircraft integrated structural health assessment – Structural health monitoring and its implementation within the European project AISHA. EU project meeting on aircraft integrated structural health assessment (AISHA), Leuven, Belgium, June (2007)

15. H. Sohn, C.R. Farrar, F.M. Hemez, D.D. Shunk, D.W. Stinemates, B.R. Nadler, A review of structural health monitoring literature: 1996–2001. Los Alamos National Laboratory Report, LA-13976-MS (2003)
16. H. Stehmeier, H. Speckmann, Comparative vacuum monitoring (CVM) monitoring of fatigue cracking in aircraft structures. 2nd European Workshop on Structural Health Monitoring, Munich/Germany, 7–9 July (2004)
17. P.W. Tan, C.A. Bigelow, J.G. Bakuckas Jr., Widespread fatigue damage assessment approach. RTO AVT Specialists' Meeting on "Life Management Techniques for Ageing Air Vehicles", Manchester, UK, 8–11 Oct (2001)
18. E. Taylor, Structural monitoring system, keeping aircraft out of the hangar. Aust. Aviat. July (2008)
19. V.K. Varadan, V.V. Varadan, Wireless remotely readable and programmable microsensors and MEMS for health monitoring of aircraft structures. *Structural Health Monitoring* 2000, (Stanford University, Palo Alto, CA, 1999), pp. 96–105
20. O.E. Wheeler, Spectrum loading and crack growth. J. Basic Eng. Trans. ASME **94**(1), 181–186 (1972)

Effects of Manufacturing-Induced Residual Stresses and Strains on Hydrogen Embrittlement of Cold Drawn Steels

J. Toribio, M. Lorenzo, D. Vergara, and V. Kharin

1 Introduction

Prestressing wires are susceptible to surface cracking, in particular of the stress corrosion origin. Environmentally assisted fracture of prestressing steels has been the subject of extensive studies on the importance of hydrogen embrittlement (HE) in material damage [1]. With regard to manufacturing factors affecting the strength and life of prestressing wires, apart of the properties of material per se, the issue of residual stresses is essential [1]. The pioneering work described in [1] established an important milestone by establishing a quantitative relationship between the level of near-surface residual stresses, represented by hypothetical stress distributions, and HE of prestressing steel wires. However, due to a lack of necessary data, the influence of realistic residual stress profiles or the effects of plastic strain on HE have not been elucidated yet. This paper goes further in the analysis, so the earlier developed model in [1] is advanced to analyze the influence of the residual stress-and-strain profiles on hydrogenation of cold drawn prestressing steel wires.

2 Background Theory of Hydrogen Induced Fracture in Metals

Hydrogen induced fracture (HIF) depends on the amount of hydrogen in prospective microstructural fracture sites in metal so that local rupture events are associated with a critical combination of the responsible stress-strain field characteristics and hydrogen concentration C over a relevant material element x_c (cell, "grain", or domain of interest), as described elsewhere [2–4]. HIF advances by hydrogen-assisted nucleation of a (micro)crack in the site of the locally worst "concentration-stress-strain" triple which may be resolved to define, for the given stress-strain field, the distribution of critical concentration of hydrogen $C_{cr} = C_{cr}(\mathbf{x})$ as a function of the material point identified by a vector $\mathbf{x}$.

J. Toribio (✉)
University of Salamanca, 49022 Zamora, Spain
e-mail: toribio@usal.es

A. Öchsner et al. (eds.), *Materials with Complex Behaviour*, Advanced Structured Materials 6, DOI 10.1007/978-3-642-12667-3_21, © Springer-Verlag Berlin Heidelberg 2010

Hydrogen from corrosive environment penetrates into the metal and is accumulated in prospective fracture sites until the evolution in time t of its concentration $C(\mathbf{x}, t)$ attains somewhere the critical level C_{cr} after a certain period of time t_f, and then rupture occurs. Therefore, the criterion to determine the fracture time t_f takes the form:

$$C(\mathbf{x}, \mathrm{t}) = C_{cr}(\mathbf{x}), \tag{1}$$

where the fracture locus $\boldsymbol{x} = \mathbf{x}_c$ (or scale) must be specified, e.g., as suggested elsewhere [1–5].

Hydrogen diffusion in metal is often the mode of transport which controls fracture time. Diffusion in metals proceeds towards maximum system entropy corresponding to uniform distribution of a given amount of specie, represented by its concentration C, over available occupation sites whose density may be characterised by the solubility factor K_S. This latter depends, firstly, on the metal lattice dilatation induced by the hydrostatic stress $\sigma = (\sigma_1+\sigma_2+\sigma_3)/3$ where $\sigma_{1,2,3}$ are the principal stresses, and secondly, on the amount of lattice imperfections (traps for hydrogen, cf. [6]) that may be associated with the equivalent plastic strain ε_p. Therefore the stress-strain affected solubility of hydrogen in metal may be expressed as follows [2–4]:

$$K_S = K_{S\varepsilon}(\varepsilon_p)\exp(\Omega\sigma) \text{ with } \Omega = \frac{V_H}{RT}, \tag{2}$$

where $K_{S\varepsilon}$ is the strain-dependent component of solubility, V_H is the partial molar volume of hydrogen in metal, R the universal gas constant, and T the absolute temperature. This leads to the stress-strain affected diffusion flux $\mathbf{J}$ [2–4]

$$\mathbf{J} = -D(\varepsilon_p)\left\{\nabla C - C\left[\Omega\nabla\sigma + \frac{\nabla K_{S\varepsilon}(\varepsilon_p)}{K_{S\varepsilon}(\varepsilon_p)}\right]\right\}, \tag{3}$$

where $D = D(\varepsilon_p)$ is the diffusion coefficient of hydrogen in metal which depends on plastic strain. Mass balance then gives the diffusion equation in terms of concentration in the form:

$$\frac{\partial C}{\partial t} = -\mathrm{div}\mathbf{J}. \tag{4}$$

Hydrogen entry into metal, i.e. the boundary condition for diffusion corresponds to the equilibrium between its environmental thermodynamic activity and concentration within metal at the entry surface Γ:

$$C(\Gamma, t) = C_\Gamma,\ C_\Gamma = C_{eq}^0 K_{S\varepsilon}\left(\varepsilon_p(\Gamma)\right)\exp\left(\Omega\sigma(\Gamma)\right), \tag{5}$$

where C_{eq}^0 is the equilibrium concentration of hydrogen in a virgin material (free of stress and strain) under given environmental conditions, which stands here as a measure of hydrogenation capacity of the environment. Finally, the initial condition, $C(\mathbf{x}, t = 0) = C_0(\mathbf{x})$, must be defined for every particular case.

3 Residual Stresses and Plastic Strains Due to Cold Drawing

To proceed with the implementation of the presented model to describe HIF of pre-stressing wires, the first step is to determine the residual stresses and plastic strains in wires generated by cold drawing. To this end, some numerical simulations of the wire drawing process have been performed recently [7]. However, focusing on the sole residual stresses, no data have been provided about the plastic strains behind them. To fill in this deficiency, a further simulation of drawing ought to be undertaken.

To start with, a one-pass cold drawing process was modelled. It consisted of passing the wire, which initial diameter was $d_0 = 12$ mm, through the die with the hole diameter $d_1 = 9.6$ mm and the inlet angle $\alpha = 7.6$, to get the wire with a final diameter $d_2 = 9.64$ mm (Fig. 1).

As a prototype wire material, the eutectoid pearlitic high-strength steel (0.74% C, 0.70% Mn, 0.20% Si, 0.015% P, 0.023% S) was considered. In order to obtain the experimental engineering stress— engineering strain curve (Fig. 2), several tensile tests were carried out. According to this curve, the Young modulus $E = 199$ GPa and the 0.2%-offset yield strength $\sigma_Y = 710$ MPa were obtained. With them, to perform finite-element simulations, the constitutive model for a wire material was chosen to be elastoplastic solid with von Mises yield surface, associated flow rule, and isotropic strain-hardening following the data shown in Fig. 2. The hard die was modelled as an elastic material with a Young's modulus of 600 GPa, corresponding to tungsten carbide.

Obviously, the axisymmetric formulation is fairly suitable for simulations. Correspondingly, a finite element model was created to model the drawing process in cylindrical coordinates (r, z), as shown in Fig. 1. A suitable length of the rod to be cold drawn was chosen to be equal to $10d_0$ in order to get rid of the

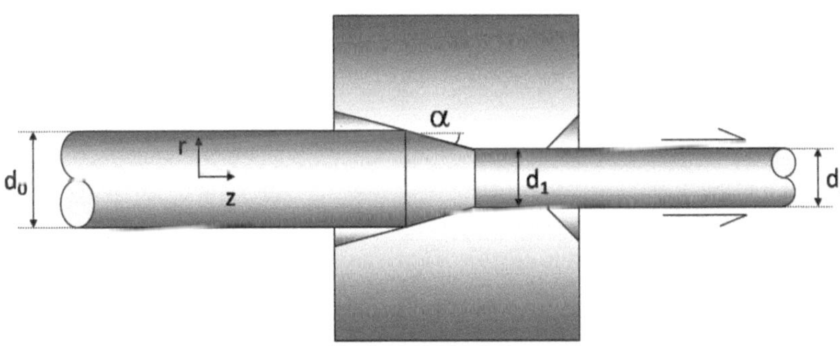

Fig. 1 Scheme of the cold drawing process

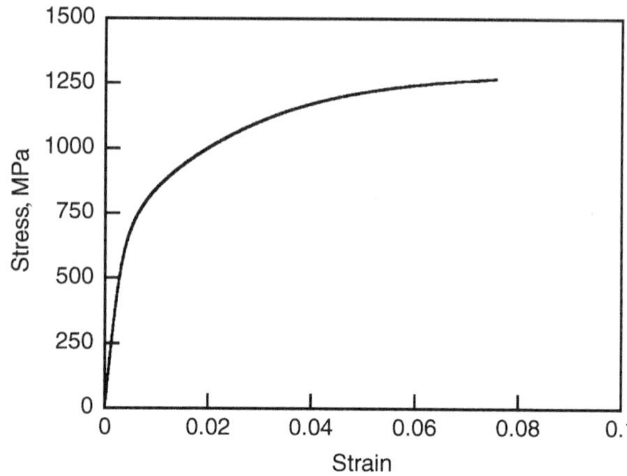

Fig. 2 Experimental stress-strain curve of the steel

end effects on the resulting stress-strain field in the wire. The boundary condition of the prescribed axial displacement was imposed on the extreme front of the rod. Elastoplastic large deformation – large strain calculations were performed using a general-purpose finite element code with updated Lagrangian formulation. Several finite element meshes formed by four-node quadrilaterals were tried till the acceptable mesh-convergence of the result was ensured. Configurations of the modeled rod before its entry into the die, within the die, and after the completion of cold drawing are presented in Fig. 3.

Focusing only on the residual hydrostatic stress σ_R and equivalent plastic strain ε_p as the only relevant variables for the analysis of the hydrogenation of cold drawn wires, Fig. 4 displays the resulting radial distributions of σ_R and ε_p along the radial coordinate $0 \leq r \leq a = d_2/2$ in the middle portion of the deformed rod, where the end effects were proved be effectively vanished and the variables of interest become

Fig. 3 Finite element simulation of the drawing process: the rod before entry into the die (*top-left*); the rod within the die (*top-right*); the cold drawn rod (*bottom*). Contour bands indicate the levels of the equivalent plastic strain

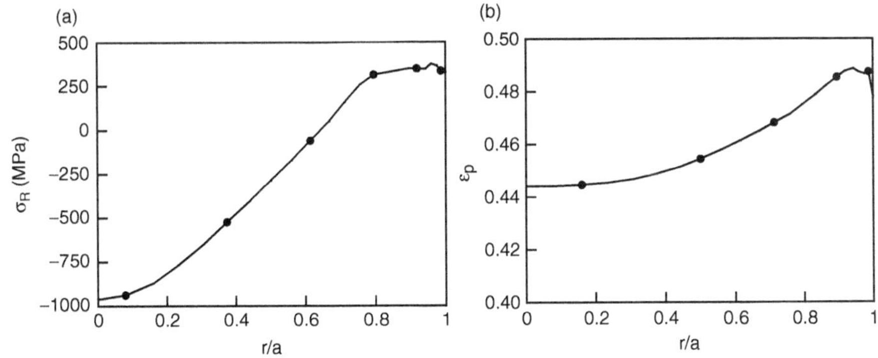

Fig. 4 Computed distributions of residual stress (**a**) and plastic strain (**b**) in cold drawn steel wire

independent on z. The obtained results are in reasonable agreement with available data of experiments and modeling [7].

4 Hydrogen Accumulation in Cold Drawn Steel Wires

Service conditions of prestressing steel wires usually combine predominantly tensile loading of round bars by uniform remotely applied tension σ_{app} and the effects of harsh environments. The *Ammonium Thiocyanate Test* (ATT) intends to reproduce these service conditions using a specific controllable environment. Obviously, cylindrical coordinates are best suited for the analysis of wire hydrogenation too. Since for long wires there are no input data depending either on the axial coordinate or on the hoop one, a one-dimensional modeling in terms of the sole spatial variable r becomes natural. A finite-element analysis of axisymmetric boundary-value problem of stress-strain affected diffusion (3), (4) and (5) was implemented basically as described elsewhere [8]. In brief, applying the Galerkin process, the same family of the element shape functions $N_e(r)$ served as trial and weighting functions in usual terms [9], and they also were used to approximate the distributions of stress $\sigma_R(r)$ and plastic strain ε_p, which are presented in Fig. 4, in the way, that

$$\sigma_R(r) = \Sigma\sigma_{Rj}N_j(r) \quad \text{and} \quad \varepsilon_p(r) = \Sigma\varepsilon_{pj}N_j(r), \tag{6}$$

where $j = 1,\ldots, M$ carries out the numbering of M nodes of the finite-element mesh. The weak form of the weighted residual statement of the problem rendered the system of ordinary differential equations with respect to the finite-element nodal concentration values $C_j(t)$ as the functions of time as follows:

$$\left[M_{ij}\right]\left\{\frac{dC_j}{dt}\right\} + \left[K_{ij}\right]\left\{C_j\right\} = \{F_i\} \quad (i,j = 1,\ldots,M), \tag{7}$$

where the components of the element matrices [...] and the vector-columns {...}, respectively, are

$$M_{ij} = \int_V N_i N_j dV, \tag{8}$$

$$K_{ij} = \int D\left(\varepsilon_{eq}^p\right)\left\{\nabla N_i \nabla N_j - \left[\left(\frac{V_H}{RT}\nabla\sigma + \frac{\nabla K_{S0}\left(\varepsilon_{eq}^p\right)}{K_{S0}\left(\varepsilon_{eq}^p\right)}\right)\cdot \nabla N_i\right] N_j\right\} dV, \tag{9}$$

$$F_i = -J_s \int_{S_f} N_i dS, \tag{10}$$

where the latter prescribes the flux of hydrogen J_S on the part S_f of the surface $S = \partial V$, whenever convenient. To solve these equations, the unconditionally stable Galerkin scheme of time-domain integration was employed [9].

Simulations were performed using linear trial functions with respect to both space and time variables. Spatial discretisation for diffusion problem was taken corresponding to the radial spacing of the nodes of the deformed finite element mesh from the simulation of the drawing process.

Although it is known [9] that strong accuracy deterioration may occur applying the Galerkin method to transport problem such as (3), (4) and (5), when a mesh-related parameter called the Peclet number increases too much, this complication has never been met in performed simulations.

The set of model parameters was assigned as follows. The temperature in the performed calculations was fixed $T = 323$ K. The partial molar volume of hydrogen for iron-based alloys, *i.e.*, steels, was well determined and considered to be fairly constant, $V_H = 2$ cm^3/mol [6]. Required experimental data on hydrogen diffusivity and solubility were not available for the steel being considered, and reasonable estimates had to be adopted. Concerning these characteristics, a wide dispersion of measurements was found in this kind of materials (*i.e.*, the same bulk composition, grade, etc.), and all these data turn out to be very sensitive to minute alterations of alloy composition, microstructure and accumulated plastic deformation, as it has been pointed out in numerous publications since long ago, cf. e.g., [6, 10, 11]. Hydrogen diffusivity in bcc-iron alloys at temperatures below 500 K may vary from 10^{-13} to 10^{-8} m^2/s [10]. Moreover, a specific feature of hydrogen behaviour, which is relevant in particular for the family two-phase steels, is the substantial anisotropy of diffusivity induced by cold-working procedures, such as rolling or drawing [11]. As a result, in heavily cold drawn pearlitic steels with highly plastically strained ferrite having high dislocation density and lamellae cementite with extensive area of highly strained interfaces, hydrogen diffusivity at ambient temperatures may be as low as 10^{-12} m^2/s and less, cf. [1]. Then, taking into account this reasoning, and adopting the simplest one from analytical forms used to fit the data about the plastic strain effect on hydrogen diffusivity and solubility in metals [12, 13], the following may be accepted for heavily cold drawn steel wires:

$$D(\varepsilon_p) = D_0 \exp(-\eta\varepsilon_p), \tag{11}$$

$$K_{S\varepsilon}(\varepsilon_p) = 1 + \beta\varepsilon_p, \tag{12}$$

where $D_0 = 3\times10^{-12}$ m^2/s is consistent with [1] for the chosen temperature, and the fitting coefficients $\eta = 2.9$ and $\beta = 4$ are taken from Ref. [13].

To discriminate the contributions of stress and plastic strain on hydrogenation of wires, the results of the computations for stress-only and for stress-strain assisted diffusion are presented in Figs. 5 and 6 in terms of dimensionless concentration $C(r, t)/C^*$ where the reference concentration C^* represents the hydrogenating capacity of environment modified by applied stress σ_{app}, i.e., the equilibrium hydrogen concentration in stressed virgin material, which within elasticity limits is $C^* = C^0_{eq}\exp(\Omega\sigma_{app}/3)$. The textbook solution [14] for diffusion in a virgin material, i.e., not taking into account the effects of residual stress-strain field on diffusion,

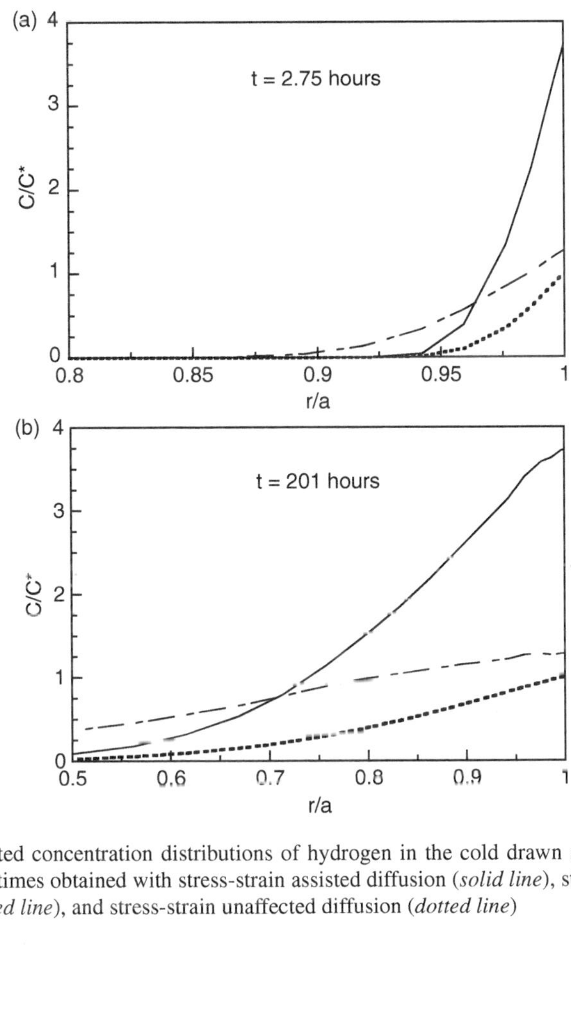

Fig. 5 Calculated concentration distributions of hydrogen in the cold drawn steel wire at indicated diffusion times obtained with stress-strain assisted diffusion (*solid line*), stress-only assisted diffusion (*dashed line*), and stress-strain unaffected diffusion (*dotted line*)

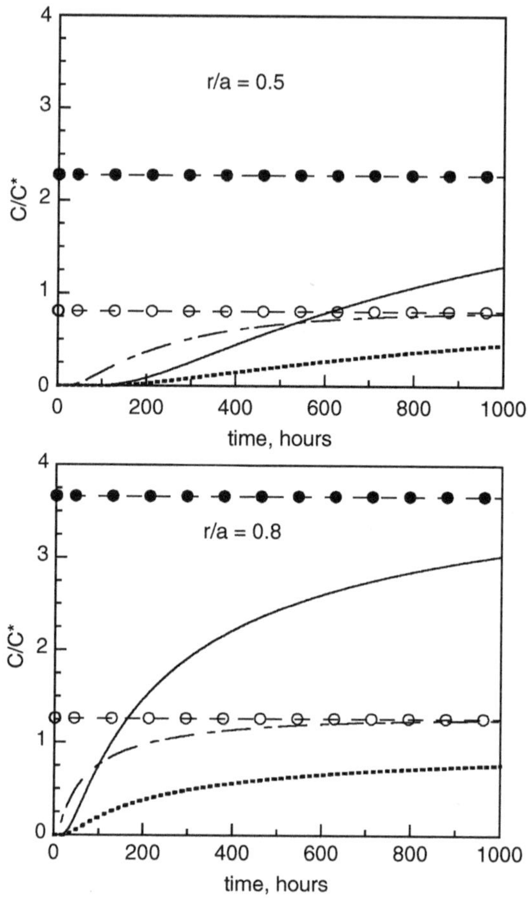

Fig. 6 Calculated concentration evolution of hydrogen at indicated depths in the cold drawn steel wire obtained with stress-strain assisted (*solid line*), stress-only assisted (*dashed line*), and stress-strain unaffected diffusion (*dotted line*). Horizontal lines marked with closed and opened circles indicate the corresponding levels of equilibrium hydrogen concentration C_∞ in stress-strain and stress-only affected cases, respectively

is also presented in Figs. 5 and 6. At long simulation times, calculated concentrations $C(r, t\rightarrow\infty)$ manifested approaching the theoretical equilibrium solutions $C_\infty(r)$ to be attained at $t\rightarrow\infty$. These latter are given by the closed-form steady-state solution of the diffusion problem [2–4], which in this case takes the form

$$C_\infty/C^* = K_S\left(\sigma_R(r), \varepsilon_p(r)\right) = K_{S\varepsilon}\left(\varepsilon_p(r)\right)\exp\left(\Omega\sigma_R(r)\right) \tag{13}$$

From the presented results, it follows that the effects of residual stress and strain on hydrogenation consist in the wire over-saturation with hydrogen in comparison with the virgin material. The distinction between stress-only and stress-strain assisted hydrogenation is that relative over-hydrogenation occurs at shorter diffusion depths from the wire surface $r = a$ and under-hydrogenation at deeper material locations in the strain-affected case, and the division between these over- and under-hydrogenated domains moves deeper into the wire as time goes on. In terms of the hydrogenation of a wire skin of definite thickness, where critical events of HIF can be expected to occur (cf. [1]), residual strain field will reduce the amount of hydrogen accumulated in this skin at short diffusion times, and augment this amount at long exposure times. This is the consequence of two kinds of effects of the stress and plastic strain fields on hydrogen diffusion. The first one is the increase of hydrogen solubility due to tensile stresses and accumulated plastic strain [cf. Eqs. (2) and (12)], which results in rising hydrogen absorption by a metal. This is what enhances the hydrogenation of cold drawn wires. On the other hand, the decrease of hydrogen diffusivity (or its mobility) in plastically strained material [cf. Eq. (11)], slows-down this process of enhanced hydrogen accumulation in the wires. Thus, in a given material element, in the HIF locus $\mathbf{x}_c$ in particular, the effect of plastic strain generated by cold drawing allows to attain substantially higher level of concentration C, especially in the wire skin where HIF initiates, but this takes notably longer time of diffusion if compared with that for stress-only assisted case.

Different consequences of the cold drawing plastic strains may be expected for hydrogen accumulation in prospective HIF sites. To exemplify the matter, let's identify the fracture nucleation site depth x_c with the dimension (location) of the critical surface (subsurface) crack, as it was suggested in the earlier model [1], and take roughly its value of 450 μm. Then, at the critical location near the point $r = a - x_c \approx 0.9a$, depending on the required critical concentration C_{cr}, would it be it lower or higher, the effect of plastic strain, can be preventing (Fig. 5, short-time plot) or promoting (Fig. 5, long-time plot) the attainment of the critical concentration C_{cr}. This would have evident consequences for the rupture time t_f in the case of stress-only dependent critical concentration $C_{cr} = C_{cr}(\sigma_{ef})$, where the effective stress σ_{ef} is the sum of the applied and residual stresses in axial (or longitudinal) direction.

Unfortunately, this is not the only factor since HIF is known to be strongly conditioned by the alloy microstructure, and this latter in prestressing steel wires is known to be strongly altered by cold drawing. Then the critical concentration must depend not only on the effective stress level, but also on the steel microstructure, *i.e.*, on the plastic strain generated by cold drawing. Thus, elucidation of the effects of cold drawing on HIF in prestressing steel wires still requires clarification of the mechanism of HIF, which must take into account the microstructural consequences of cold drawing on rupture events.

Nevertheless, the presented theory may be employed to interpret the consequences of various treatments, which affect the residual stress-strain fields, on workability of prestressing wires under the risk of HIF from the point of view of their influence on the (residual) stress-and-strain profiles, and as a consequence, on the stress-strain assisted diffusion of hydrogen towards prospective rupture sites.

It also may be useful to reveal the situations in potential HIF loci (i.e., the levels of stress, strain and hydrogenation therein), which would be beneficial for the understanding of the mechanisms of HIF in wires.

5 Conclusions

A theory to predict the lives of high strength pearlitic steel wires under conditions of hydrogen assisted fracture in harsh environments is presented, which is based on the consideration of stress-and-strain affected diffusion of hydrogen in metal and on the criterion of hydrogen induced local fracture. It associates the effects of various manufacturing procedures on the serviceability of prestressing wires with the influence of the residual stress-strain profiles on the diffusion of hydrogen towards prospective rupture sites, apart from the role of the stress-strain field as the relevant mechanical factor of fracture.

To carry out predictions of HIF in wires with the use of this theory, the knowledge of the whole distributions of plastic strains and triaxial residual stresses generated in wires by cold drawing is necessary. Numerical modelling is the feasible way to determine them throughout the entire wire diameter. The paper presents the results of simulations of the cold drawing process, that give the desired inhomogeneous distributions of the triaxial residual stresses and plastic strains, whose magnitudes and gradients are considered the influencing factors of metal hydrogenation within the adopted theoretical framework.

Numerical modelling of the stress-strain assisted hydrogen diffusion in cold drawn wires was performed. The evolutions of hydrogen concentration in wires were calculated. The relevance of residual stresses and plastic strains for hydrogen accumulation in prospective rupture sites in wires was manifested. This way, this paper emphasizes the roles of *both* residual stresses and plastic strains (generated by cold drawing) in the performance of prestressing steel wires under hydrogen environment.

The proposed theory and modelling seem to be a promising tool for further development of life-prediction procedures for prestressing steel wires under HIF conditions on the basis of reduced testing, and to account for various residual stress and strain profiles induced by surface treatments. The theory may be employed to foresee the consequences of various manufacturing procedures and treatments, which affect the residual stress-strain fields, on the workability of prestressing wires under the risk of HIF from the point of view of their influence on the (residual) stress-and-strain profiles, and as a consequence, on the stress-strain assisted diffusion of hydrogen towards prospective rupture sites.

Acknowledgments The authors wish to acknowledge the financial support provided by the following Spanish Institutions: Ministry for Science and Technology (MCYT; Grant MAT2002-01831), Ministry for Education and Science (MEC; Grant BIA2005-08965), Ministry for Science and Innovation (MCINN; Grant BIA2008-06810), *Junta de Castilla y León* (JCyL; Grants SA067A05, SA111A07 and SA039A08).

References

1. J. Toribio, M. Elices, Influence of residual stresses on hydrogen embrittlement susceptibility of prestressing steels. Int. J. Solids Struct. **28**, 791–803 (1991)
2. J. Toribio, V. Kharin, K-dominance condition in hydrogen assisted cracking: The role of the far field. Fatigue Fract. Eng. Mater. Struct. **20**, 729–745 (1997)
3. J. Toribio, V. Kharin, Evaluation of hydrogen assisted cracking: the meaning and significance of the fracture mechanics approach. Nucl. Eng. Des. **182**, 149–163 (1998)
4. J. Toribio, V. Kharin, The effect of history on hydrogen assisted cracking: 1. Coupling of hydrogenation and crack growth. Int. J. Fract. **88**, 233–245 (1998)
5. V.S. Kharin, Crack growth in deformed metals under the action of hydrogen. Soviet Mater. Sci. **23**, 348–357 (1987)
6. J.P. Hirth, Effects of hydrogen on the properties of iron and steel. Metall. Trans. **11**A, 861–890 (1980)
7. M. Elices, Influence of residual stresses in the performance of cold drawn pearlitic wires. J. Mater. Sci. **39**, 3889–3899 (2004)
8. J. Toribio, V. Kharin, Role of fatigue crack closure stresses in hydrogen assisted cracking. in *Advances in Fatigue Crack Closure Measurement and Analysis*, ed. by R.C. McClung, J.C. Newman, vol. 2. (ASTM STP 1343, Philadelphia, 1999), pp. 440–458
9. O.C. Zienkiewicz, R.L. Taylor, *The Finite Element Method*, vols. 1–2. (McGraw-Hill, London, 1989)
10. B.A. Kolachev, *Hydrogen Brittleness of Metals* (Metallurgia, Moscow, 1985)
11. H.L. Lee, S.L.I. Chan, Hydrogen embrittlement of AISI 4130 steel with an alternate ferrite/pearlite banded structure. Mater. Sci. Eng. A**142**, 193–201 (1991)
12. J. Lufrano, P. Sofronis, Hydrogen transport and large strain elastoplasticity near a notch in alloy X-750. Eng. Fract. Mech. **59**, 827–845 (1998)
13. M.S. Astiz, J.A. Álvarez, F. Gutiérrez-Solana, Modelo numérico para analizar el efecto del hidrógeno sobre los procesos de fisuración dúctil. Anal. Mech. Fract. **15**, 79–84 (1998) (in Spanish)
14. P. Shewmon, *Diffusion in Solids* (The Minerals, Metals and Materials Society, Warrendale, PA, 1989)

Hybrid Bonding of Advanced High Strength Steels in the Lightweight Body Shell Design for the Automobile Manufacturing

G. Weber, T. Bschorr, H. Cramer, O. Hahn, M. Rethmeier, and H. Thommes

1 Introduction

In the lightweight body shell mass production of automobiles, the application of hybrid bonding techniques becomes more and more important in the future. One of these hybrid bonding techniques is the combination of adhesive bonding and resistance spot welding called weldbonding. This contribution is based on some special results of a research project which was carried out by the German Welding Institute SLV Munich and the University of Paderborn [1]. The weldbonding procedure affords many advantages, e. g. improved crash performance, fatigue behaviour and corrosion resistance. Therefore, this technology is state of the art in many branches joining metal sheets. Especially for newly developed high strength multiphase steels, also called advanced high strength steels (AHSS), this joining technique has more advantages than other joining procedures for thin steel sheets. The application of AHSS materials in conjunction with economically efficient and reliable joining processes helps saving costs and conserving resources (weight reduction, energy minimization) and provides at the same time consistent or improved safety of the passenger cell (crash optimization). In this context, the Ultra Light Steel Auto Body design (ULSAB) and New Steel Body (NSB) design are referred to [2, 3]. For more information on AHSS see [4–6]. There are many contributions on the resistance spot weldability of AHSS, the process reliability of the resistance spot welding process and the mechanical behaviour of spot welded AHSS and some other steel grades [7–12]. Furthermore, a lot of important publications have been written about the weldbonding process and the properties of weldbonded joints [13–23]. But papers on weldbonded AHSS do not exist [24]. In this contribution, statements will be given about the process reliability of the weldbonding process and the mechanical properties of the hybrid bonded joints for different adhesives and different combinations of a mild and some high strength steels.

G. Weber (✉)
Federal Institute for Materials Research and Testing (BAM), Berlin, Germany
e-mail: gert.weber@bam.de

A. Öchsner et al. (eds.), *Materials with Complex Behaviour*, Advanced Structured Materials 6, DOI 10.1007/978-3-642-12667-3_22,

The results are based on the application of a rubber-based and two epoxy resin-based adhesive systems. It will be shown that both the process reliability of the hybrid bonding process and the mechanical and metallurgical behaviour of the hybrid bonded joints are influenced by the combinations of base metals, the applied adhesive system and the choice of the joining parameters. Especially the mechanical behaviour of the weldbonded joints will be studied under quasi-static and impact loads. An interpretation of the metallurgical behaviour of the weldbonded joints will be given, too. Furthermore, a comparison of the fatigue behaviour of resistance spot welded, adhesive bonded and weldbonded joints will be carried out. The presentation of these results may be interpreted as widening of weldbonding knowledge.

2 Conditions for the Application of the Weldbonding Process

2.1 Welding Equipment

The weldbonded joints were realized by a medium frequency direct current (MF-DC) spot welding gun. The advantage of the medium frequency technology with an operation frequency of 1,000 Hz consists in the possibility of a welding current regulation which enables a very fast reaction with respect to any variation in the weldbonding process. Figure 1 shows the principle of the weldbonding process when welding with a MF-DC spot welding gun. The used equipment was a stiff spot welding gun with short arms and a servo electric drive. The choice of the welding parameters electrode force F_e and welding time t_w depends on the applied base metals. The electrode forces for the mild steel are 2.5 kN and 3.5 kN, for the high strength steels 3.5 kN and 4.5 kN and for the Martensitic steel MS-W1200 4.5 kN and 5.5 kN. Depending on the strength of the base metals welding times from 260 ms up to 400 ms are chosen [1].

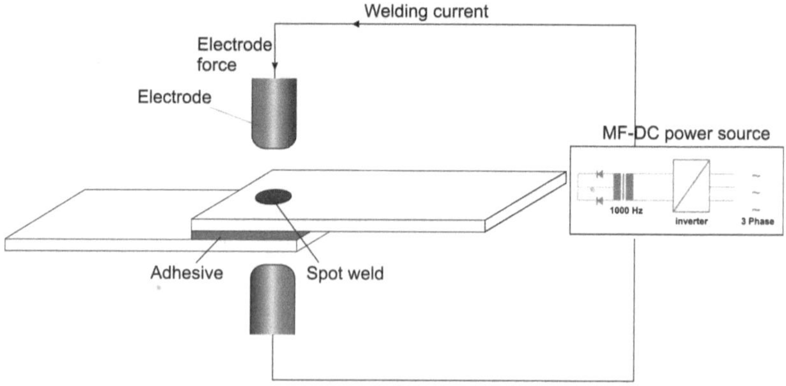

Fig. 1 Principle of weldbonding and resistance spot welding

2.2 Applied Adhesives

The realized weldbonding processes described in this contribution are based on three one-component hot-curing adhesives. Especially a rubber-based adhesive (Terostat 5194), an epoxy resin-based adhesive (Betamate 1480) and an epoxy resin PUR-based adhesive (SikaPower 498) were applied ([1]; see Table 1).

The viscous behaviour of these adhesives is different. The adhesive Terostat 5194 can be defined as an adhesive of low viscosity, the adhesives Betamate 1480 and the SikaPower 498 can be defined as adhesives of medium viscosity. For instance, Fig. 2 shows the shear viscosity depending on the shear stress. These results were generated by the application of the shear stress slope to the adhesive specimen. Both, the viscous behaviour of the adhesives and their yield point in the non-cured condition were determined. It can be seen from Fig. 2 that the SikaPower 498 adhesive, despite its higher viscosity, exhibits a lower yield point compared to Betamate 1480. Terostat 5194 has the lowest viscosity and the lowest yield point. The shear viscosities of the two adhesives Betamate 1480 and SikaPower-498 align to one another over all temperature ranges considered at shear stresses between 2.3 MPa and 5.0 MPa. Any difference in their squeezability may therefore not be expected to occur during the application of the electrode force (i.e. when the bond gap has nearly reached its final geometry). By application of a force controlled amplitude sweep to

Table 1 Applied adhesives

Type of adhesive	Base	Viscosity
Terostat 5194 (A1)	Rubber-based	Low
Betamate 1480 (A2)	Epoxy resin-based	Medium
SikaPower 498 (A3)	Epoxy resin PUR-based	Medium

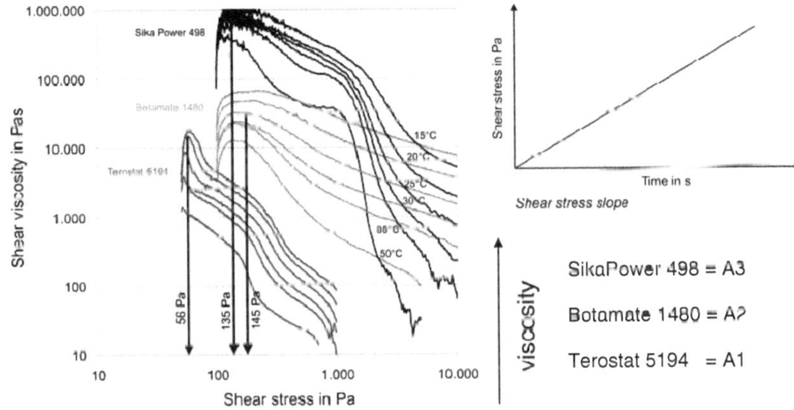

Fig. 2 Shear viscosity versus shear stress – determination of the flow limits

the adhesive specimen the viscoelastic behaviour of the non-cured adhesives was examined, too [1]. This was accomplished by identifying the characteristics of the complex viscosity [1].

2.3 Advanced High Strength Steels

The newly developed so called Advanced High Strength Steels (AHSS) show superior properties in comparison with traditional high strength steels, for example carbon steels and other conventional grades (see Fig. 3; [5, 6]). Due to the chemical composition and the manufacturing process, AHSS offer an optimal solution as regards high strength and good formability. The reason for such generally contradictory behaviour is the combination of different microstructures, like ferrite, austenite, bainite and martensite. The ULSAB (UltraLight Steel AutoBody) project has demonstrated that the excellent properties of AHSS in combination with new manufacturing processes and innovative design leads to a significant reduction in the weight of body shells in automobile production and, accordingly, to a reduction of CO_2 emissions [2].

2.3.1 Grades of Advanced High Strength Steels

AHSS are normally classified into four types, i.e. dual phase steels (DP), TRIP steels, complex phase steels (CP) and martensitic steels (MS). With the aid of a temperature specific manufacturing process and thanks to different chemical compositions the amount as well as the dispersion of different microstructures (ferrite, martensite, etc.) and the mechanical properties of AHSS can considerably be varied

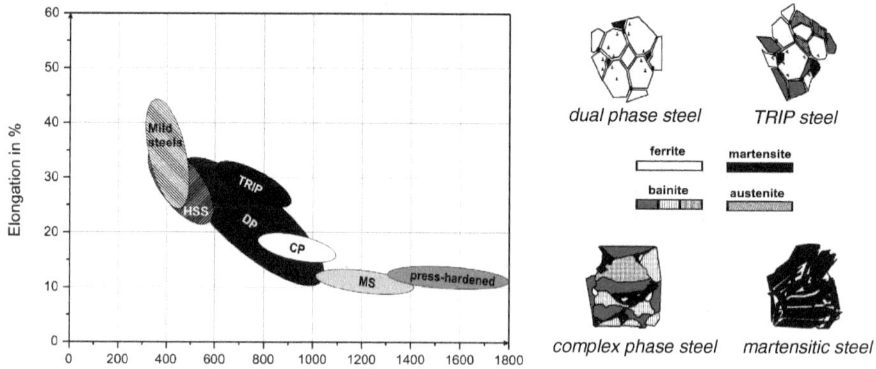

Fig. 3 Formability versus strength for conventional and advanced high strength steels (AHSS)

according to user requirements. The different microstructures of AHSS can be characterized as follows [3–6]:

Dual Phase Steels (DP)

Dual phase steels have a ferritic matrix (ferrite content varying between 85 and 90%) with inclusion of martensite islands. The mechanical properties of DP steel can be influenced by changing the amount of martensite. Generally, increasing the martensite content also increases the mechanical strength. An advantage of DP steels is the combination of high strength with high elongation when compared to conventional high strength steels (see Fig. 3).

TRIP Steels

TRIP steels possess a complex microstructure dominated by ferrite (70–85%) with residual austenite (up to 15%) and additions of martensite and bainite. The strength/ductility balance is increased by strain induced austenite to martensite transformation (TRIP effect). Steels with TRIP effect attain high uniform elongation values at high tensile strength levels because of their very strong work hardening.

Complex Phase Steels (CP)

CP steels have a fine complex microstructure of bainite and with islands of retained austenite and inclusions of ferrite and martensite. CP steels offer higher yield strength in comparison with TRIP steels, however, with simultaneous decrease of formability.

Martensitic Steels (MS)

MS steels offer a predominantly martensitic microstructure with minor quantities of ferrite and bainite. In consequence of the martensitic microstructure, the mechanical strength reaches up to 1,400 MPa and more, but with strongly restricted formability, which is an essential factor to be considered in practical application in the automobile industry. Table 2 shows typical mechanical properties of selected AHSS, which are used for the following interpretations.

Press-Hardened Steels

Press-hardened steels such as UsiBor or Ultraform use boron as an alloying element. Boron acts as hardening agent during thermomechanical treatments and provides a material with excellent hardness and high strength. A quenching treatment leads to precipitation of boron carbide at the grain boundaries as well as boron segregation. The suppressing of austenite to ferrite transformation caused by this segregation phenomenon leads to increasing hardenability. Furthermore, substitutional solid solution elements such as Mn influence the strength after quenching.

Table 2 Mechanical properties of the applied steel grades

Grade definition by future standardization	Sheet thickness in mm	Surface	Yield strength in MPa	Tensile strength In MPa	Fracture strain A_{20}
DC04ZE	1.50	ZE50/50 + KSP (Granocoat)	195	323	49
HCT780T	1.25	ZE75/75	696	879	29
HCT600XD	1.30	Z100	398	625	26
MS-W1200	1.55	–	948	1,216	12
22MnB5 (UsiBor)	1.50	AlSi	1,282	1,471	11

3 Process Reliability of the Weldbonding Process

3.1 Welding Current Ranges and Process Reliability

The introduction of the weldbonding process reliability is based on the definition of the resistance spot welding process reliability [7–12]. Thus, to ensure process reliability during weldbonding under production conditions, the knowledge of the welding current range (WCR) is of basic importance. Different WCR are obtained depending on the selected welding parameters. The primary welding parameters of resistance spot welding are the root mean square value (r.m.s.-value) of welding current I, the stationary electrode force F_e and the welding time t_w. These parameters are the primary joining parameters of the weldbonding process, too.

In resistance spot welding, the WCR is usually defined on the basis of requirements imposed on the r.m.s.-value of welding current during spot welding. A particular requirement is that the r.m.s.-value of welding current must be kept within certain limits set by quality demands placed on the spot weld diameter d_p.

The lower limit of the r.m.s.-value of welding current is determined by a minimum spot weld diameter, whereas the upper limit of the welding current is given by the physics of the spot welding process and the weldbonding process respectively. Commonly used lower quality limits are spot weld diameters of $d_p = 3.5\sqrt{t}$ or $d_p = 4\sqrt{t}$, whereby t is the sheet thickness. These lower quality limits are then also referred to as $3.5\sqrt{t}$-limit or $4\sqrt{t}$-limit. The situation is completely different with the upper quality limits. The maximum admissible upper quality limit is usually referred to as splash limit I_{SL}. This limit constitutes a stability limit of the resistance spot welding process and accordingly of the weldbonding process. The splash limit is the very quality limit at which a spot weld can still be performed without the occurrence of a splash. In order to ensure that at and below this limit no splashes do in fact occur, it is necessary in setting this limit to take into account that it varies within a certain scatter band. The variation of the upper quality limit depends on the welding/joining parameters electrode force and welding time, on the material to be joined and its coating, on the applied adhesive system, on the electrode cap types, on the applied current form as well as on the electrical and mechanical machine properties of the spot welding unit. The spot weld diameter, and hence the upper and lower quality limits depend on the test procedure, and the fracture type must always be indicated in welding range determinations. The WCR for AHSS, by definition, is the setting range of the welding current. Particularly in the definition of the welding range, a welding current difference according to

$$\Delta I = I_U - I_L = I_{SL} - I_{4\sqrt{t}} \tag{1}$$

is usually assumed, whereby the lower (I_L) and the upper quality limits (I_U) are set by the $4\sqrt{t}$ –limit $I_L = I_{4\sqrt{t}}$ and by the splash limit $I_U = I_{SL}$. The representation of the difference ΔI of the r.m.s.-value of welding current, according to Eq. (1), between the upper and the lower quality limit as a function of the (stationary) electrode force F_e and of the welding time t_w

$$\Delta I = f(F_e\ t_w) \tag{2}$$

is referred to as three-dimensional weldability lobe. From the three-dimensional weldability lobe according to this definition, the classical two-dimensional weldability lobes can be derived as special representations [7–10], i.e.

$$\Delta I = f(F_e = \text{const.}, t_w) \tag{3}$$

$$\Delta I = f(F_e\ t_w = \text{const.}). \tag{4}$$

Furthermore, welding current ranges can be described by extended weldability lobes [12]. The process reliability of the spot welding and the weldbonding process depends on the size of the WCR. Roughly speaking, the process reliability will be higher if the WCR becomes wider for the chosen welding parameters. The WCR is determined by the welding parameters electrode force F_e and welding time t_w and is affected by a lot of influencing factors [7–11]. Thus, the process reliability depends on the selection of welding parameters and on these additional influencing factors. Important influencing factors are the shape and material of the electrode caps, the base metal and the base metal combinations, the coating of the steel sheets and the static and dynamic mechanical machine properties of the welding equipment. Additionally the process reliability of the weldbonding process is influenced by the properties of the applied adhesive system.

3.2 Welding Current Ranges of the Weldbonding Process

The size of the welding current range is a measure for the process reliability of the weldbonding process (Fig. 4). In order to show the weldbonding process reliability for different weldbonded AHSS, the corresponding weldability lobes were first set up. The applied adhesives are given by Table 1. The mechanical properties of the tested base metals are given by Table 2. For the tested base metals DC04ZE-KSP, HCT600XD, HCT780Z + ZE50/50, MS-W1200 and 22MnB5 (UsiBor) and

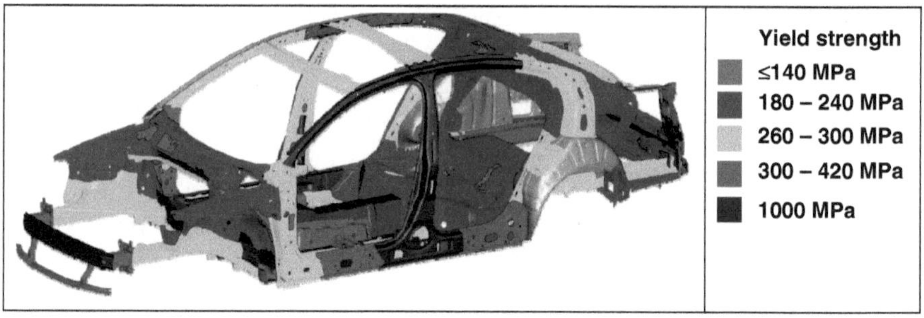

Fig. 4 Application of steels with different yield strengths in the car body shell design

the applied adhesives Terostat 5194 (A1), Betamate 1480 (A2) and SikaPower 498 (A3) the lower and upper quality limits I_L and I_U of the welding current I are given by Table 3. The welding current differences corresponding to eq. (2) are given by Table 4. In the following two-dimensional weldability lobes in accordance with the special representation (4) with constant welding time (t_W = const.) will be used. Depending on the tested base metal different constant welding times (from 260 ms up to 400 ms) are used. Therefore, according to the representation (4), i.e. $\Delta I = f(F_e, t_w = \text{const.})$ only the influence of the welding parameter electrode force F_e on the welding current ranges ΔI for the different combinations of base metals and for the three applied adhesives represented in Table 1 are investigated. The corresponding weldability lobes for these combinations are given by the Figs. 5, 6, 7, 8 and 9 It can be seen that in the case of weldbonding (RSW + A_1, RSW + A_2, RSW + A_3 with RSW = Resistance Spot Welding) both quality limits I_L and I_U

Table 3 Welding current limits for identical base metals and different adhesives

		Joining technique							
		Weldbonding						RSW	
Base metal combinations and thickness		Terostat 5194		Betamate 1480		SikaPower 498		–	
Welding parameters – Electrode force F_e – Welding time t_w	F_e in kN	Welding current limits in kA							
		I_L	I_U	I_L	I_U	I_L	I_U	I_L	I_U
DC04ZE-KSP (1 mm)	2.5	6.6	7.2	6.8	9.3	6.7	8.6	6.4	8.6
t_w = 300 ms	3.5	7.1	8.7	7.3	9.8	7.3	9.4	6.8	9.0
HCT600XD (1.3 mm)	3.5	6.6	8.5	7.1	8.3	7.1	9.6	6.8	8.0
t_w = 260 ms	4.5	7.1	9.6	7.5	9.5	8.1	10.1	7.2	9.8
HCT780T (1.3 mm)	3.5	5.4	7.3	5.3	7.0	5.5	7.3	5.2	7.0
t_w = 260 ms	4.5	5.6	7.5	5.8	8.1	6.2	7.8	5.7	8.2
MS-W1200 (1.5 mm)	4.5	6.4	7.9	6.3	8.6	6.2	8.5	6.4	8.8
t_w = 300 ms	5.5	6.6	8.3	6.6	8.8	6.7	9.6	6.8	9.8
22MnB5 (2.0 mm)	3.5	5.8	8.5	5.7	7.5	5.9	7.8	5.6	8.6
t_w = 400 ms	4.5	6.4	8.7	6.0	9.0	6.1	8.5	6.1	9.2

Table 4 Welding current ranges for identical base metals and different adhesives

		Joining technique			
		Weldbonding			RSW
Base metal combinations and thickness		Terostat 5194	Betamate 1480	SikaPower 498	–
Welding Parameters: – Electrode Force F_e – Welding Time t_w	F_e in kN	Welding current ranges ΔI in kA	ΔI in kA	ΔI in kA	ΔI in kA
DC04ZE-KSP (1 mm)	2.5	0.6	2.5	1.9	2.2
$t_w = 300$ ms	3.5	1.6	2.5	2.1	2.2
HCT600XD (1.3 mm)	3.5	1.9	1.2	2.5	1.2
$t_w = 260$ ms	4.5	2.5	2.0	2.0	2.6
HCT780T (1.3 mm)	3.5	1.9	1.7	1.8	1.8
$t_w = 260$ ms	4.5	1.9	2.3	1.6	2.5
MS-W1200 (1.5 mm)	4.5	1.5	2.3	2.3	2.4
$t_w = 300$ ms	5.5	1.7	2.2	2.9	3.0
22MnB5 (2.0 mm)	3.5	2.7	1.9	1.9	3.0
$t_w = 400$ ms	4.5	2.3	3.0	2.4	3.1

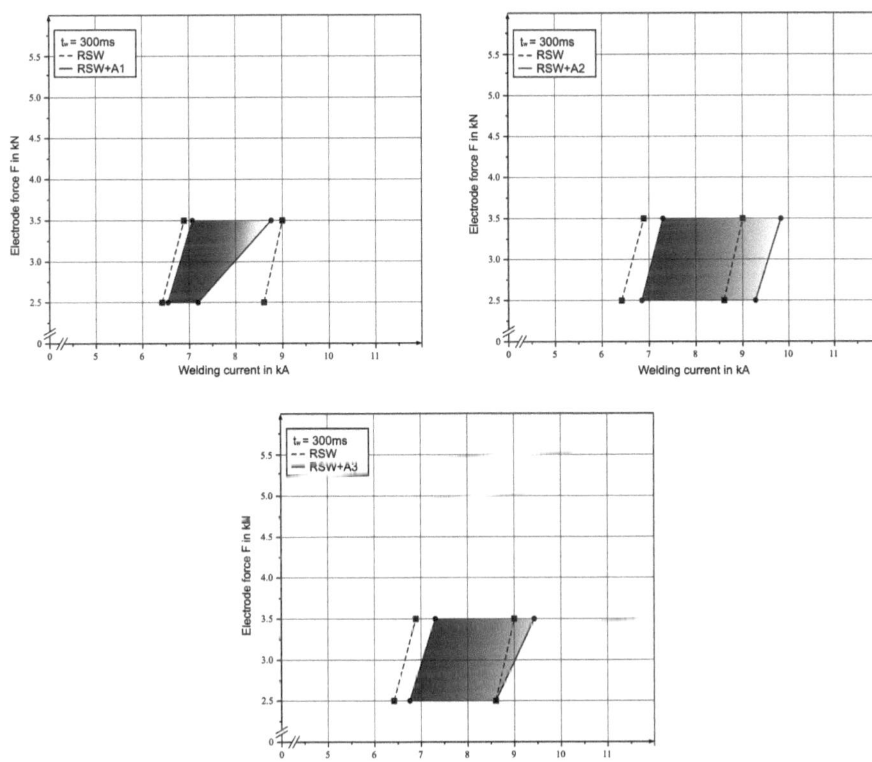

Fig. 5 Weldability lobes for the base metal DC04ZE-KSP (mild steel) and the different adhesives Terostat 5194 (A1), Betamate 1480 (A2) and SikaPower498 (A3)

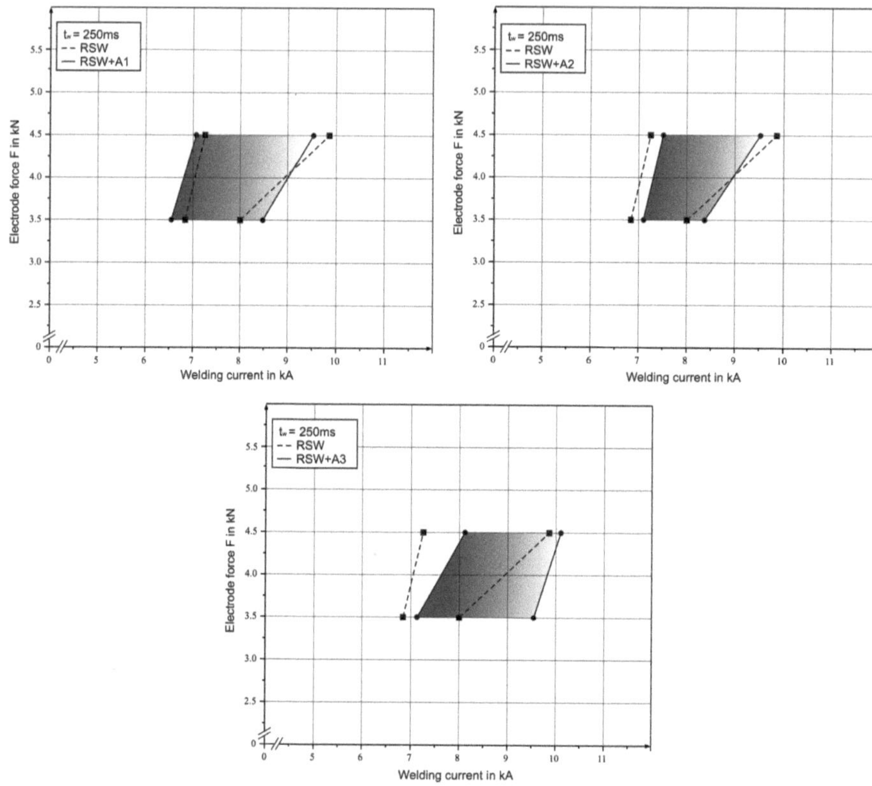

Fig. 6 Weldability lobes for the base metal HCT 600 XD (dual phase steel) and the different adhesives Terostat 5194 (A1), Betamate 1480 (A2) and SikaPower498 (A3)

increase with the increasing of electrode force F_e. Furthermore, the figures show that for all tested base metals and each applied adhesives these quality limits shift towards higher current values. This means that the energy input of the weldbonding process is higher than in the case of resistance spot welding. Clearly, roughly speaking the electrical energy needed for spot welding and weldbonding is proportional to the second power I^2 of the welding current. At the beginning of the weldbonding process there is a higher energy input than for the resistance spot welding process due to the influence of the used adhesive. The reason for this fact is the shunt behaviour of the electrical resistance, which is strongly increased by the insulating effect of the adhesive between the two metal sheets at the beginning of weldbonding process [1].

An influence of the viscosity of the applied adhesives can be seen for the low strength base metal DC04ZE-KSP and for the high strength Martensitic steel MS-W1200. In these cases the low-viscosity adhesive Terostat 5194 (A1) produce smaller welding current ranges than the medium-viscosity adhesives Betamate 1480 (A2) and SikaPower 498 (A3). But for the other three types of tested high strength

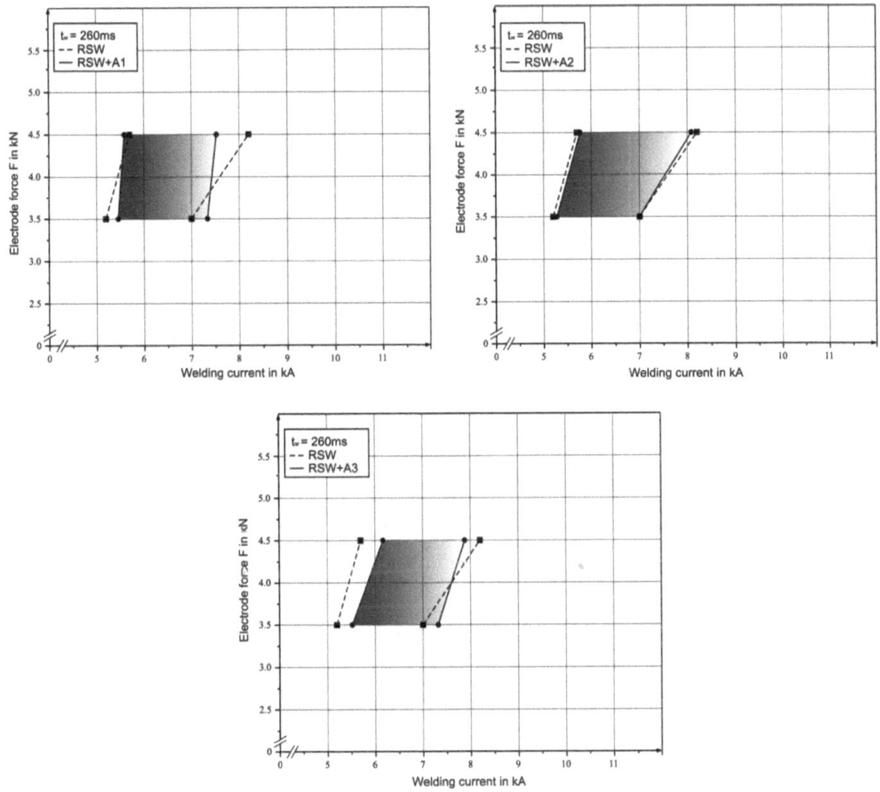

Fig. 7 Weldability lobes for the base metal HCT 700 T (TRIP steel) and the different adhesives Terostat 5194 (A1), Betamate 1480 (A2) and SikaPower498 (A3)

steel grades such an influence of the viscosity on the welding current ranges and consequently on the process reliability cannot be observed.

In summary, one can say that the influence of the tested low- and medium-viscosity adhesives for all weldbonded base metals of the same type on the welding current ranges is not significant in comparison to the spot welding process. Exemplary, the behaviour of the two base metal combinations DC04ZE-KSP/HCT600XD and HCT600XD/22MnB5 will be discussed. The lower and upper quality limits I_L and I_U for the welding current I are given by Table 5. The corresponding welding current ranges are given by Table 6 and Fig. 10 shows the welding current ranges for the base metal combination DC04ZE-KSP/HCT600XD depending on the electrode force F_e. The welding current ranges for the weldbonding process become significantly smaller than in the case of resistance spot welding. Interesting in this case is the different behaviour of lower and upper welding current limits for the low-viscosity and the two medium-viscosity adhesives. For the low-viscosity adhesive the lower quality limits are nearly the same and the upper limit changes towards lower welding currents. Compared with

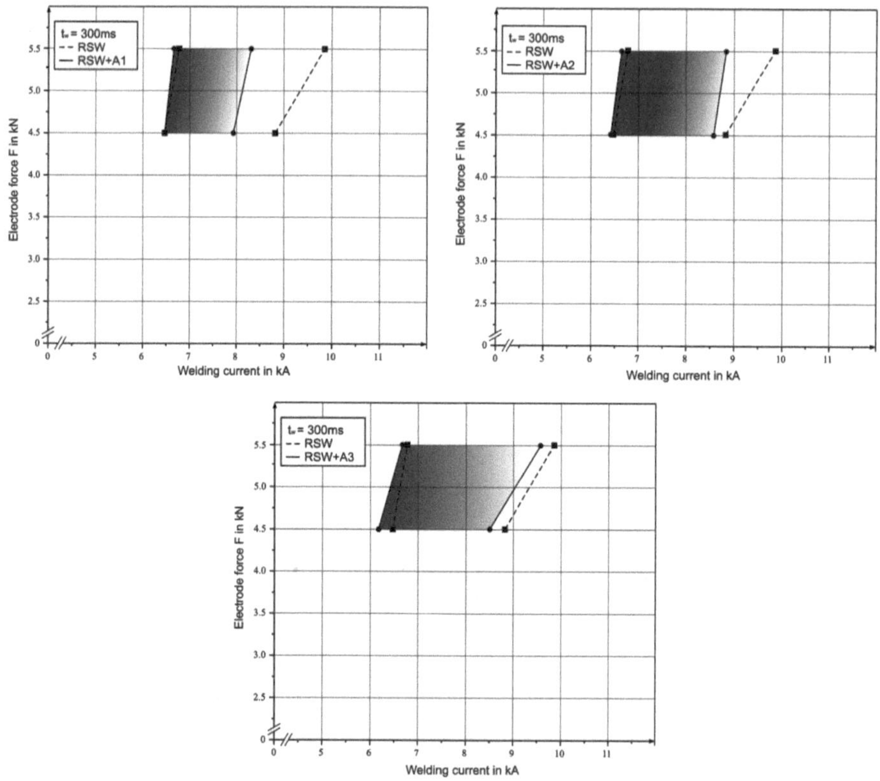

Fig. 8 Weldability lobes for the base metal MS-W1200 (martensitic steel) and the different adhesives Terostat 5194 (A1), Betamate 1480 (A2) and SikaPower498 (A3)

the quite different behaviour of the quality limits for the weldbonding process with respect to the medium-viscosity adhesives, this fact might be interpreted as an influence of the viscosity of the applied adhesives. Interesting in this connection is the significantly lower energy input for realizing the upper quality limits in the case of the low-viscosity adhesive Terostat 5194 (A1). A similar behaviour of the welding current ranges cannot be stated in the case of weldbonding the base metal combination of dual phase steel HCT600XD and press hardened steel 22MnB5 (UsiBor) (see Fig. 11). But analogously, for the first regarded base metal combinations the behaviour of the welding current ranges for the weldbonding process with the low-viscosity adhesive Terostat 5194 (A1) is quite different, too. This fact may also to be interpreted as an influence of the viscosity of the used adhesive.

Summarized it may be stated that the viscosity of the adhesives has a significant influence, when combinations of base metals are weldbonded.

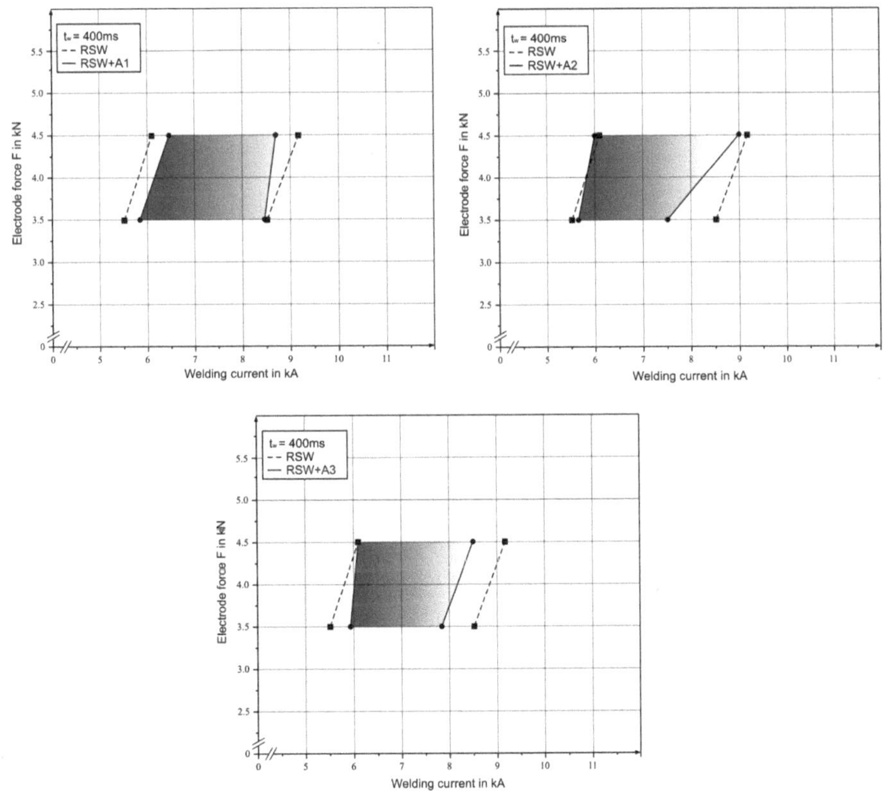

Fig. 9 Weldability lobes for the base metal 22 MnB5 (UsiBor, press-hardened steel) and the different adhesives Terostat 5194 (A1), Betamate 1480 (A2) and SikaPower498 (A3)

Table 5 Welding current limits for base metal combinations and different adhesives

Base metal combinations and thickness		Joining technique: Weldbonding – Terostat 5194		Betamate 1480		SikaPower 498		RSW –	
Welding parameters: – Electrode force F_e – Welding time t_w	F_e in kN	Welding current limits in kA: I_L	I_U	I_L	I_U	I_L	I_U	I_L	I_U
DC04ZE (0.8 mm)	2.5	7.2	8.7	8.3	10.1	8.3	10.0	7.4	9.7
HCT600XD (1.3 mm)	3.5	8.0	9.5	8.9	10.5	9.0	10.8	8.0	10.5
$t_w = 200$ ms									
HCT600XD (1.3 mm)	3.5	6.7	7.9	6.6	8.4	6.8	8.4	6.5	8.4
22MnB5 (2.0 mm)	4.5	7.1	8.2	7.4	8.7	7.2	8.7	6.9	8.7
$t_w = 300$ ms									

Table 6 Welding current ranges for base metal combinations and different adhesives

		Joining technique			
		Weldbonding			RSW
Base metal combinations and thickness		Terostat 5194	Betamate 1480	SikaPower 498	–
Welding parameters – Electrode force *Fe* – Welding time t_w	F_e in kN	Welding current ranges ΔI in kA	ΔI in kA	ΔI in kA	ΔI in kA
DC04ZE (0.8 mm) / HCT600XD	2.5	1.5	1.8	1.7	2.3
(1.3 mm) $t_w = 200$ ms	3.5	1.5	1.6	1.8	2.5
HCT600XD (1.3 mm) / 22MnB5	3.5	1.2	1.8	1.6	1.9
(2.0 mm) $t_w = 300$ ms	4.5	1.1	1.3	1.5	1.8

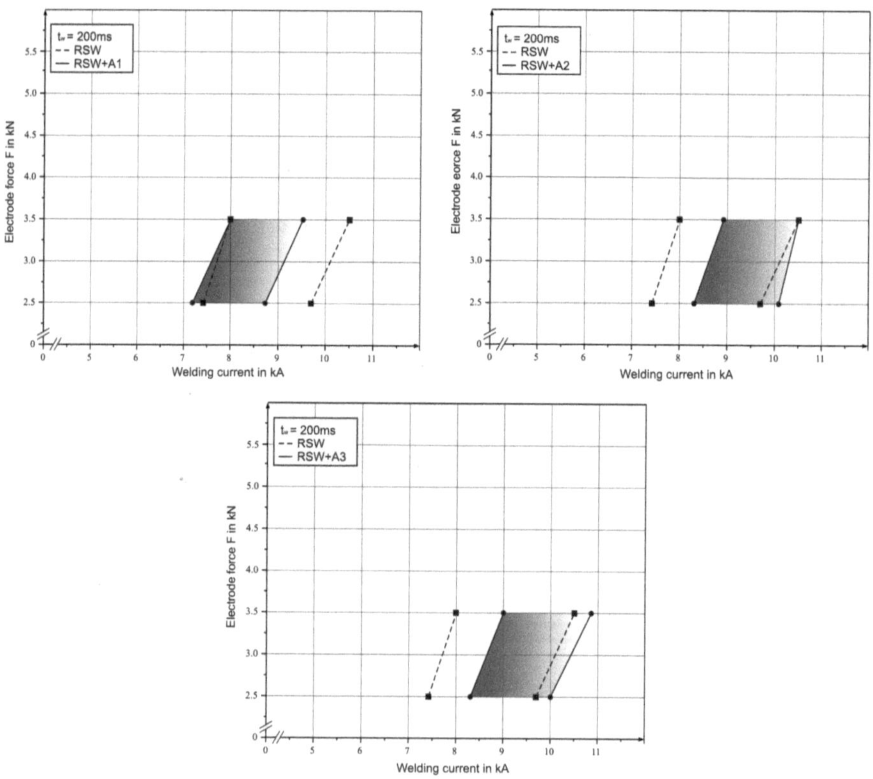

Fig. 10 Weldability lobes for the base metal combination DCO4ZE-KSP/HCT600XD and the different adhesives Terostat 5194 (A1), Betamate 1480 (A2) and SikaPower498 (A3)

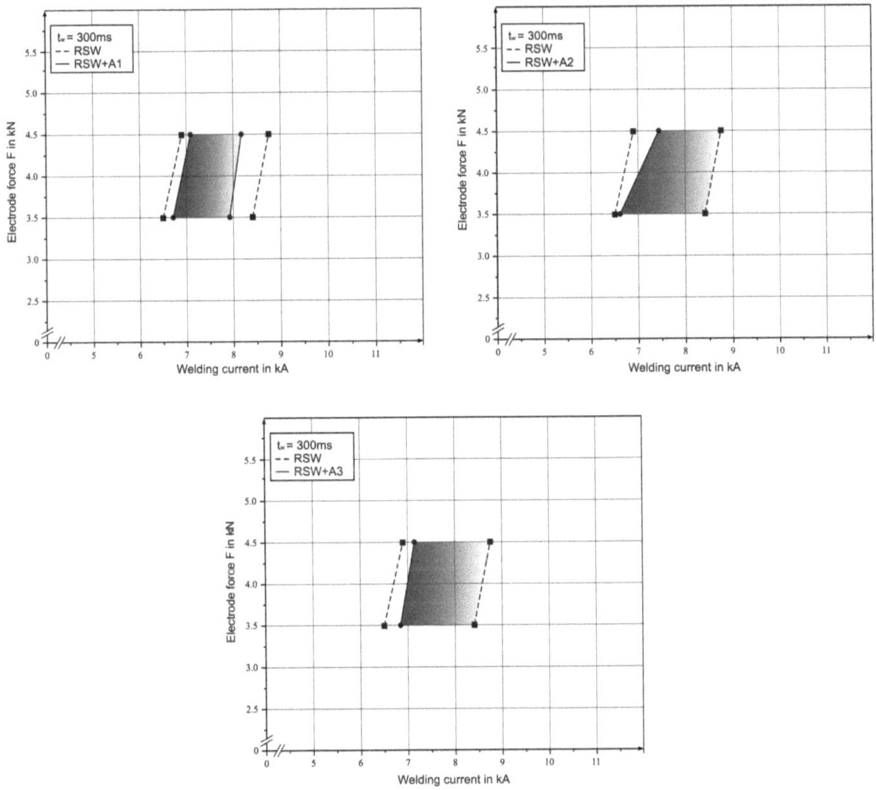

Fig. 11 Weldability lobe for the base metal combination HTC600XD/22MnB5 and the different adhesives Terostat 5194 (A1), Betamate 1480 (A2) and SikaPower498 (A3)

4 Mechanical Properties of Weldbonded Sheets of AHSS

In this Section, a comparison of some selected mechanical properties of weldbonded and spot welded sheets of AHSS and low strength reference sheets of mild steel is carried out. For the characteristic data of the tested steel grades see Table 1. The most important mechanical properties of weldbonded and spot welded joints are the behaviour under mechanical static and impact loads. Furthermore, the fatigue behaviour of weldbonded joints in comparison with spot welded joints is of main importance.

In the following, some selected results on the tensile-shear strength under static and impact load and on the fatigue strength under shear load for weldbonded and spot welded joints of some different high and low strength steels will be given and commented. Furthermore, the absorbed energy of the materials and its combinations under static and impact load will be given, too. All results are based on the same types of adhesives as used in Sect. 3 for weldability lobes construction showing the process reliability of the weldbonding process.

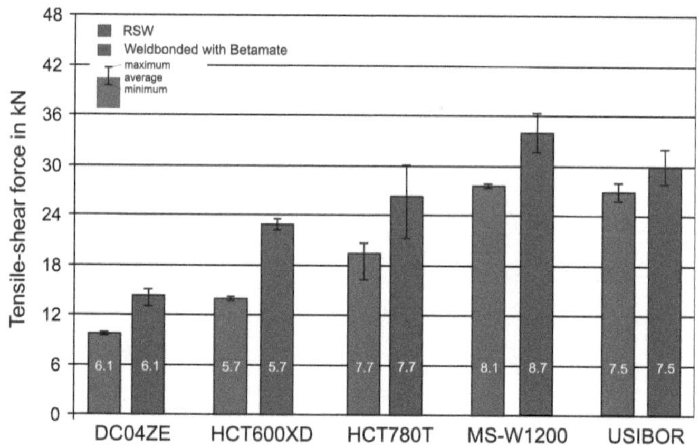

Fig. 12 Influence of the base metal strength on the strength of weldbonded (for Betamate 1480 (A2)) and spot welded joints for same base metals, with weld diameters (*in columns*)

The difference between the influence of the base metal strength on the spot welded and the weldbonded joints for the medium-viscosity adhesive Betamate 1480 (A2) is given by the Fig. 12 and Fig. 13. On the other hand, a significant increase of the tensile-shear force values is observed for the application of the medium-viscosity adhesives Betamate 1480 (A2) and SikaPower 498 (A3) (see Fig. 14). The influence of the viscosity is stronger in the case of impact load. The absorbed energy of the spot welded and the weldbonded base metal combination HCT780T/HCT800XD with the applied adhesives Terostat 5194 (A1), Betamate 1480 (A2) and SikaPower 498 (A3) under static (a) and impact shear load

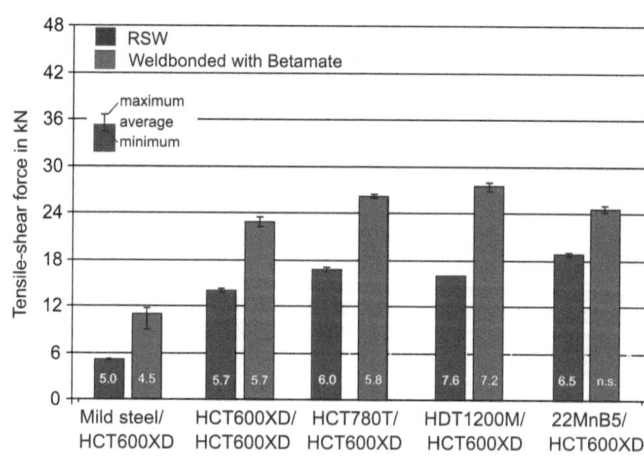

Fig. 13 Influence of the base metal strength on the strength of weldbonded (for Betamate 1480 (A2)) and spot welded joints for the base metal combinations, with weld diameters (*in columns*)

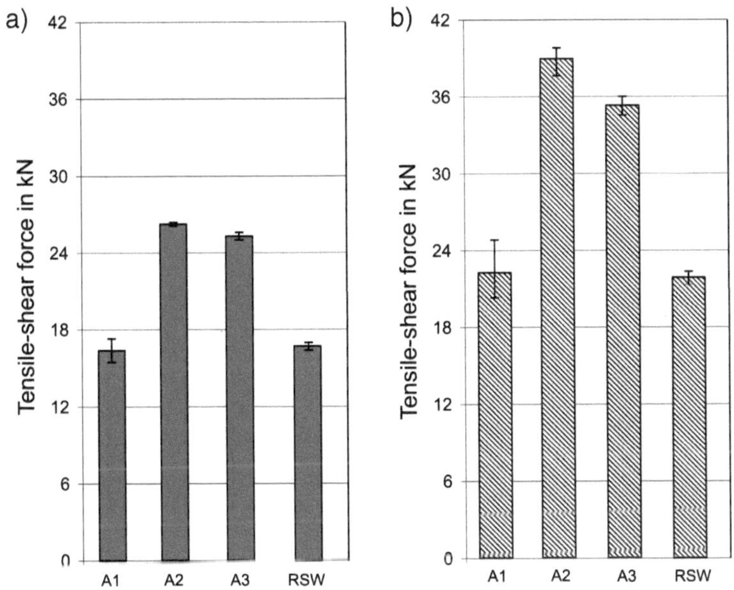

Fig. 14 Tensile-shear force of the base metal combination HCT780T/HCT600XD and the different adhesives Terostat 5194 (A1), Betamate 1480 (A2) and SikaPower498 (A3) under static **(a)** impact shear load **(b)**

is illustrated by the Fig. 15. Here, an analogous effect occurs. The absorbed energy of the weldbonded joints with the medium-viscosity adhesives Betamate 1480 (A2) and SikaPower 498 (A3) is significantly higher than for the application of the low-viscosity adhesive Terostat 5194 (A1). This means that there exists a better crash behaviour when the weldbonding is carried out with adhesives of higher viscosities. Figure 12 shows the influences for the same base metals and Fig. 13 shows the influence for base metal combinations with the dual phase steel (AHSS) HCT600XD. In all cases with the same base metals and the base metal combinations, an advantage is discovered for the weldbonded joints in comparison to the spot welded joints. This tendency becomes stronger for base metal combinations. Figure 14 shows the tensile-shear strength for weldbonded and spot welded joints of the base metal combination HCT780T/HCT600XD for the applied adhesives Terostat 5194 (A1), Betamate 1480 (A2) and SikaPower 498 (A3) under static (a) and impact shear load. It can be seen that both under static and impact load the tensile-shear force values of the weldbonded joints for the application of the low-viscosity adhesive Terostat 5194 (A1) are nearly identical to those of the spot welded joints. It can be stated that the tensile-shear strength behaviour and the crash behaviour of the weldbonded joints and static under impact shear load are influenced by the viscosity.

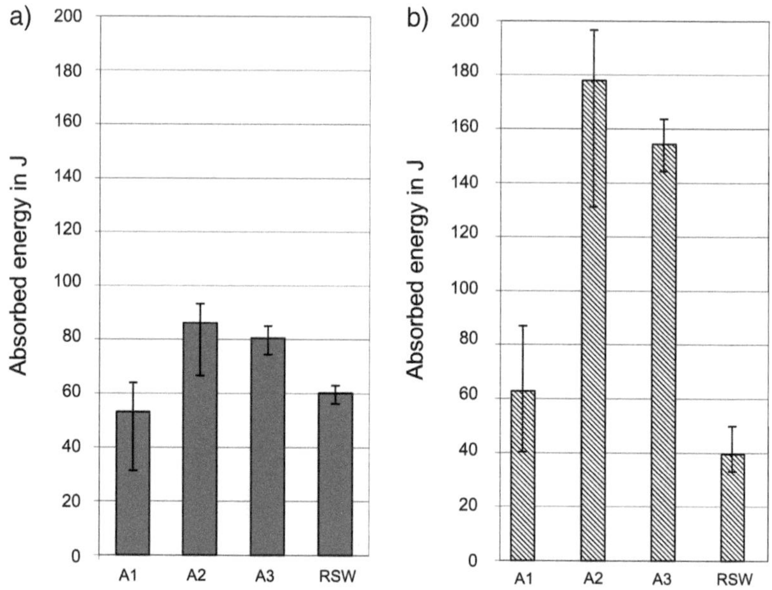

Fig. 15 Absorbed energy of the base metal combination HCT780T/HCT600XD and the different adhesives Terostat 5194 (A1), Betamate 1480 (A2) and SikaPower498 (A3) under static **(a)** impact shear load **(b)**

In order to analyze the fatigue behaviour of weldbonded joints Wöhler tests were carried out by a dynamic testing machine under a force-controlled regime with a load ratio of R = 0.1. This was done for shear specimens using tensile-fatigue test procedures according to EN ISO 14324. The applied test stop criterion was the total fracture of the tested specimens. In this connection plug failures of fractures in the material were the dominating failure types. Exemplarily the fatigue behaviour under tensile-shear load for the base metal combination HCT600XD/22MnB5 will be discussed. A comparison of the S/N-curves for spot welded joints with weld diameter $d_p = 6.7$ mm and adhesives bonded joints with the medium-viscosity adhesives Betamate 1480 (A2) and SikaPower 498 (A3) is given by Fig. 16. It can be seen that in both cases of weldbonded joints there is an advantage with respect to the spot welded joints. Figure 17 shows the S/N-curves for spot welded joints with weld diameter $d_p = 6.7$ mm and weldbonded joints with weld diameter $d_p = 6.3$ mm for the adhesive Betamate 1480 (A2). There is an advantage of weldbonded joints with adhesive Betamate 1480 (A2) with respect to the spot welded joints. The S/N-curves for spot welded joints with weld diameter $d_p = 6.7$ mm and weldbonded joints with weld diameter $d_p = 6.5$ mm for the adhesive SikaPower 498 (A3) are given by Fig. 17. Figure shows an advantage of the weldbonded joints with respect to the spot welded joints, too. In summary, weldbonded joints have a significant advantage over spot welded joints (Fig. 18).

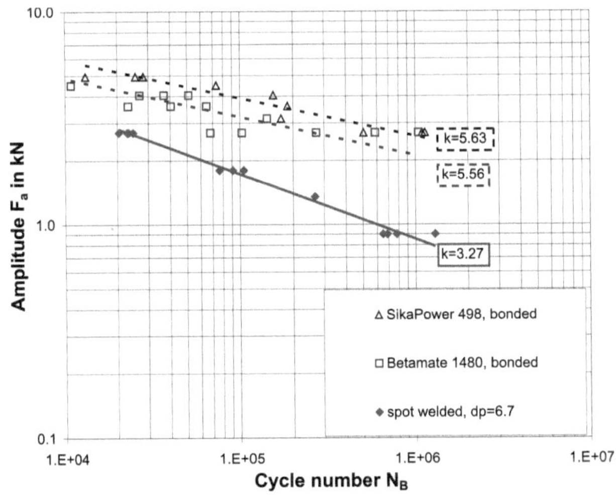

Fig. 16 Fatigue behaviour (S/N curves) of the base metal combination HCT600XD/22MnB5 for spot welded (d_p = 6.7 mm) and adhesive bonded joints with Betamate 1480 (A2) and SikaPower498 (A3)

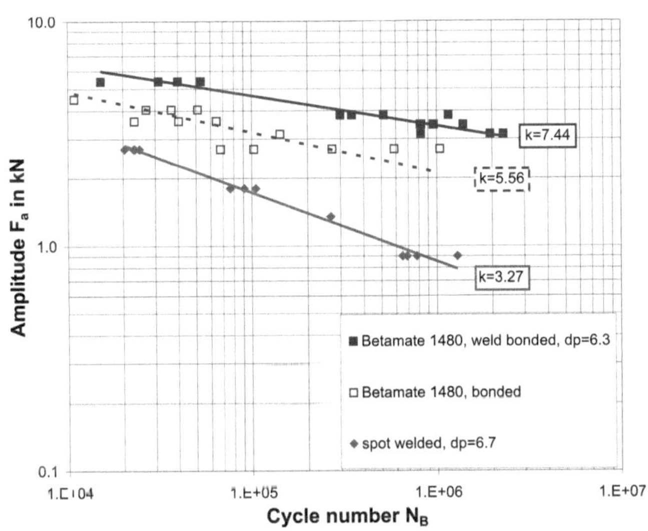

Fig. 17 Fatigue behaviour (S/N-curves) of the base metal combination HCT600XD/22MnB5 for spot welded (d_p = 6.7 mm), adhesive bonded and weldbonded (d_p = 6.3 mm) joints with Betamate 1480 (A2)

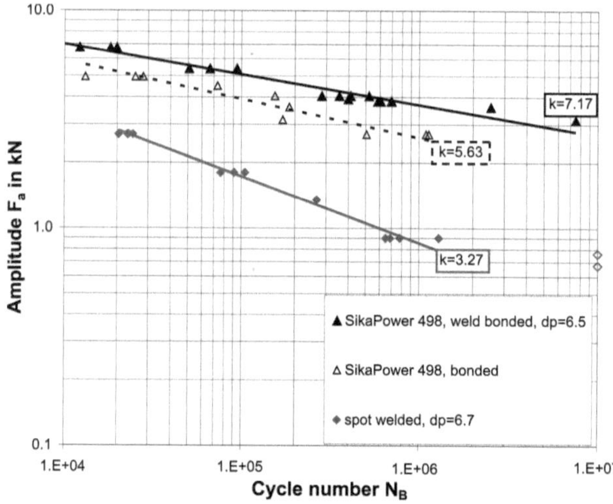

Fig. 18 Fatigue behaviour (S/N-curves) of the base metal combination HCT600XD/22MnB5 for spot welded (d_p = 6.7 mm), adhesive bonded and weldbonded (d_p = 6.5 mm) joints with SikaPower498 (A3)

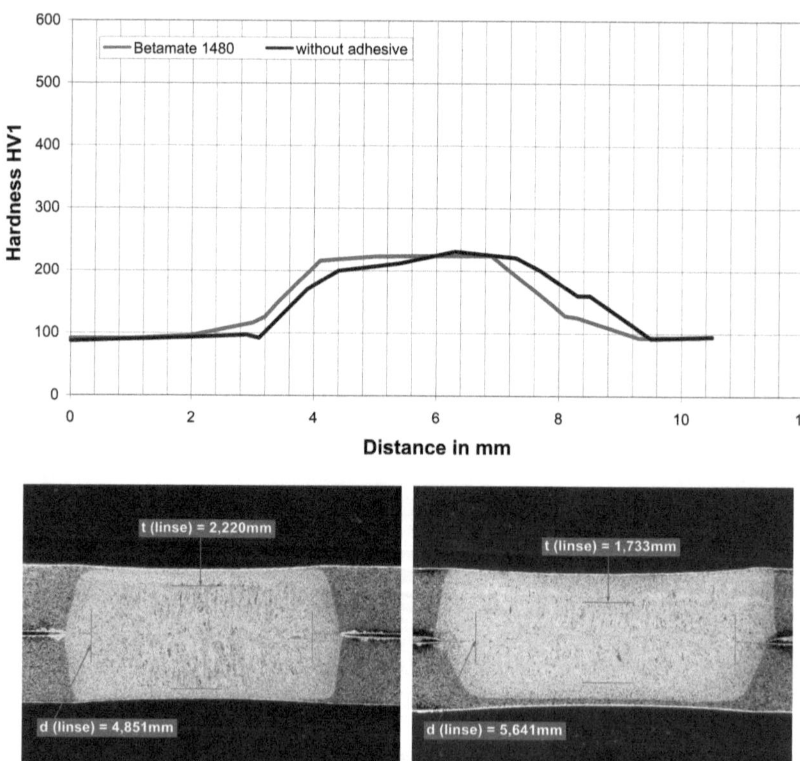

Fig. 19 Weld nugget structures and the hardness curves of spot welded and weldbonded base metal DC04ZE-KSP for Betamate 1480 (A2)

5 Metallurgical and Fracture Behaviour of Weldbonded Joints

For the mild steel DC04ZE-KSP, the advanced high strength steel (AHSS) HCT 780T and the press hardened steel 22MnB5 (UsiBor) the metallurgical structures of the weld nuggets for resistance spot welded and weldbonded joints with the medium-viscosity adhesive Betamate 1480 (A2) are shown by the Figs. 19, 20 and 21. It can be seen that there is no difference between the spot welded and weldbonded nugget structures. Similar results are given for the application of other base metals and its combinations for the adhesives A1 and A3 [1]. These results are consistent with the hardness curves, also given by the Figs. 19, 20 and 21. There are no significant differences between the hardness curves for the weldbonded steels mentioned above and the applied adhesive Betamate 1480 (A2). Same results are given for the application of the other adhesives Terostat 5194 (A1) and SikaPower 498 (A3) for the weldbonding of the investigated base metals and its combinations [1]. The reason for these facts is the similarity of the electrical energy input after the beginning phases of the weldbonding and the resistance spot welding process. The higher total electrical resistance between the steel sheets during the beginning

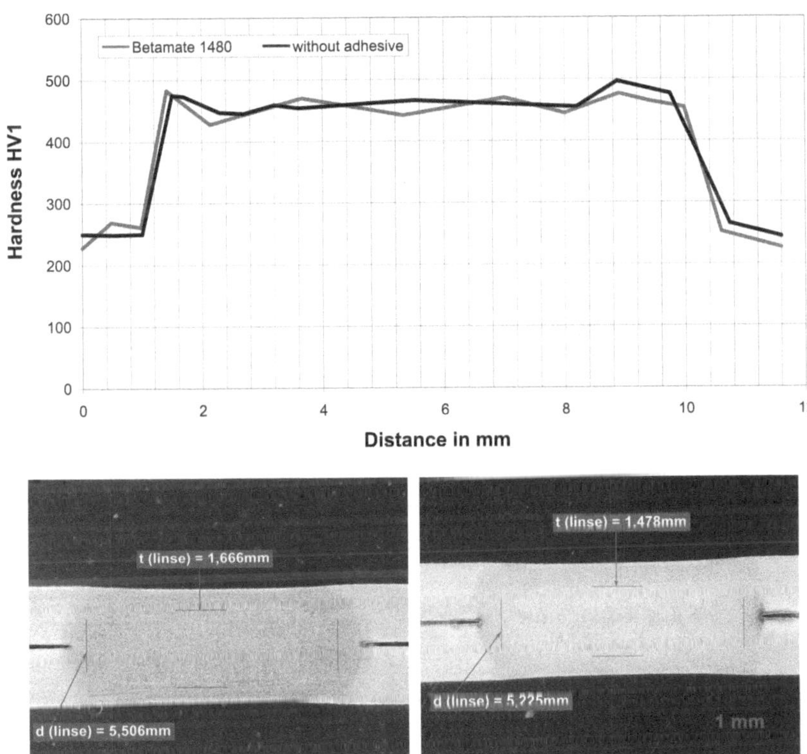

Fig. 20 Weld nugget structures and the hardness curves of spot welded and weldbonded base metal HCT780T for Betamate 1480 (A2)

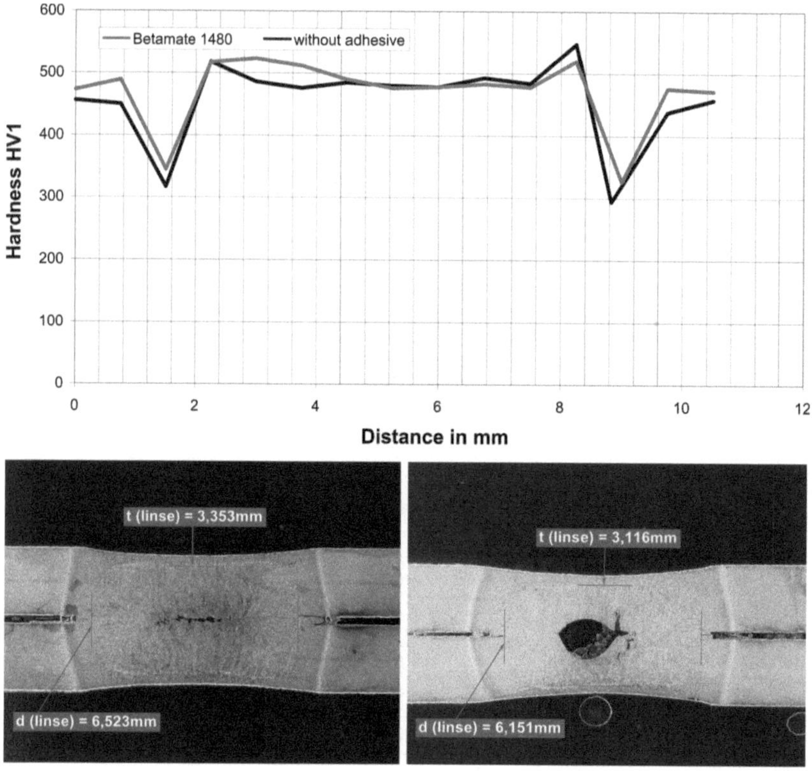

Fig. 21 Weld nugget structures and the hardness curves of spot welded and weldbonded base metal 22MnB5 (UsiBor) for Betamate 1480 (A2)

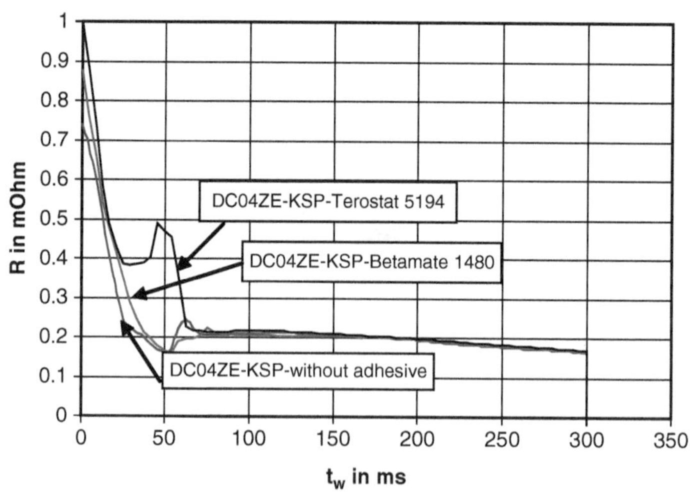

Fig. 22 Electrical resistance R between the steel sheets during the welding time t_w for spot welding and weldbonding with the adhesives Terostat 5194 (A1) and Betamate 1480 (A2) for DC04ZE-KSP

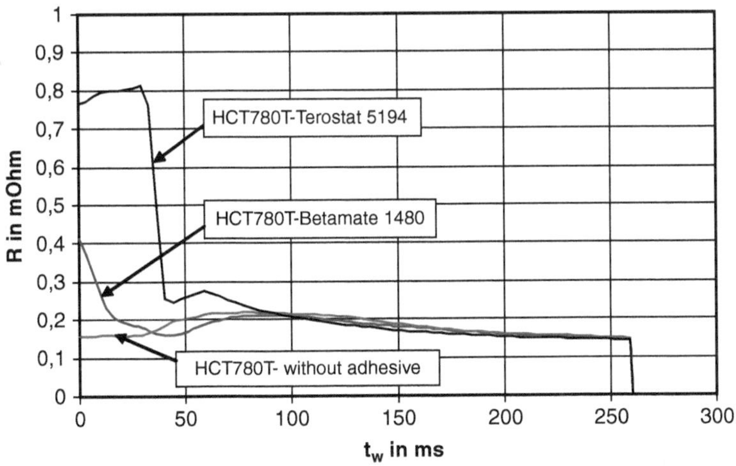

Fig. 23 Electrical resistance R between the steel sheets during the welding time t_W for spot welding and weldbonding with the adhesives Terostat 5194 (A1) and Betamate 1480 (A2) for HCT780T

of the weldbonding process is attributed to the electrical insulating behaviour of the applied adhesives. After the break down of the constant resistance, the calculated total electrical resistance between the steel sheets achieves comparable values for the weldbonding and spot welding process. Figures 22, 23 and 24 show exemplary the electrical resistance between the steel sheets during the welding time for the weldbonding and the spot welding process for the base metals DC04ZE-KSP,

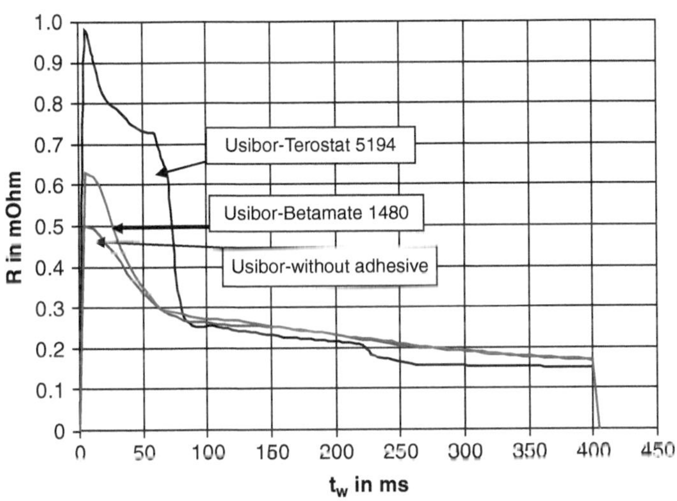

Fig. 24 Electrical resistance R between the steel sheets during the welding time t_W for spot welding and weldbonding with the adhesives Terostat 5194 (A1) and Betamate 1480 (A2) for 22MnB5 (UsiBor)

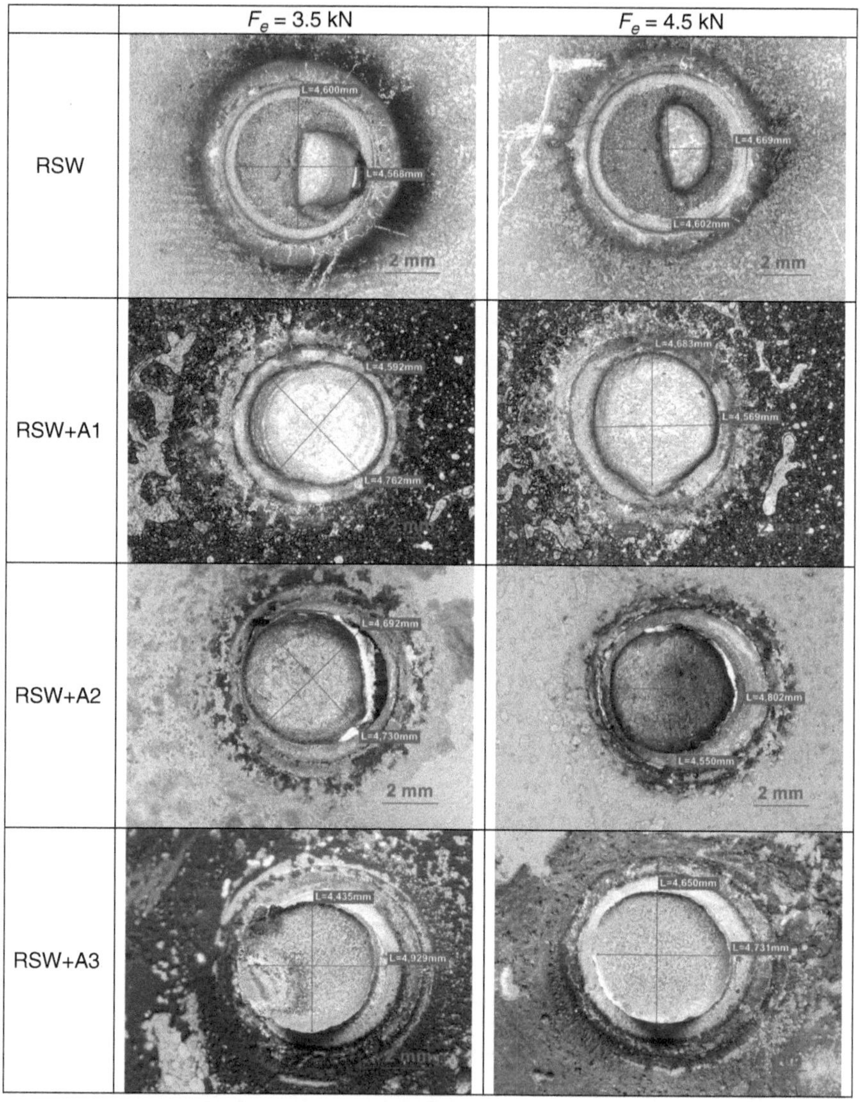

Fig. 25 Fracture behaviour of weldbonded and spot welded joints of the base metal HCT780T (TRIP steel) for the adhesives Terostat 5194 (A1), Betamate 1480 (A2) and SikaPower498 (A3). Results for the lower quality limit and different electrode forces $F_e = 3.5/4.5$ kN

HCT780T and 22MnB5 (UsiBor). It can be seen that the low-viscosity adhesive Terostat 5194 (A1) leads to higher deviations from the electrical resistance for the spot welding process than the medium-viscosity adhesive Betamate 1480 (A2) for the above mentioned base metals during the beginning phase (process time < 80 m sec). But after the beginning phase, the values of the electrical resistances

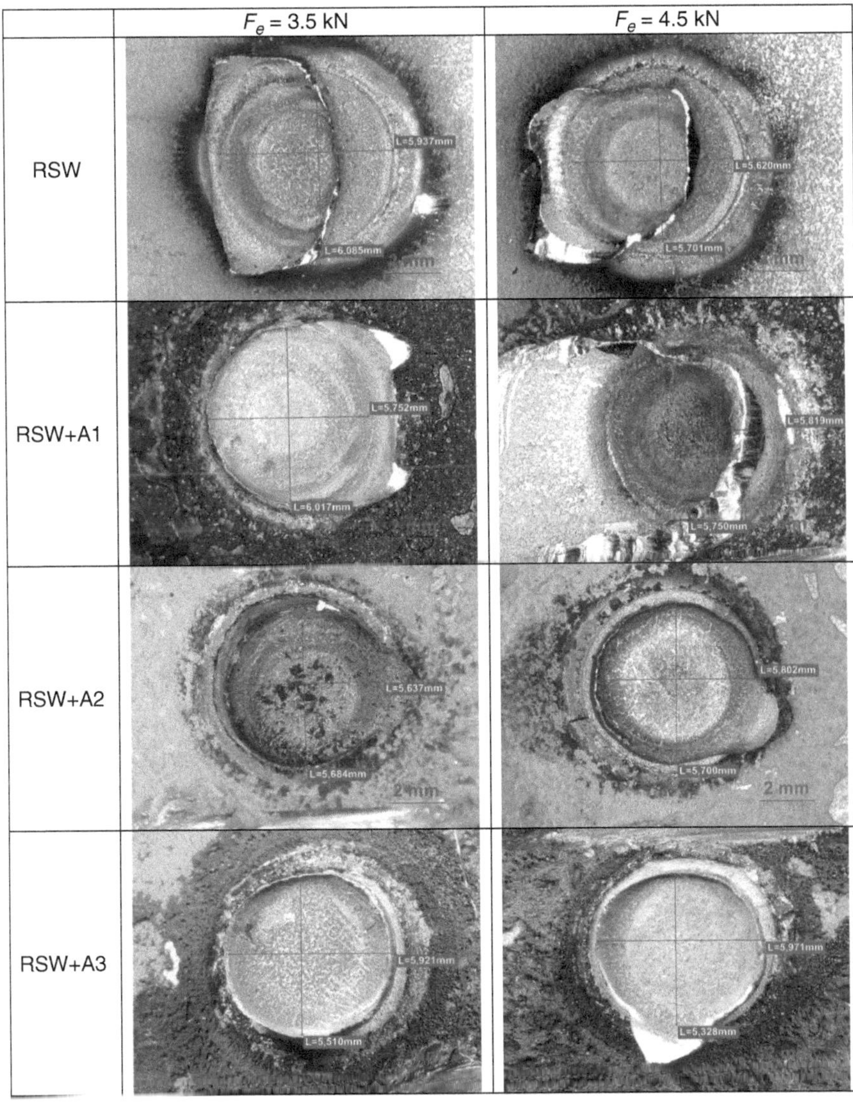

Fig. 26 Fracture behaviour of weldbonded and spot welded joints of the base metal HCT780T (TRIP steel) for the adhesives Terostat 5194 (A1), Betamate 1480 (A2) and SikaPower498 (A3). Results for the upper quality limit and different electrode forces $F_e = 3.5/4.5$ kN

between the steel sheets are independent of the applied adhesives, i.e. nearly identical in all cases. Therefore, the identical metallurgical behaviour in form of the nugget structures and the hardness curves given above can be interpreted as a result of the similar electrical behaviour of the weldbonding and the resistance spot welding process after the beginning phase. A comparison of the fracture behaviour of weldbonded and spot welded joints of the base metals HCT780T (TRIP Steel) and

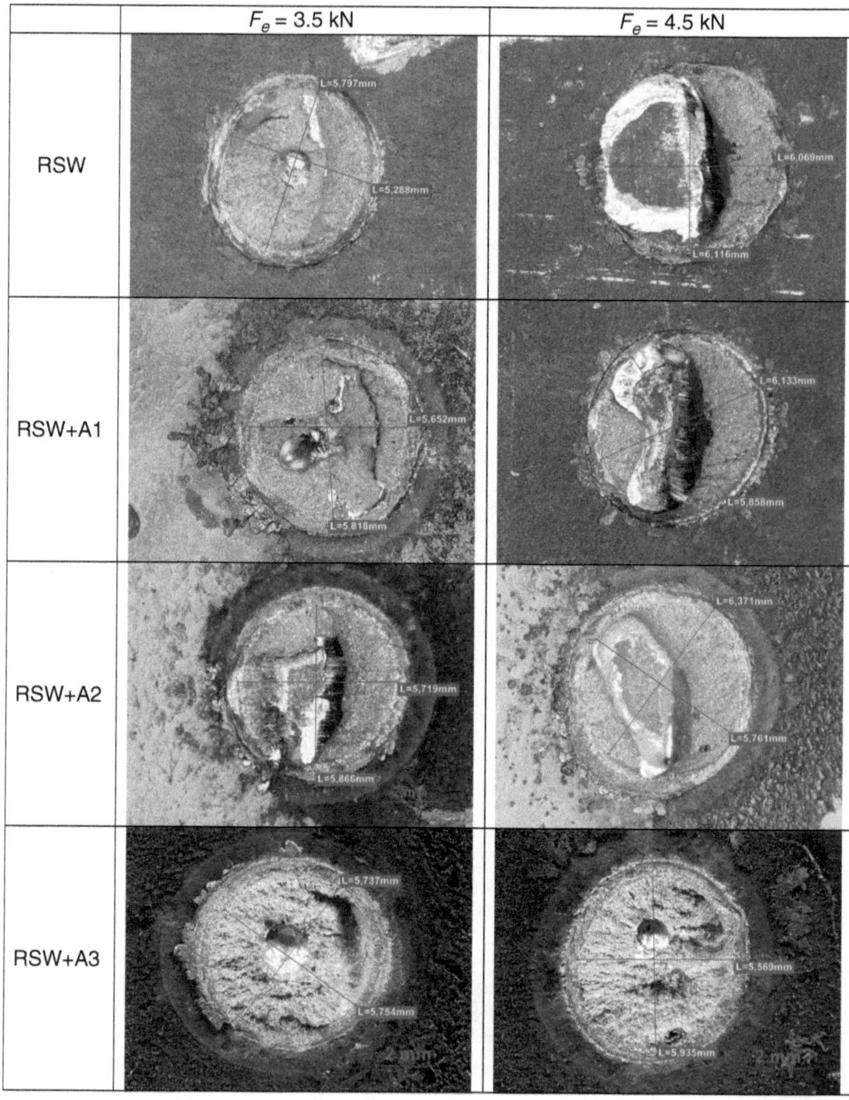

Fig. 27 Fracture behaviour of weldbonded and spot welded joints of the base metal 22MnB5 (UsiBor) for the adhesives Terostat 5194 (A1), Betamate 1480 (A2) and SikaPower498 (A3). Results for the lower quality limit and different electrode forces $F_e = 3.5/4.5$ kN

22MnB5 (UsiBor) for different electrode forces (F_e = 3.5/4.5 kN) is given by the Figs. 25, 26, 27 and 28. The applied adhesives in these cases are the low- and medium-viscosity adhesives A1, A2 and A3 mentioned above. It can be seen that there is no influence of the applied adhesives and, consequently, of the viscosity on the failure type. The fracture behaviour of the press hardened steel 22MnB5 (UsiBor) is very irregular, Figs. 27 and 28. But the fracture behaviour of the

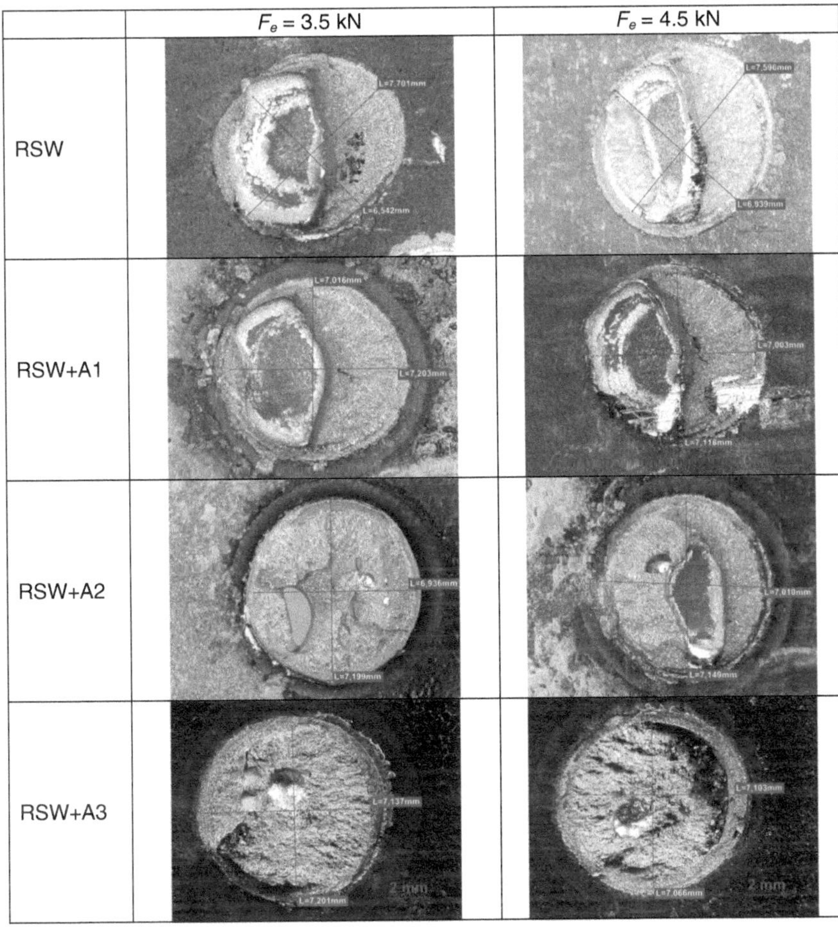

Fig. 28 Fracture behaviour of weldbonded and spot welded joints of the base metal 22MnB5 (UsiBor) for the adhesives Terostat 5194 (A1), Betamate 1480 (A2) and SikaPower498 (A3). Results for the upper quality limit d_{pmax} and different electrode forces $F_e = 3.5/4.5$ kN

weldbonded TRIP steel HCT780T shows compared with the spot welded joints a tendency towards a plug failure independently of the applied adhesives, Figs. 25 and 26.

6 Summary

The well known advantages of the weldbonding process for mild steels of lower yield strength are an obvious truth for the application of high strength steels, too. Disadvantages due to smaller sizes of welding current ranges and consequently significantly lower process reliabilities do not occur. Clearly, in almost all cases the welding current ranges and therefore the process reliability of the weldbonding process become smaller in comparison with the resistance welding process. But it is possible, by changing the welding parameters electrode force and welding time, to alter the width of the welding current ranges. The influence of the applied adhesives of low and medium viscosities on the welding current ranges is nearly identical for weldbonding the same base metals. Significant differences do not occur. Only for weldbonding base metal combinations it can be stated that there is an influence of the viscosity on the welding current ranges. But there is a disadvantage with respect to the welding current ranges of the weldbonding process, too. The welding current for the lower and upper quality limits change towards higher current values. Because of the dependence of the electrical energy to the second power of the welding current this means a higher energy needs in comparison with the resistance spot welding process. The welding current ranges for the weldbonding process become significantly smaller in comparison with the resistance spot welding process for the base metal combination of a mild steel with a dual phase steel (AHSS), when the low-viscosity adhesive is applied.

For weldbonding the combination of a mild steel with a dual phase steel (AHSS) or with an extremely high strength press hardened steel, respectively, there occurs a quite different behaviour of the lower and upper quality limits for the different applied adhesives. These facts may be interpreted as an influence of the viscosity of adhesives.

Further advantages of the weldbonding process are given by the mechanical behaviour. For all cases of weldbonded base metals and their combinations, higher values of the tensile-shear force under static and impact load are found for the applied different adhesives in comparison with the spot welded joints. The absorbed energy is higher for weldbonded than for spot welded joints. This means that the crash behaviour of the weldbonded joints becomes better as in the case of spot welded joints. There is a difference between the influences on the tensile-shear strength and the crash behaviour of the weldbonded joints for adhesives of low and medium viscosity. Furthermore, there is an advantage regarding the fatigue behaviour of weldbonded joints. The metallurgical behaviour of weldbonded and spot welded joints is nearly the same. There is no difference in the hardness behaviour of weldbonded and spot welded joints. Furthermore, the metallurgical structure of the weldbonded and spot welded joints is almost equal. The fracture behaviour of the weldbonded and spot welded joints depends on the joined base metals and their combinations and on the applied adhesives. It is quite different for the tested combinations. Only in the case of the advanced high strength TRIP steel there occurs a tendency towards the plug failure behaviour of the weldbonded joints compared with the spot welded joints. In summary, one can say that the contribution has shown that the influence of the applied adhesives and their viscosities on the process reliability and the energy input of the weldbonding process depends on the weldbonded base metal and its combinations. This dependence is also true for the static strength and fatigue strength of the weldbonded joints. Furthermore, there exits a different fracture behaviour of the weldbonded joints influenced by the applied adhesives, but no influence of the adhesives on the metallurgical behaviour was observed.

The results of this chapter are based on a research project entitled "Investigations into the combined spot welding and adhesive bonding of higher-strength steel sheets with new hot-curing and cold-curing adhesive systems" (No. 14476 N) of the Research Association for Steel Application (FOSTA). The project was financed by the German Ministry of Economic Affairs and Technology (BMWi) via the Federation of Industrial Research Associations (AiF) in the programme for the promotion of joint industrial research (IGF) and was carried out at the Laboratory of Materials Engineering and Joining Technology of the University of Paderborn and the German Welding Institute and expertly accompanied and supported by FOSTA (No. P 704). We would like to express our gratitude for this. Our thanks goes to the members of the industrial support circle for the test materials and services made available in order to carry out this research project as well as for the support and the constructive discussions.

References

1. H. Cramer, T. Bschorr, O. Hahn, H. Thommes, et al. Final Report for FOSTA-No. P704/10 AiF-No. 14476 N (Forschungsvereinigung Stahlanwendung e. V., Düsseldorf 2009)
2. N. N. (1998) Ultra Light Steel Auto Body Final Report. American Iron and Steel Institute, Washington, DC
3. B. Osburg, L. Patberg, A. Grüneklee et al. New Steel Body – Sicherer und wirtschaftlicher Karosserieleichtbau mit Stahl. Automobiltechnische Zeitschrift ATZ 3 **106**, 190–194 (2004)
4. N. N. (2004) in *Proceeding of the Conference "State-of-the-Art Multi-Phase Steels"*, 23/24 Sept, Berlin
5. J.M. Demeri, Forming of advanced high strength steels. ASM Handbook Volume 14B Metal Working Sheet Forming, 530–538 (2006)
6. S. Maggi, M. Murgia, Introduction to the metallurgic characteristics of advanced high strength steels for automobile applications. Weld. Int. **22**(9), 610–618, Sept (2008)
7. G. Weber, S. Göklü, Resistance spot welding of advanced high strength steels – Influence of welding parameters and electrode cap type Proceedings of the International Conference of the International Institute of Welding (IIW), Prague, 14–15 July, 135–155 (2005)
8. G. Weber, S. Göklü, Schweißeignung und Prozesssicherheit beim Widerstandspunktschweißen hoch fester Mehrphasenstähle im Rohkarosserie-Leichtbau EURO-JOIN 5, Wien:1 V9/1–30 (2004)
9. G. Weber, S. Göklü, Resistance Spot Welding of Uncoated and Zinc Coated Advanced High Strength Steels (AHSS) Weldability and Process Reliability Influence of Welding Parameters International Institute of Welding (IIW) IIW-Doc. No. III-1347-05, Prague (2005)
10. G. Weber, S. Göklü, Resistance Spot Welding of Uncoated and Zinc Coated Advanced High Strength Steels (AHSS) – Weldability and Process Reliability – Influence of Welding Parameters Welding in the World **50**(3/4), 3–12, Roissy (2006)
11. G. Weber, S. Göklü, M. Rethmeier, Influence of the Type of Electrode Caps on the Welding Current Ranges and the Process Reliability in Resistance Spot Welding IIW Doc. No. III-1484-08, 61th Annual Assembly of IIW, Graz (2008)
12. G. Weber, M. Rethmeier, S. Brauser, Extended Weldability Lobes in Resistance Spot Welding of Advanced High Strength Steels (AHSS). In Proceedings of EUROJOIN, Vol 7, Lido de Venezia, Italy (2009)

13. S.M. Darwish, Weldbonding strengthens and balances the stresses in spot-welded dissimilar thickness joints. J. Mater. Proc. Technol. **134**, 352–362 (2003)
14. S.M. Darwish, A. Al-Samhan, Design rationale of weldbonded joints Int. J. Adhes. Adhes. **24**, 367–377 (2004)
15. S.M. Darwish, A. Al-Samhan, Thermal stresses developed in weldbonded joints. J. Mater. Proc. Technol. **153–154**, 971–977 (2004)
16. O. Hahn, H. Roland, B. Wender, Berechnung der maximalen Spannungs-konzentration in Punktschweißverbindungen mit geschlossenen Berechnungsverfahren Schweißen und Schneiden Heft **1**, 24–27 (1983)
17. O. Hahn, A. Peetz, Eigenschaften und Wirtschaftlichkeit kombiniert gefügter Blechverbindungen Konferenz-Einzelbericht DVS-Berichte Band 1, Düsseldorf, 272–277 (1995)
18. D. Knoll, Punktschweißen und Kleben im Karosserie-Rohbau Adhäsion Heft **12**, 25–31 (1991)
19. G. Kötting, G. Schmid, Widerstandspunktschweißkleben geklebter und beschichteter Karosseriebleche: Verbindungseigenschaften Technologie und Gefahr-stoffemission Konferenz-Einzelbericht DVS-Berichte Band **165**, 17–20 (1995)
20. A. Al-Samhan, S.M.H. Darwish, Finite element modelling of weldbonded joints. J. Mater. Proc. Technol. **142**, 587–598 (2003)
21. A. Al-Samhan, S.M.H. Darwish, Strength prediction of weldbonded joints. Int. J. Adhes. Adhes. **23**, 23–28 (2003)
22. I.O. Santosa, W. Zhangb, V.M. Goncalvesa, et al. Weldbonding of stainless steel, Lissabon (2004)
23. G. Schmid, M. Korte, U. Walther, Punktschweißkleben im Automobilbau Konferenz-Einzelbericht DVS-Berichte Band 213 Düsseldorf, 53–58 (2001)
24. G. Weber, S. Brauser, L.-A. Pepke, M. Rethmeier, Description of welding current ranges by extended weldability lobes in resistance spot welding. In *Annual Meeting SC-Auto of the International Institute of Welding (IIW)*, Wolfsburg, Germany (2009)